Lanthanide and Actinide Chemistry

Inorganic Chemistry

A Wiley Series of Advanced Textbooks

Editorial Board

Previously Published Books In this Series

Chemical Bonds: A Dialog
Author: J. K. Burdett

Bioinorganic Chemistry: Inorganic Elements in the Chemistry
of Life – An Introduction and Guide
Author: W. Kaim

Synthesis of Organometallic Compounds: A Practical Guide
Edited by: S. Komiya

Main Group Chemistry Second Edition
Author: A. G. Massey

Inorganic Structural Chemistry
Author: U. Muller

Stereochemistry of Coordination Compounds
Author: A. Von Zelewsky

Lanthanide and Actinide Chemistry
Author: S. A. Cotton

Email (for orders and customer service enquiries): cs-books@wiley.co.uk
Visit our Home Page on www.wileyeurope.com or www.wiley.com

Reprinted with corrections May 2007

Other Wiley Editorial Offices

John Wiley & Sons Inc., 111 River Street, Hoboken, NJ 07030, USA

Jossey-Bass, 989 Market Street, San Francisco, CA 94103-1741, USA

Wiley-VCH Verlag GmbH, Boschstr. 12, D-69469 Weinheim, Germany

John Wiley & Sons Australia Ltd, 42 McDougall Street, Milton, Queensland 4064, Australia

John Wiley & Sons (Asia) Pte Ltd, 2 Clementi Loop #02-01, Jin Xing Distripark, Singapore 129809

John Wiley & Sons Canada Ltd, 22 Worcester Road, Etobicoke, Ontario, Canada M9W 1L1

Wiley also publishes its books in a variety of electronic formats. Some content that appears in print may not be available in electronic books.

Library of Congress Cataloging-in-Publication Data

Cotton, Simon, Dr.
      Lanthanide and actinide chemistry / Simon Cotton.
         p.   cm. – (Inorganic chemistry)
      Includes bibliographical references and index.
      ISBN-13: 978-0-470-01005-1 (acid-free paper)
      ISBN-10: 0-470-01005-3 (acid-free paper)
      ISBN-13: 978-0-470-01006-8 (pbk. : acid-free paper)
      ISBN-10: 0-470-01006-1 (pbk. : acid-free paper)
      1. Rare earth metals.   2. Actinide elements.   I. Title.
   II. Inorganic chemistry (John Wiley & Sons)
   QD172.R2C68 2006
   546′.41 dc22   2005025297

British Library Cataloguing in Publication Data

A catalogue record for this book is available from the British Library

ISBN-13 9-78-0-470-01005-1 (Cloth)   9-78-0-470-01006-8 (Paper)
ISBN-10 0-470-01005-3 (Cloth)   0-470-01006-1 (Paper)

Typeset in 10/12pt Times by TechBooks, New Delhi, India

Lanthanide and Actinide Chemistry

Simon Cotton

Uppingham School, Uppingham, Rutland, UK

John Wiley & Sons, Ltd

In memory of Ray and Derek Cotton, my parents.

Remember that it was of your parents you were born;
how can you repay what they have given to you?

(Ecclesiasticus 7.28 RSV)

also in memory of María de los Ángeles Santiago Hernández,
a lovely lady and devout Catholic, who died far too young.

and to Lisa.

Dr Simon Cotton obtained his PhD at Imperial College London. After postdoctoral research and teaching appointments at Queen Mary College, London, and the University of East Anglia, he has taught chemistry in several different schools, and has been at Uppingham School since 1996. From 1984 until 1997, he was Editor of Lanthanide and Actinide Compounds for the *Dictionary of Organometallic Compounds* and the *Dictionary of Inorganic Compounds*. He authored the account of Lanthanide Coordination Chemistry for the 2nd edition of *Comprehensive Coordination Chemistry* (Pergamon) as well as the accounts of Lanthanide Inorganic and Coordination Chemistry for both the 1st and 2nd editions of the *Encyclopedia of Inorganic Chemistry* (Wiley).

His previous books are:

S.A. Cotton and F.A. Hart, "*The Heavy Transition Elements*", Macmillan, 1975.
D.J. Cardin, S.A. Cotton, M. Green and J.A. Labinger, "*Organometallic Compounds of the Lanthanides, Actinides and Early Transition Metals*", Chapman and Hall, 1985.
S.A. Cotton, "*Lanthanides and Actinides*", Macmillan, 1991.
S.A. Cotton, "*Chemistry of Precious Metals*", Blackie, 1997.

Contents

Preface

This book is aimed at providing a sound introduction to the chemistry of the lanthanides, actinides and transactinides to undergraduate students. I hope that it will also be of value to teachers of these courses. Whilst not being anything resembling a comprehensive monograph, it does attempt to give a factual basis to the area, and the reader can use a fairly comprehensive bibliography to range further.

Since I wrote a previous book in this area (1991), the reader may wonder why on earth I have bothered again. The world of f-block chemistry has moved on. It is one of active and important research, with names like Bünzli, Evans, Ephritikhine, Lappert, Marks and Parker familiar world-wide (I am conscious of names omitted). Not only have several more elements been synthesized (and claims made for others), but lanthanides and their compounds are routinely employed in many areas of synthetic organic chemistry; gadolinium compounds find routine application in MRI scans; and there are other spectroscopic applications, notably in luminescence. Whilst some areas are hardly changed, at this level at least (e.g. actinide magnetism and spectroscopy), a lot more compounds have been described, accounting for the length of the chapters on coordination and organometallic chemistry. I have tried to spell out the energetics of lanthanide chemistry in more detail, whilst I have provided some end-of-chapter questions, of variable difficulty, which may prove useful for tutorials. I have supplied most, but not all, of the answers to these (my answers, which are not always definitive).

It is a pleasure to thank all those who have contributed to the book: Professor Derek Woollins, for much encouragement at different stages of the project; Professor James Anderson, for many valuable comments on Chapter 8; Martyn Berry, who supplied valuable comment on early versions of several chapters; to Professors Michel Ephritikhine, Allan White and Jack Harrowfield, and Dr J.A.G. Williams, and many others, for exchanging e-mails, correspondence and ideas. I'm very grateful to Dr Mary P. Neu for much information on plutonium. The staff of the Libraries of the Chemistry Department of Cambridge University and of the Royal Society of Chemistry, as well as the British Library, have been quite indispensable in helping with access to the primary literature. I would also wish to thank a number of friends – once again Dr Alan Hart, who got me interested in lanthanides in the first place; Professor James Anderson (again), Dr Andrew Platt, Dr John Fawcett, and Professor Paul Raithby, for continued research collaboration and obtaining spectra and structures from unpromising crystals, so that I have kept a toe-hold in the area. Over the last 8 years, a number of Uppingham 6th form students have contributed to my efforts in lanthanide coordination chemistry – John Bower, Oliver Noy, Rachel How, Vilius Franckevicius, Leon Catallo, Franz Niecknig, Victoria Fisher, Alex Tait and Joanna Harris. Finally, thanks are most certainly due to Dom Paul-Emmanuel Clénet and the Benedictine community of the Abbey of Bec, for continued hospitality during several Augusts when I have been compiling the book.

Simon Cotton

1 Introduction to the Lanthanides

By the end of this chapter you should be able to:

- understand that lanthanides differ in their properties from the s- and d-block metals;
- recall characteristic properties of these elements;
- appreciate reasons for their positioning in the Periodic Table;
- understand how the size of the lanthanide ions affects certain properties and how this can be used in the extraction and separation of the elements;
- understand how to obtain pure samples of individual Ln^{3+} ions.

1.1 Introduction

Lanthanide chemistry started in Scandinavia. In 1794 Johann Gadolin succeeded in obtaining an 'earth'(oxide) from a black mineral subsequently known as gadolinite; he called the earth yttria. Soon afterwards, M.H. Klaproth, J.J. Berzelius and W. Hisinger obtained ceria, another earth, from cerite. However, it was not until 1839–1843 that the Swede C.G. Mosander first separated these earths into their component oxides; thus ceria was resolved into the oxides of cerium and lanthanum and a mixed oxide 'didymia' (a mixture of the oxides of the metals from Pr through Gd). The original yttria was similarly separated into substances called erbia, terbia, and yttria (though some 40 years later, the first two names were to be reversed!). This kind of confusion was made worse by the fact that the newly discovered means of spectroscopic analysis permitted misidentifications, so that around 70 'new' elements were erroneously claimed in the course of the century.

Nor was Mendeleev's revolutionary Periodic Table a help. When he first published his Periodic Table in 1869, he was able to include only lanthanum, cerium, didymium (now known to have been a mixture of Pr and Nd), another mixture in the form of erbia, and yttrium; unreliable information about atomic mass made correct positioning of these elements in the table difficult. Some had not yet been isolated as elements. There was no way of predicting how many of these elements there would be until Henry Moseley (1887–1915) analysed the X-ray spectra of elements and gave meaning to the concept of atomic number. He showed that there were 15 elements from lanthanum to lutetium (which had only been identified in 1907). The discovery of radioactive promethium had to wait until after World War 2.

It was the pronounced similarity of the lanthanides to each other, especially each to its neighbours (a consequence of their general adoption of the +3 oxidation state in aqueous solution), that caused their classification and eventual separation to be an extremely difficult undertaking.

Lanthanide and Actinide Chemistry   S. Cotton
© 2006 John Wiley & Sons, Ltd.

Subsequently it was not until the work of Bohr and of Moseley that it was known precisely how many of these elements there were. Most current versions of the Periodic Table place lanthanum under scandium and yttrium.

1.2 Characteristics of the Lanthanides

The lanthanides exhibit a number of features in their chemistry that differentiate them from the d-block metals. The reactivity of the elements is greater than that of the transition metals, akin to the Group II metals:

1. A very wide range of coordination numbers (generally 6–12, but numbers of 2, 3 or 4 are known).
2. Coordination geometries are determined by ligand steric factors rather than crystal field effects.
3. They form labile 'ionic' complexes that undergo facile exchange of ligand.
4. The 4f orbitals in the Ln^{3+} ion do not participate directly in bonding, being well shielded by the $5s^2$ and $5p^6$ orbitals. Their spectroscopic and magnetic properties are thus largely uninfluenced by the ligand.
5. Small crystal-field splittings and very sharp electronic spectra in comparison with the d-block metals.
6. They prefer anionic ligands with donor atoms of rather high electronegativity (e.g. O, F).
7. They readily form hydrated complexes (on account of the high hydration energy of the small Ln^{3+} ion) and this can cause uncertainty in assigning coordination numbers.
8. Insoluble hydroxides precipitate at neutral pH unless complexing agents are present.
9. The chemistry is largely that of one (3+) oxidation state (certainly in aqueous solution).
10. They do not form $Ln=O$ or $Ln\equiv N$ multiple bonds of the type known for many transition metals and certain actinides.
11. Unlike the transition metals, they do not form stable carbonyls and have (virtually) no chemistry in the 0 oxidation state.

1.3 The Occurrence and Abundance of the Lanthanides

Table 1.1 presents the abundance of the lanthanides in the earth's crust and in the solar system as a whole. (Although not in the same units, the values in each list are internally consistent.)

Two patterns emerge from these data. First, that the lighter lanthanides are more abundant than the heavier ones; secondly, that the elements with even atomic number are more abundant than those with odd atomic number. Overall, cerium, the most abundant lanthanide

Table 1.1 Abundance of the lanthanides

	La	Ce	Pr	Nd	Pm	Sm	Eu	Gd	Tb	Dy	Ho	Er	Tm	Yb	Lu	Y
Crust (ppm)	35	66	9.1	40	0.0	7	2.1	6.1	1.2	4.5	1.3	3.5	0.5	3.1	0.8	31
Solar System (with respect to 10^7 atoms Si)	4.5	1.2	1.7	8.5	0.0	2.5	1.0	3.3	0.6	3.9	0.9	2.5	0.4	2.4	0.4	40.0

on earth, has a similar crustal concentration to the lighter Ni and Cu, whilst even Tm and Lu, the rarest lanthanides, are more abundant than Bi, Ag or the platinum metals.

The abundances are a consequence of how the elements were synthesized by atomic fusion in the cores of stars with heavy elements only made in supernovae. Synthesis of heavier nuclei requires higher temperature and pressures and so gets progressively harder as the atomic number increases. The odd/even alternation (often referred to as the Oddo–Harkins rule) is again general, and reflects the facts that elements with odd mass numbers have larger nuclear capture cross sections and are more likely to take up another neutron, so elements with odd atomic number (and hence odd mass number) are less common than those with even mass number. Even-atomic-number nuclei are more stable when formed.

1.4 Lanthanide Ores

Principal sources (Table 1.2) are the following:

Bastnasite $LnFCO_3$; Monazite $(Ln, Th)PO_4$ (richer in earlier lanthanides); Xenotime $(Y, Ln)PO_4$ (richer in later lanthanides). In addition to these, there are Chinese rare earth reserves which amount to over 70% of the known world total, mainly in the form of the ionic ores from southern provinces. These Chinese ion-absorption ores, weathered granites with lanthanides adsorbed onto the surface of aluminium silicates, are in some cases low in cerium and rich in the heavier lanthanides (Longnan) whilst the Xunwu deposits are rich in the lighter metals; the small particle size makes them easy to mine. The Chinese ores have made them a leading player in lanthanide chemistry.

Table 1.2 Typical abundance of the lanthanides in ores[a]

%	La	Ce	Pr	Nd	Pm	Sm	Eu	Gd	Tb	Dy	Ho	Er	Tm	Yb	Lu	Y
Monazite	20	43	4.5	16	0	3	0.1	1.5	0.05	0.6	0.05	0.2	0.02	0.1	0.02	2.5
Bastnasite	33.2	49.1	4.3	12	0	0.8	0.12	0.17	*160*	*310*	*50*	*35*	*8*	*6*	*1*	0.1
Xenotime	0.5	5	0.7	2.2	0	1.9	0.2	4	1	8.6	2	5.4	0.9	6.2	0.4	60.0

[a] Bold values are in ppm.

1.5 Extracting and Separating the Lanthanides

These two processes are not necessarily coterminous. Whilst electronic, optical and magnetic applications require individual pure lanthanides, the greatest quantity of lanthanides is used as mixtures, e.g. in mischmetal or oxide catalysts.

1.5.1 Extraction

After initial concentration by crushing, grinding and froth flotation, bastnasite is treated with 10% HCl to remove calcite, by which time the mixture contains around 70% lanthanide oxides. This is roasted to oxidize the cerium content to Ce^{IV}; on further extraction with HCl, the Ce remains as CeO_2, whilst the lanthanides in the (+3) state dissolve as a solution of the chlorides.

Monazite is usually treated with NaOH at 150 °C to remove phosphate as Na_3PO_4, leaving a mixture of the hydrated oxides, which are dissolved in boiling HCl at pH 3.5, separating the lanthanides from insoluble ThO_2. Sulfuric acid can also be used to dissolve the lanthanides.

1.5.2   Separating the Lanthanides

These can be divided into four types: chemical separations, fractional crystallization, ion-exchange methods and solvent extraction. Of these, only the last-named is used on a commercial scale (apart from initial separation of cerium). Chemical separations rely on using stabilities of unusual oxidation states; thus Eu^{2+} is the only ion in that oxidation state formed on reduction by zinc amalgam and can then be precipitated as $EuSO_4$ (note the similarity with heavier Group 2 metals). Repeated (and tedious) fractional crystallization, which made use of slight solubility differences between the salts of neighbouring lanthanides, such as the bromates $Ln(BrO_3)_3.9H_2O$, ethyl sulfates and double nitrates, were once the only possible way of obtaining pure lanthanides, as with the 15 000 recrystallizations carried out by the American C. James to get pure thulium bromate (1911) (Figure 1.1 indicates the principle of this method).

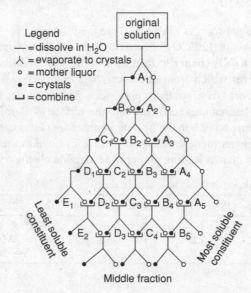

Figure 1.1
Diagrammatic representation of the system of fractional crystallization used to separate salts of the rare-earth elements (reproduced with permission from D.M. Yost, H. Russell and C.S. Garner, The Rare Earth Elements and their Compounds, John Wiley, 1947.)

Ion-exchange chromatography is not of real commercial importance for large-scale production, but historically it was the method by which fast high-purity separation of the lanthanides first became feasible. As radioactive lanthanide isotopes are important fission products of the fission of ^{235}U and therefore need to be separated from uranium, and because the actinides after plutonium tend to resemble the lanthanides, the development of the technique followed on the Manhattan project. It was found that if Ln^{3+} ions were adsorbed at the top of a cation-exchange resin, then treated with a complexing agent such as buffered citric acid, then the cations tended to be eluted in reverse atomic number order (Figure 1.2a); the anionic ligand binds most strongly to the heaviest (and smallest) cation, which has the highest charge density. A disadvantage of this approach when scaled up to high concentration is that the peaks tend to overlap (Figure 1.2b).

It was subsequently found that amine polycarboxylates such as $EDTA^{4-}$ gave stronger complexes and much better separations. In practice, some Cu^{2+} ions ('retainer') are added to prevent precipitation of either the free acid H_4EDTA or the lanthanide complex

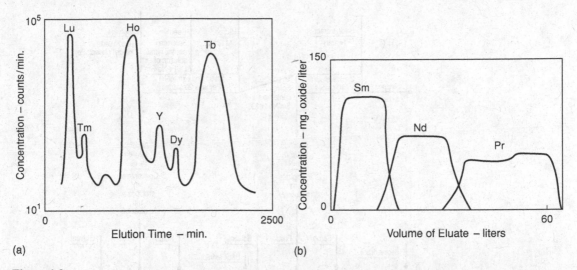

Figure 1.2
(a) Cation-exchange chromatography of lanthanides, (b) overlap of peaks at high concentration. (a) Tracer-scale elution with 5% citrate at pH 3.20 (redrawn from B.H. Ketelle and G.E. Boyd, *J. Am. Chem. Soc.*, 1947, **69**, 2800). (b) Macro-scale elution with 0.1% citrate at pH 5.30 (redrawn from F.H. Spedding, E.I. Fulmer, J.E. Powell, and T.A. Butler, *J. Am. Chem. Soc.*, 1950, **72**, 2354). Reprinted with permission of the American Chemical Society ©1978.

HLn(EDTA).xH$_2$O on the resin. The major disadvantage of this method is that it is a slow process for large-scale separations.

Solvent extraction has come to be used for the initial stage of the separation process, to give material with up to 99.9% purity. In 1949, it was found that Ce^{4+} could readily be separated from Ln^{3+} ions by extraction from a solution in nitric acid into tributyl phosphate [(BuO)$_3$PO]. Subsequently the process was extended to separating the lanthanides, using a non-polar organic solvent such as kerosene and an extractant such as (BuO)$_3$PO or bis (2-ethylhexyl)phosphinic acid [[C$_4$H$_9$CH(C$_2$H$_5$)]$_2$P=O(OH)] to extract the lanthanides from aqueous nitrate solutions. The heavier lanthanides form complexes which are more soluble in the aqueous layer. After the two immiscible solvents have been agitated together and separated, the organic layer is treated with acid and the lanthanide extracted. The solvent is recycled and the aqueous layer put through further stages.

For a lanthanide Ln$_A$ distributed between two phases, a distribution coefficient D_A is defined:

$$D_A = [\text{Ln}_A \text{ in organic phase}] / [\text{Ln}_A \text{ in aqueous phase}]$$

For two lanthanides Ln$_A$ and Ln$_B$ in a mixture being separated, a separation factor β_B^A can be defined, where

$$\beta_B^A = D_A/D_B$$

β is very close to unity for two adjacent lanthanides in the Periodic Table (obviously, the larger β is, the better the separation).

In practice this process is run using an automated continuous counter-current circuit in which the organic solvent flows in the opposite direction to the aqueous layer containing the lanthanides. An equilibrium is set up between the lanthanide ions in the aqueous phase and the organic layer, with there tending to be a relative enhancement of the concentration of the heavier lanthanides in the organic layer. Because the separation between adjacent

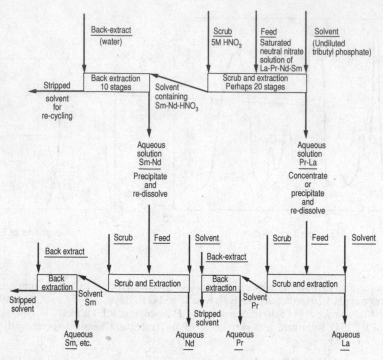

Figure 1.3
Schematic diagram of lanthanon separation by solvent extraction. From R.J. Callow, *The Rare Earth Industry*, Pergamon, 1966; reproduced by permission.

lanthanides in each exchange is relatively slight, over a thousand exchanges are used (see Figure 1.3). This method affords lanthanides of purity up to the 99.9% purity level and is thus well suited to large-scale separation, the products being suited to ordinary chemical use. However, for electronic or spectroscopic use ('phosphor grade') 99.999% purity is necessary, and currently ion-exchange is used for final purification to these levels. The desired lanthanides are precipitated as the oxalate or hydroxide and converted into the oxides (the standard starting material for many syntheses) by thermal decomposition.

Various other separation methods have been described, one recent one involving the use of supercritical carbon dioxide at 40 °C and 100 atm to convert the lanthanides into their carbonates whilst the quadrivalent metals (e.g. Th and Ce) remain as their oxides.

1.6   The Position of the Lanthanides in the Periodic Table

As already mentioned, neither Mendeleev nor his successors could 'place' the lanthanides in the Periodic Table. Not only was there no recognizable atomic theory until many years afterwards, but, more relevant to how groupings of elements were made in those days, there was no comparable block of elements for making comparisons. The lanthanides were *sui generis*. The problem was solved by the combined (but separate) efforts of Moseley and Bohr, the former showing that La–Lu was composed of 15 elements with atomic numbers from 57 to 71, whilst the latter concluded that the fourth quantum shell could accommodate 32 electrons, and that the lanthanides were associated with placing electrons into the 4f orbitals.

The Periodic Table places elements in atomic number order, with the lanthanides falling between barium (56) and hafnium (72). For reasons of space, most present-day Periodic Tables are presented with Groups IIA and IVB (2 and 4) separated only by the Group IIIB (3) elements. Normally La (and Ac) are grouped with Sc and Y, but arguments have been advanced for an alternative format, in which Lu (and Lr) are grouped with Sc and Y (see e.g. W.B. Jensen, *J. Chem. Educ.*, 1982, **59**, 634) on the grounds that trends in properties (e.g. atomic radius, I.E., melting point) in the block Sc-Y-Lu parallel those in the Group Ti-Zr-Hf rather closely, and that there are resemblances in the structures of certain binary compounds. Certainly on size grounds, Lu resembles Y and Sc (it is intermediate in size between them) rather more than does La, owing to the effects of the 'lanthanide contraction'. The resemblances between Sc and Lu are, however, by no means complete.

1.7 The Lanthanide Contraction

The basic concept is that there is a decrease in radius of the lanthanide ion Ln^{3+} on crossing the series from La to Lu. This is caused by the poor screening of the 4f electrons. This causes neighbouring lanthanides to have similar, but not identical, properties, and is discussed in more detail in Section 2.4.

Question 1.1 Using the information you have been given in Section 1.2, draw up a table comparing (in three columns) the characteristic features of the s-block metals (use group 1 as typical) and the d-block transition metals.

Answer 1.1 see Table 1.3 for one such comparison.

Table 1.3 Comparison of 4f, 3d and Group I metals

	4f	3d	Group I
Electron configurations of ions	Variable	Variable	Noble gas
Stable oxidation states	Usually +3	Variable	1
Coordination numbers in complexes	Commonly 8–10	Usually 6	Often 4–6
Coordination polyhedra in complexes	Minimise repulsion	Directional	Minimise repulsion
Trends in coordination numbers	Often constant in block	Often constant in block	Increase down group
Donor atoms in complexes	'Hard' preferred	'Hard' and 'soft'	'Hard' preferred
Hydration energy	High	Usually moderate	Low
Ligand exchange reactions	Usually fast	Fast and slow	Fast
Magnetic properties of ions	Independent of environment	Depends on environment and ligand field	None
Electronic spectra of ions	Sharp lines	Broad lines	None
Crystal field effects in complexes	Weak	Strong	None
Organometallic compounds	Usually ionic, some with covalent character	Covalently bonded	Ionically bonded
Organometallics in low oxidation states	Few	Common	None
Multiply bonded atoms in complexes	None	Common	None

2 The Lanthanides – Principles and Energetics

By the end of this chapter you should be able to:

- recognise the difference between f-orbitals and other types of orbitals;
- understand that they are responsible for the particular properties of the lanthanides;
- give the electron configurations of the lanthanide elements and Ln^{3+} ions;
- explain the reason for the lanthanide contraction;
- understand the effect of the lanthanide contraction upon properties of the lanthanides and subsequent elements;
- explain patterns in properties such as ionization and hydration energies;
- recall that lanthanides behave similarly when there is no change in the 4f electron population, but that they differ when the change involves a change in the number of 4f electrons;
- relate the stability of oxidation states to the ionization energies;
- calculate enthalpy changes for the formation of the aqua ions and of the lanthanide halides and relate these to the stability of particular compounds.

2.1 Electron Configurations of the Lanthanides and f Orbitals

The lanthanides (and actinides) are those in which the 4f (and 5f) orbitals are gradually filled. At lanthanum, the 5d subshell is lower in energy than 4f, so lanthanum has the electron configuration [Xe] $6s^2 5d^1$ (Table 2.1).

As more protons are added to the nucleus, the 4f orbitals contract rapidly and become more stable than the 5d (as the 4f orbitals penetrate the 'xenon core' more) (see Figure 2.1), so that Ce has the electron configuration [Xe] $6s^2 5d^1 4f^1$ and the trend continues with Pr having the arrangement [Xe] $6s^2 4f^3$. This pattern continues for the metals Nd–Eu, all of which have configurations [Xe] $6s^2 4f^n$ (n = 4–7) After europium, the stability of the half-filled f subshell is such that the next electron is added to the 5d orbital, Gd being [Xe] $6s^2 5d^1 4f^7$; at terbium, however, the earlier pattern is resumed, with Tb having the configuration [Xe] $6s^2 4f^9$, and succeeding elements to ytterbium being [Xe] $6s^2 4f^n$ (n = 10–14). The last lanthanide, lutetium, where the 4f subshell is now filled, is predictably [Xe] $6s^2 5d^1 4f^{14}$.

Lanthanide and Actinide Chemistry   S. Cotton
© 2006 John Wiley & Sons, Ltd.

Table 2.1   Electron configurations of the lanthanides and their common ions

	Atom	Ln^{3+}	Ln^{4+}	Ln^{2+}
La	[Xe] $5d^1$ $6s^2$	[Xe]		
Ce	[Xe] $4f^1$ $5d^1$ $6s^2$	[Xe] $4f^1$	[Xe]	
Pr	[Xe] $4f^3$ $6s^2$	[Xe] $4f^2$	[Xe] $4f^1$	
Nd	[Xe] $4f^4$ $6s^2$	[Xe] $4f^3$	[Xe] $4f^2$	[Xe] $4f^4$
Pm	[Xe] $4f^5$ $6s^2$	[Xe] $4f^4$		
Sm	[Xe] $4f^6$ $6s^2$	[Xe] $4f^5$		[Xe] $4f^6$
Eu	[Xe] $4f^7$ $6s^2$	[Xe] $4f^6$		[Xe] $4f^7$
Gd	[Xe] $4f^7$ $5d^1$ $6s^2$	[Xe] $4f^7$		
Tb	[Xe] $4f^9$ $6s^2$	[Xe] $4f^8$	[Xe] $4f^7$	
Dy	[Xe] $4f^{10}$ $6s^2$	[Xe] $4f^9$	[Xe] $4f^8$	[Xe] $4f^{10}$
Ho	[Xe] $4f^{11}$ $6s^2$	[Xe] $4f^{10}$		
Er	[Xe] $4f^{12}$ $6s^2$	[Xe] $4f^{11}$		
Tm	[Xe] $4f^{13}$ $6s^2$	[Xe] $4f^{12}$		[Xe] $4f^{13}$
Yb	[Xe] $4f^{14}$ $6s^2$	[Xe] $4f^{13}$		[Xe] $4f^{14}$
Lu	[Xe] $4f^{14}$ $5d^1$ $6s^2$	[Xe] $4f^{14}$		
Y	[Kr] $4d^1$ $5s^2$	[Kr]		

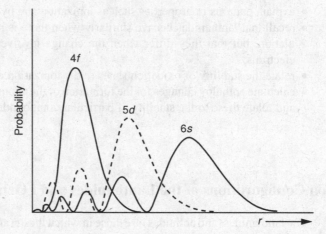

Figure 2.1
The radial part of the hydrogenic wave functions for the 4f, 5d and 6s orbitals of cerium (after H.G. Friedman *et al. J. Chem. Educ.* 1964, **41**, 357). Reproduced by permission of the American Chemical Society © 1964.

2.2   What do f Orbitals Look Like?

They are generally represented in one of two ways, either as a cubic set, or as a general set, depending upon which way the orbitals are combined. The cubic set comprises f_{xyz}; $f_{z(x2-y2)}$, $f_{z(y2-z2)}$ and $f_{y(z2-x2)}$; f_{z3}, f_{x3} and f_{y3}.

The general set, more useful in non-cubic environments, uses a different combination: f_{z3}; f_{xz2} and f_{yz2}; f_{xyz}; $f_{z(x2-y2)}$, $f_{x(x2-3y2)}$ and $f_{y(3x2-y2)}$; Figure 2.2 shows the general set.

2.3   How f Orbitals affect Properties of the Lanthanides

The 4f orbitals penetrate the xenon core appreciably. Because of this, they cannot overlap with ligand orbitals and therefore do not participate significantly in bonding. As a result of

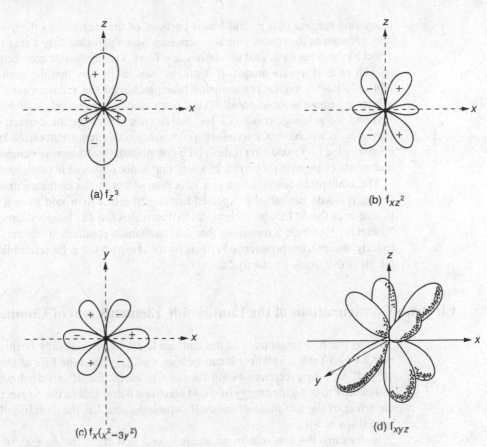

(a) f_{z^3}

(b) f_{xz^2}

(c) $f_{x(x^2-3y^2)}$

(d) f_{xyz}

Figure 2.2
(a) f_{z^3}, (f_{x^3} and f_{y^3} are similar, extending along the x- and y-axes repectively); (b) f_{xz^2}, (f_{yz^2} is similar, produced by a 90° rotation about the z-axis); (c) $f_{x(x^2-3y^2)}$, $f_{y(3x^2-y^2)}$ is similar, formed by a 90° clockwise rotation round the z-axis); (d) f_{xyz}, ($f_{xz^2-y^2}$), $f_{y(z^2-y^2)}$ and $f_{z(x^2-y^2)}$ are produced by a 45° rotation about the x, y and z-axes respectively). The cubic set comprises f_{x^3}, f_{y^3}, f_{z^3}, f_{xyz}, $f_{x(z^2-y^2)}$ $f_{y(z^2-x^2)}$ and $f_{z(x^2-y^2)}$; the general set is made of f_{z^3}, f_{xz^2}, f_{yz^2}, f_{xyz}, $f_{z(x^2-y^2)}$, $f_{x(x^2-3y^2)}$ and $f_{y(3x^2-y^2)}$. (Reproduces with permission from S.A. Cotton, *Lanthanides and Actinides*, Macmillan, 1991).

their isolation from the influence of the ligands, crystal-field effects are very small (and can be regarded as a perturbation on the free-ion states) and thus electronic spectra and magnetic properties are essentially unaffected by environment. The ability to form π bonds is also absent, and thus there are none of the M=O or M≡N bonds found for transition metals (or, indeed, certain early actinides). The organometallic chemistry is appreciably different from that of transition metals, too.

2.4   The Lanthanide Contraction

As the series La–Lu is traversed, there is a decrease in both the atomic radii and in the radii of the Ln^{3+} ions, more markedly at the start of the series. The 4f electrons are 'inside' the 5s and 5p electrons and are core-like in their behaviour, being shielded from the ligands, thus taking no part in bonding, and having spectroscopic and magnetic properties largely independent of environment. The 5s and 5p orbitals penetrate the 4f subshell and are not shielded from increasing nuclear charge, and hence because of the increasing effective nuclear charge they contract as the atomic number increases. Some part (but only a small fraction) of this effect has also been ascribed to relativistic effects. The lanthanide contraction is sometimes spoken of as if it were unique. It is not, at least in the way that the term is usually used. Not only does a similar phenomenon take place with the actinides (and here relativistic effects are much more responsible) but contractions are similarly noticed on crossing the first and second long periods (Li–Ne; Na–Ar) not to mention the d-block transition series. However, as will be seen, because of a combination of circumstances, the lanthanides adopt primarily the (+3) oxidation state in their compounds, and therefore demonstrate the steady and subtle changes in properties in a way that is not observed in other blocks of elements.

The lanthanide contraction has a knock-on effect in the elements in the 5d transition series. It would naturally be expected that the 5d elements would show a similar increase in size over the 4d transition elements to that which the 4d elements demonstrate over the 3d metals. However, it transpires that the 'lanthanide contraction' cancels this out, almost exactly, and this has pronounced effects on the chemistry, e.g. Pd resembling Pt rather than Ni, Hf is extremely similar to Zr.

2.5   Electron Configurations of the Lanthanide Elements and of Common Ions

The electronic arrangements of the lanthanide atoms have already been mentioned (Section 2.1 and Table 2.1) where it can be seen that in general the ECs of the atoms are [Xe] $4f^n$ $5d^0$ $6s^2$, the exceptions being La and Ce, where the 4f orbitals have not contracted sufficiently to bring the energy of the 4f electrons below that of the 5d electrons; Gd, where the effect of the half-filled 4f subshell dominates; and Lu, the 4f subshell having already been filled at Yb.

In forming the ions, electrons are removed first from the 6s and 5d orbitals (rather reminiscent of the case of the transition metals, where they are removed from 4s before they are taken from 3d), so that all the Ln^{3+} ions have [Xe] $4f^n$ arrangements.

2.6   Patterns in Ionization Energies

The values of the first four ionization energies for the lanthanides (and yttrium) are listed in Table 2.2.

As usual, for a particular element, $I_4 > I_3 > I_2 > I_1$, as the electron being removed is being taken from an ion with an increasingly positive charge, affording greater electrostatic attraction. Yttrium has greater ionization energies than the lanthanides as there is one fewer filled shell, and decreased distance effects outweigh the effect of the reduced nuclear charge.

There is in general a tendency for ionization energies to increase on crossing the series but it is irregular. The low I_3 values for gadolinium and lutetium, where the one electron removed comes from a d orbital, not an f orbital, and the high I_3 value for Eu and Yb show some correlation with the stabilizing effects of half-filled and filled f sub shells. Results can be presented diagrammatically as cumulative ionization energies (Figure 2.3). These diagrams demonstrate at a glance the effect of the alternately high and low values of I_3 for the elements Eu and Gd, influencing the respective stabilities of the +2 and +3 states of these elements, similarly illustrated by the neighbours Yb and Lu. The low I_4 value for Ce (and to some extent for Pr and Tb) can be correlated with the accessibility of the Ce^{4+} ion.

Table 2.2 Ionization Energies (kJ/mole)

	I_1	I_2	I_3	I_4	$I_1 + I_2$	$I_1 + I_2 + I_3$	$I_1 + I_2 + I_3 + I_4$
La	538	1067	1850	4819	1605	3455	8274
Ce	527	1047	1949	3547	1574	3523	7070
Pr	523	1018	2086	3761	1541	3627	7388
Nd	529	1035	2130	3899	1564	3694	7593
Pm	536	1052	2150	3970	1588	3738	7708
Sm	543	1068	2260	3990	1611	3871	7990
Eu	546	1085	2404	4110	1631	4035	8145
Gd	593	1167	1990	4250	1760	3750	8000
Tb	564	1112	2114	3839	1676	3790	7629
Dy	572	1126	2200	4001	1698	3898	7899
Ho	581	1139	2204	4110	1720	3924	8034
Er	589	1151	2194	4115	1740	3934	8049
Tm	597	1163	2285	4119	1760	4045	8164
Yb	603	1176	2415	4220	1779	4194	8414
Lu	523	1340	2033	4360	1863	3896	8256
Y	616	1181	1980	5963	1797	3777	9740

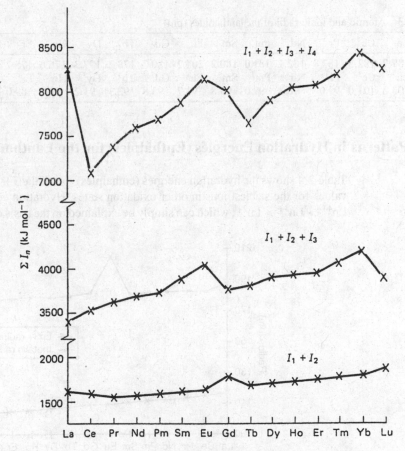

Figure 2.3
Cumulative ionization energies across the lanthanide Series (reproduced by permission of Macmillan from S.A. Cotton, *Lanthanides and Actinides*, Macmillan, 1991).

2.7   Atomic and Ionic Radii

These are listed in Table 2.3 and shown in Figure 2.4. It will be seen that the atomic radii exhibit a smooth trend across the series with the exception of the elements europium and ytterbium. Otherwise the lanthanides have atomic radii intermediate between those of barium in Group 2A and hafnium in Group 4A, as expected if they are represented as Ln^{3+} $(e^-)_3$. Because the screening ability of the f electrons is poor, the effective nuclear charge experienced by the outer electrons increases with increasing atomic number, so that the atomic radius would be expected to decrease, as is observed. Eu and Yb are exceptions to this; because of the tendency of these elements to adopt the (+2) state, they have the structure $[Ln^{2+}(e^-)_2]$ with consequently greater radii, rather similar to barium. In contrast, the ionic radii of the Ln^{3+} *ions* exhibit a smooth decrease as the series is crossed.

The patterns in radii exemplify a principle enunciated by D.A. Johnson: 'The lanthanide elements behave similarly in reactions in which the 4f electrons are conserved, and very differently in reactions in which the number of 4f electrons change' (*J. Chem. Educ.*, 1980, **57**, 475).

Table 2.3   Atomic and ionic radii of the lanthanides (pm)

Ba	La	Ce	Pr	Nd	Pm	Sm	Eu	Gd	Tb	Dy	Ho	Er	Tm	Yb	Lu	Hf
217.3	187.7	182.5	182.8	182.1	181.0	180.2	204.2	180.2	178.2	177.3	176.6	175.7	174.6	194.0	173.4	156.4
	La^{3+}	Ce^{3+}	Pr^{3+}	Nd^{3+}	Pm^{3+}	Sm^{3+}	Eu^{3+}	Gd^{3+}	Tb^{3+}	Dy^{3+}	Ho^{3+}	Er^{3+}	Tm^{3+}	Yb^{3+}	Lu^{3+}	Y^{3+}
	103.2	101.0	99.0	98.3	97.0	95.8	94.7	93.8	92.3	91.2	90.1	89.0	88.0	86.8	86.1	90.0

2.8   Patterns in Hydration Energies (Enthalpies) for the Lanthanide Ions

Table 2.4 shows the hydration energies (enthalpies) for all the 3+ lanthanide ions, and also values for the stablest ions in other oxidation states. Hydration energies fall into a pattern $Ln^{4+} > Ln^{3+} > Ln^{2+}$, which can simply be explained on the basis of electrostatic attraction,

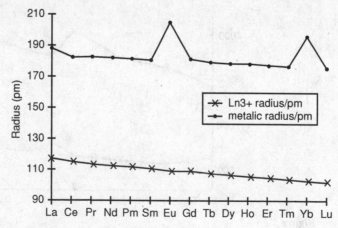

Figure 2.4
Metallic and ionic radii across the lanthanide series.

since the ions with a larger charge have a greater charge density. The hydration energies for the Ln^{3+} ions show a pattern of smooth increase with increasing atomic number, as the ions becomes smaller and their attraction for water molecules increases. This is another example of the principle enunciated by Johnson.

2.9 Enthalpy Changes for the Formation of Simple Lanthanide Compounds

Observed patterns of chemical behaviour can sometimes appear confusing or sometimes just be encapsulated in rules [e.g. Ce has a stable (+4) oxidation state]. They can frequently be explained in terms of the energy changes involved in the processes.

2.9.1 Stability of Tetrahalides

Among the fluorides, lanthanum forms only LaF_3, whilst cerium forms CeF_3 and CeF_4. (Tetrafluorides are also known for Pr and Tb, see Section 3.4). Why do neighbouring metals behave so differently? First, examine the energetics of formation of LaF_3 using a Born–Haber cycle.

	ΔH (kJ/mol)
$La(s) \rightarrow La(g)$	$+402$
$La(g) \rightarrow La^{3+}(g) + 3\,e^-$	$+538 + 1067 + 1850$
$3/2\,F_2(g) \rightarrow 3F\,(g)$	$+252$
$3F(g) + 3\,e^- \rightarrow 3F^-(g)$	-984
$La^{3+}(g) + 3F^-(g) \rightarrow LaF_3$	-4857

Thus $La(s) + 3/2\,F_2(g) \rightarrow LaF_3$

$\Delta H = +402 + (538 + 1067 + 1850) + 252 - 984 - 4857 = -1732 \text{ kJ/mol}$

Calculated and observed enthalpies of formation for LnX_n ($n = 2$–4) are given in Table 2.5.

The same method can be used to calculate ΔH_f for LaF_4, making the assumption that the lattice energy is the same as that for CeF_4 (-8391 kJ/mol), and that I_4 for lanthanum is $+4819$ kJ/mol. In this case, $\Delta H_f(LaF_4) = -691$ kJ/mol. Similarly, using the same method, ΔH_f for LaF_2 can be calculated as $\Delta H = -831$ kJ/mol (assuming that the lattice energy is the same as for BaF_2 (-2350 kJ/mol). This poses the question: if ΔH_f for LaF_2 and for LaF_4 are -831 and -691 kJ/mol respectively, why can't these compounds be isolated?

Apart from the oversimplification of using ΔH rather than ΔG values as an index of stability, what the preceding calculations have done is to indicate that these compounds are stable with respect to the elements, and *no other decomposition pathways have been considered.*

Table 2.4 Enthalpies of hydration of the lanthanide ions (values given as $-\Delta H$ hydr/kj mol^{-1}

La^{3+}	Ce^{3+}	Pr^{3+}	Nd^{3+}	Pm^{3+}	Sm^{3+}	Eu^{3+}	Gd^{3+}	Tb^{3+}	Dy^{3+}	Ho^{3+}	Er^{3+}	Tm^{3+}	Yb^{3+}	Lu^{3+}	Y^{3+}
3278	3326	3373	3403	3427	3449	3501	3517	3559	3567	3623	3637	3664	3706	3722	3583
	Ce^{4+}				Sm^{2+}	Eu^{2+}							Yb^{2+}		
	6309				1444	1458							1594		

Table 2.5 Enthalpies of formation of lanthanide halides[a]

	LnF$_2$	LnF$_3$	LnF$_4$	LnCl$_2$	LnCl$_3$	LnCl$_4$	LnBr$_2$	LnBr$_3$	LnI$_2$	LnI$_3$
La	880	*1732*	600	520	*1073*	−480	430	*907*	320	*699*
Ce	950	*1733*	*1946*	580	*1058*	820	490	890	380	*686*
Pr	1050	*1712*	1690	700	*1059*	630	610	*891*	490	*678*
Nd	1050	*1661*	1500	*707*	*1042*	490	630	*873*	510	*665*
Pm	1080	1700	1500	720	1040	430	620	850	510	640
Sm	*1160*	*1669*	1450	820	*1040*	400	720	*857*	600	*640*
Eu	*1188*	*1584*	1290	*824*	*1062*	230	760	*799*	630	*540*
Gd	870	*1699*	1310	500	*937*	180	400	*829*	290	*619*
Tb	980	*1707*	*1742*	600	*1008*	600	500	850	390	*598*
Dy	1050	*1678*	1540	*693*	*1007*	420	590	*831*	470	*603*
Ho	1040	*1698*	1500	660	*990*	350	580	830	450	*594*
Er	1020	*1699*	1510	640	*995*	360	560	*836*	420	*586*
Tm	1090	*1689*	1500	*709*	*995*	360	640	840	500	*582*
Yb	*1172*	*1570*	1250	*799*	*960*	250	710	800	580	*561*
Lu	790	*1640*	1210	420	*986*	160	330	850	190	*556*

[a] Values are quoted as $-\Delta H_f$ (kJ/mol). Experimental values in bold. Values taken from D.W. Smith, *J. Chem. Educ.*, 1986, **63**, 228. Although values calculated in the text for lanthanum and cerium halides differ slightly, older values are retained here for consistency.

LaF$_4$ might decompose thus :

$$\text{LaF}_4 \rightarrow \text{LaF}_3 + \tfrac{1}{2} \text{F}_2$$

Applying Hess's Law to this in the form

$$\Delta H_{\text{reaction}} = \Sigma \Delta H_f(\text{products}) - \Sigma \Delta H_f(\text{reactants})$$
$$\Delta H_{\text{reaction}} = (-1732 + 0) - (-691) = -1041 \text{ kJ/mol}$$

This decomposition is thus thermodynamically favourable (especially as it would be favoured on entropy grounds too, with the formation of fluorine gas)

LaF$_2$ might decompose by disproportionation:

$$3\text{LaF}_2 \rightarrow 2\text{LaF}_3 + \text{La}$$

Using ΔH_f for LaF$_2$ and for LaF$_3$ (−831 and −1732 kJ/mol respectively), ΔH for this decomposition reaction can be calculated as −971 kJ/mol. This is again a very exothermic process, indicating that LaF$_2$ is likely to be unstable (this is analogous to the reason for the non-existence of MgCl, as disproportionation into Mg and MgCl$_2$ is favoured, as the reader may be aware). The reason for this is that, although I_3 for lanthanum is large (and endothermic), it is more than compensated for by the higher lattice energy for LaF$_3$ compared with the value for LaF$_2$.

Since both CeF$_3$ and CeF$_4$ are isolable, what makes the difference here? If similar calculations are carried out (assuming lattice energies for CeF$_3$ and CeF$_4$ of −4915 and −8391 kJ/mol respectively, and an enthalpy of atomization of 398 kJ/mol for Ce (see Table 2.6), and using the same enthalpy of atomization and electron affinity for fluorine as in the lanthanum examples), ΔH_f for CeF$_3$ can be calculated as −1726 kJ/mol and ΔH_f for CeF$_4$ as −1899 kJ/mol.

The discrepancy between ΔH_f for CeF$_3$ and CeF$_4$ is much smaller than is the case for lanthanum. In fact, for the decomposition reaction

$$\text{CeF}_4 \rightarrow \text{CeF}_3 + \tfrac{1}{2} \text{F}_2$$

Table 2.6 Enthalpies of atomization of the lanthanides (kJ/mol)

Ba	La	Ce	Pr	Nd	Pm	Sm	Eu	Gd	Tb	Dy	Ho	Er	Tm	Yb	Lu	Hf
150.9	402.1	398	357	328		164.8	176	301	391	293	303	280	247	159	428.0	570.7

$\Delta H = -1726 - (-1899) = +173$ kJ/mol, making the decomposition of CeF_4 relatively unfavourable. CeF_4 is stable, certainly at ambient temperatures.

If the values of the parameters used in the calculation are compared, the determining factor is the much lower value of I_4 for cerium (3547 kJ/mol compared with 4819 kJ/mol for La) due to the fact that the fourth electron is being removed from a different shell (nearer the nucleus) in the case of lanthanum.

Other tetrahalides do not exist. Thus, though both $CeCl_3$ and salts of the $[CeCl_6]^{2-}$ ion can be isolated, $CeCl_4$ cannot be made. The reasons for this are those that enable fluorine to support high oxidation states (see Question 2.4). Similar factors indicate that tetrabromides and tetraiodides are much less likely to be isolated.

2.9.2 Stability of Dihalides

The stability of the dihalides can be explained in a similar way. As already noted, LaF_2 does not exist. When might dihalides be expected?

One way in which dihalides can decompose is disproportionation

$$3\,LnX_2 \rightarrow 2\,LnX_3 + Ln$$

though a number of dihalides (particularly dichlorides) have been made by the reverse reproportionation, by heating the mixture to a high temperature and rapidly quenching (Section 3.5.1.)

$$2\,LnX_3 + Ln \rightarrow 3\,LnX_2$$

It can be shown, using Hess's Law, that the disproportionation will be exothermic unless:

$$\Delta H_f\,LnX_3/\Delta H_f\,LnX_2 < 1.5$$

The disproportionation can be broken down into individual components (Figure 2.5)

$$\Delta H_1 = \Delta H_2 + \Delta H_3 + \Delta H_4$$
$$\Delta H_1 = [-3\Delta H_{latt}(LnX_2)] + \{2\,I_3(Ln) - [I_1(Ln) + I_2(Ln)]\}$$
$$+ [-\Delta H_{at}(Ln) + 2\,\Delta H_{latt}(LnX_3)]$$
$$\Delta H_1 = \{2\,I_3(Ln) - [I_1(Ln) + I_2(Ln)]\} + 2\,\Delta H_{latt}(LnX_3) - 3\,\Delta H_{latt}(LnX_2) - \Delta H_{at}(Ln)$$

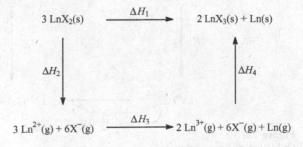

Figure 2.5
Disproportion of lanthanide dihalides.

These equations can be used qualitatively, first to suggest for which lanthanides the halides LnX_2 are most likely to be stable. For the disproportionation process to be more likely to be endothermic (i.e. stabilizing the +2 state), the preceding equation suggests that high values of I_3 are favourable. Study of Table 2.2 shows that this is more likely to be associated with Eu, Sm and Yb. Iodide is the halide most often found in low oxidation state halides. In the case of LnI_2, the large size of the iodide ion will reduce the lattice energy for both LnI_2 and LnI_3 so that the *difference* in lattice enthalpy will become less significant, favouring LnI_2. The dihalide will also be favoured by a low $\Delta H_{at}(Ln)$, again associated with the lanthanides most often found in the +2 state, Eu and Yb (see Table 2.6).

Applying the $\Delta H_f\ LnX_3/\Delta H_f\ LnX_2 <1.5$ criterion, and using data in Table 2.5, the halides LnX_2 are most likely to be stable are predicted to be LnF_2 (Ln = Sm, Eu, Yb); $LnCl_2$(Ln = Nd, Pm, Sm, Eu, Dy, Tm, Yb); LnY_2 (Y = Br, I; Ln = Pr, Nd, Pm, Sm, Eu, Dy, Ho, Er, Tm, Yb).

The known dihalides are listed in Table 3.2. There is a reasonably good correlation, given that the Pm dihalides have not been investigated on account of promethium's short half-life. Some dihalides listed are 'metallic' and are not covered by this argument.

2.9.3   Stability of Aqua Ions

Since Ln^{3+}(aq) is the most stable aqua ion, then both of the following processes are favoured.

$$Ln^{2+}(aq) + H^+(aq) \rightarrow Ln^{3+}(aq) + 1/2\ H_2(g)$$
$$\text{and}\ 2\ Ln^{4+}(aq) + H_2O(aq) \rightarrow 2\ Ln^{3+}(aq) + 2\ H^+(aq) + 1/2\ O_2(g)$$

In other words, Ln^{2+} ions tend to reduce water and Ln^{4+} ions tend to oxidize it. We can examine the stability of the Ln^{2+} ion using a treatment similar to the one just employed (Figure 2.6), with

$$\Delta H_{ox}(Ln^{2+}) = I_3 + \left\{[\Delta H_{hydr}[Ln^{3+}(aq)] - \Delta H_{hydr}[Ln^{2+}(aq)]\right\} - 439\ kJ/mol.$$

When is it most likely that Ln^{2+}(aq) ions will be stable? For the first of the two reactions above to be favoured, the single factor that will help make ΔH positive is a high value of I_3. Less important would be the size of the ions, as this could affect the hydration enthalpies; the difference between the hydration enthalpies will be less, the larger the lanthanide ions. Substituting into the above equation, we can investigate the relative stabilities of La^{2+}(aq) and Eu^{2+}(aq), making use of ionization energies from

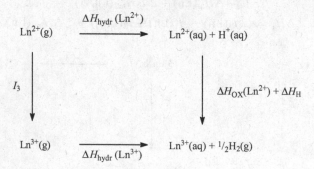

Figure 2.6
Oxidation of Ln^{2+} (aq).

Table 2.2 and enthalpies of hydration found in Table 2.4, also assuming ΔH_{hydr} [(La^{2+}(aq)] $= -1327$ kJ/mol:

For La^{2+}:

$$\Delta H_{ox}(La^{2+}) = I_3 + \{\Delta H_{hydr}[La^{3+}(aq)] - \Delta H_{hydr}[La^{2+}(aq)]\} - 439$$
$$= 1850 + [-3278 - (-1327)] - 439$$
$$= -540 \text{ kJ/mol}$$

For Eu^{2+}:

$$\Delta H_{ox}(Eu^{2+}) = I_3 + \{\Delta H_{hydr}[Eu^{3+}(aq)] - \Delta H_{hydr}[Eu^{2+}(aq)]\} - 439$$
$$= 2404 + [-3501 - (-1458)] - 439$$
$$= -78 \text{ kJ/mol}$$

The large exothermic value for ΔH_{ox} (La^{2+}) indicates that it is not likely to exist in aqueous solution. The Eu^{2+}(aq) ion is known to have a short lifetime in water, even though ΔH is negative for the oxidation process, so the activation energy for oxidation may be rather high.

2.10 Patterns in Redox Potentials

Known and estimated values are listed in Table 2.7. The values for the reduction potential for Ln^{3+} + 3 e$^-$ → Ln are very consistent, with slight irregularities at Eu and Yb. The potential largely depends upon three processes:

$$Ln(s) \rightarrow Ln(g) \quad \Delta H_{at}(Ln)$$
$$Ln(g) \rightarrow Ln^{3+}(g) + 3 \text{ e}^- \quad I_1 + I_2 + I_3$$
$$Ln^{3+}(g) \rightarrow Ln^{3+}(aq) \quad \Delta H_{hydr}(Ln^{3+})$$

The first two of these are endothermic and the third exothermic; overall ΔH is the difference between two large quantities. It remains fairly constant across the series, apart from Eu and Yb, with values of 608 (La); 712 (Eu); 630 (Gd); 613 (Tm); 644 (Yb) and 593 (Lu) kJ/mol being representative, the values for Eu and Yb resulting from the high I_3 values. The very negative value for the reduction potential is expected for such reactive metals (and also reflects the difficulty in isolating them). The potentials for the Ln^{3+} + e$^-$ → Ln^{2+} process reflect the stability of the +2 state. Since I_3 relates to ΔH for the process, and $\Delta G = -n$FE, a relationship between these is unsurprising. Similarly, the only potential for the Ln^{4+} + e$^-$ → Ln^{3+} process within reasonable range is that for Ce, and indicates that Ce^{4+} is the only ion in this state likely to be encountered in aqueous solution.

Question 2.1 What is the trend in atomic radii, and that in ionic radii, for the lanthanides? What are the exceptions to this, and why?
Answer 2.1 The structure of metals is usually described as one in which metal ions are surrounded by a 'sea' of delocalised outer-shell electrons. The greater the number of loosely held electrons, the stronger the metallic bonding and the smaller the atomic radius. If the

Table 2.7 Redox potentials of the lanthanide ions (v)^a

	La	Ce	Pr	Nd	Pm	Sm	Eu	Gd	Tb	Dy	Ho	Er	Tm	Yb	Lu	Y
$Ln^{3+} + 3e \rightarrow Ln$	−2.37	−2.34	−2.35	−2.32	−2.29	−2.30	−1.99	−2.29	−2.30	−2.29	−2.33	−2.31	−2.31	−2.22	−2.30	−2.37
$Ln^{3+} + e \rightarrow Ln^{2+}$	(−3.1)	(−3.2)	(−2.7)	−2.6^b	(−2.6)	−1.55	−0.34	(−3.9)	(−3.7)	−2.5^b	(−2.9)	(−3.1)	−2.3^b	−1.05		
$Ln^{4+} + e \rightarrow Ln^{3+}$		1.70	(3.4)	(4.6)	(4.9)	(5.2)	(6.4)	(7.9)	(3.3)	(5.0)	(6.2)	(6.1)	(6.1)	(7.1)	(8.5)	

^{a} Values in parentheses are estimated.
^{b} = in THF.

lanthanides are represented as Ln^{3+} $(e^-)_3$, then the atomic radii would be expected to fall between those of barium $[Ba^{2+}(e^-)_2]$ and hafnium $[Hf^{4+}(e^-)_4]$, as is generally observed.

Question 2.2 Using Hess's Law and the ΔH_f for LaF_2 $(-831$ kJ/mol) and for $LaF_3(-1732$ kJ/mol), calculate ΔH for this decomposition reaction.

$$3\,LaF_2 \rightarrow 2\,LaF_3 + La$$

Answer 2.2

$$\Delta H_{reaction} = \Sigma\,\Delta H_f(products) - \Sigma\,\Delta H_f(reactants)$$
$$\Delta H_{reaction} = 2\,(-1732) - 3\,(-831) = -971\ kJ/mol$$

Question 2.3 Assuming lattice energies for CeF_3 and CeF_4 of -4915 and -8391 kJ/mol respectively, and using the same enthalpy of atomization and electron affinity for fluorine as in the lanthanum examples (Section 2.9.1.), calculate ΔH_f for CeF_3 and CeF_4. Take an enthalpy of atomization of 398 kJ/mol for cerium (Table 2.6). Use ionization energies from Table 2.2.

Answer 2.3

	ΔH(kJ/mol)
$Ce(s) \rightarrow Ce\,(g)$	$+398$
$Ce(g) \rightarrow Ce^{3+}(g) + 3\ e^-$	$+527 + 1047 + 1949$
$3/2\ F_2(g) \rightarrow 3F\,(g)$	$+252$
$3F(g) + 3\ e^- \rightarrow 3F^-\,(g)$	-984
$Ce^{3+}(g) + 3F^-(g) \rightarrow CeF_3(s)$	-4915

Thus $Ce(s) + 3/2\ F_2(g) \rightarrow CeF_3(s)$

$\Delta H = +398 + (527 + 1047 + 1949) + 252 - 984 - 4915 = -1726\ kJ/mol$

	ΔH(kJ/mol)
$Ce(s) \rightarrow Ce(g)$	$+398$
$Ce(g) \rightarrow Ce^{4+}(g) + 4\ e^-$	$+527 + 1047 + 1949 + 3547$
$2F_2(g) \rightarrow 4F\,(g)$	$+336$
$4F(g) + 4\ e^- \rightarrow 4F^-\,(g)$	-1312
$Ce^{4+}(g) + 4F^-(g) \rightarrow CeF_4(s)$	-8391

Thus $Ce(s) + 2F_2(g) \rightarrow CeF_4(s)$

$\Delta H = +398 + (527 + 1047 + 1949 + 3547) + 336 - 1312 - 8391 = -1899\ kJ/mol$

Question 2.4 Unlike CeF_4, $CeCl_4$ does not exist, though $CeCl_3$ does. Suggest why this might be. Values of ΔH_f for $CeCl_3$ and $CeCl_4$ are -1058 and -820 (calculated) kJ/mol, respectively.

Answer 2.4 Fluorine is well known to promote high oxidation states. Factors associated with this are the high lattice energies associated with the small fluoride ion, along with the very small F–F bond energy (due to non-bonding electron pair repulsions) as well as high bond energies involving fluorine (not relevant in this case). Because of the larger size of the Cl^- ion, there is going to be much less difference between the lattice energies of $CeCl_3$ and

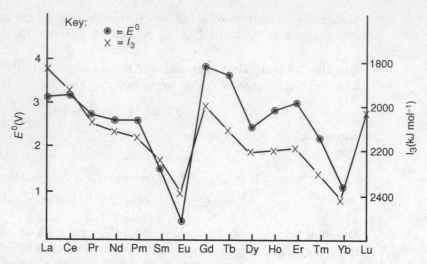

Figure 2.7
Strong correlation between I_3 and $E°$ for $Ln^{3+} + e \rightarrow Ln^{2+}$ (reproduced by permission of Macmillan from S.A. Cotton, *Lanthanides and Actinides*, 1991, p. 26.)

$CeCl_4$, and this is probably the determining factor. The higher $\Delta H_{at}(Cl)$ of 121.5 kJ/mol also mitigates against the formation of $CeCl_4$.

Question 2.5 Applying the $\Delta H_f LnX_3 / \Delta H_f LnX_2 < 1.5$ criterion for stability of dihalides, and using data in Table 2.5, predict which the halides LnX_2 are most likely to be exist.
Answer 2.5 As in section 2.9.2 $LnF_2(Ln = Sm, Eu, Yb)$; $LnCl_2(Ln = Nd, Pm, Sm, Eu, Dy, Tm, Yb)$; $LnY_2(Y = Br, I; Ln = Pr, Nd, Pm, Sm, Eu, Dy, Ho, Er, Tm, Yb)$ are predicted to be stable, in reasonably good agreement with the currently known facts.

Exercise 2.6 Using the same horizontal axis (atomic numbers 57–71), plot values of (a) the third ionization enthalpy I_3 and (b) the $Ln^{3+} + e \rightarrow Ln^{2+}$ reduction potential on the y axis (choose appropriate scales). Comment.
Answer 2.6 There is a strong correlation between them, not surprisingly! See Figure 2.7.

3 The Lanthanide Elements and Simple Binary Compounds

By the end of this chapter you should be able to:

- know how to prepare lanthanide metals and simple binary compounds such as the halides, oxides, and hydrides;
- apply the principle of the lanthanide contraction to explain patterns in co-ordination number in these compounds;
- apply knowledge gained in the study of Chapter 2 to the compounds in unusual oxidation states;
- understand the uses of the metals and certain compounds in applications such as hydrogen storage and in superconductors.

3.1 Introduction

This chapter discusses the synthesis of the lanthanide metals, their properties, reactions, and uses. It also examines some of the most important binary compounds of the lanthanides, particularly the halides, which well illustrate patterns and trends in the lanthanide series.

3.2 The Elements

3.2.1 Properties

The lanthanides are rather soft reactive silvery solids with a metallic appearance, which tend to tarnish on exposure to air. They react slowly with cold water and rapidly in dilute acid. They ignite in oxygen at around 150–200 °C; similar reactions occur with the halogens, whilst they react on heating with many nonmetals such as hydrogen, sulfur, carbon, and nitrogen (above 1000 °C).

The metals are relatively high-melting and -boiling. Their physical properties usually show smooth transition across the series, except that discontinuities are often observed for the metals that have a stable +2 state, europium and ytterbium. Thus the atomic radii of europium and ytterbium are about 0.2 Å greater than might be predicted by interpolation from values for the flanking lanthanides (Figure 3.1).

Similarly, Sm, Eu, and Yb have boiling points that are lower than those of the neighbouring metals (Figure 3.2).

Lanthanide and Actinide Chemistry   S. Cotton
© 2006 John Wiley & Sons, Ltd.

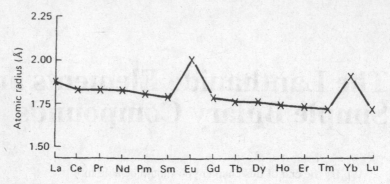

Figure 3.1
The atomic radii of the lanthanide metals (reproduced with permission from S.A. Cotton, *Lanthanides and Actinides*, Macmillan, 1991).

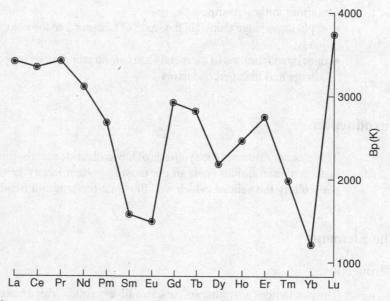

Figure 3.2
The boiling points of the lanthanide metals (reproduced with permission from S.A. Cotton, *Lanthanides and Actinides*, Macmillan, 1991).

Assuming that one can represent the structure of a metal as a lattice of metal ions permeated by a sea of electrons, then metals like lanthanum can be shown as $(Ln^{3+}) (e^-)_3$; however, metals based upon divalent ions (like Eu and Yb) would be $(Ln^{2+}) (e^-)_2$. The ions with the (+3) charge have a smaller radius, as the higher charge draws in the electrons more closely, and the stronger attraction means that it takes more energy to boil them (similarly, they would be predicted to have higher conductivities).

3.2.2 Synthesis

The metals have similar reactivities to magnesium. This means that they cannot be extracted by methods like carbon reduction of the oxides; in practice, metallothermic reduction of the lanthanide fluoride or chloride with calcium at around 1450 °C is used. The product is an alloy of (excess of) calcium and the lanthanide, from which the calcium can be distilled.

$$2\,LnF_3 + 3\,Ca \rightarrow 2\,Ln + 3\,CaF_2$$

The reduction is carried out under an atmosphere of argon, not nitrogen.

The above method is not suitable for obtaining the metals with a stable +2 state, which are only reduced as far as the difluoride (Eu, Yb, Sm). The lanthanide can be removed by distillation.

$$2\,La + M_2O_3 \rightleftharpoons 2\,M + La_2O_3$$

The 'divalent' metals Sm, Eu, and Yb have boiling points of 1791, 1597 and 1193 °C respectively, much lower than that of La (3457 °C), so that on heating they are distilled off, their volatility meaning that their removal from the mixture will displace the equilibrium to the right, so the reaction will proceed to completion.

3.2.3 Alloys and Uses of the Metals

Mischmetal is the lanthanide alloy with the longest history. It is a mixture of the lighter lanthanides, cerium in particular, which is manufactured from an 'unseparated' mixture of the oxides. This is first converted into a mixture of the anhydrous chlorides, which is then electrolysed using a graphite anode and iron cathode at around 820 °C to afford the mixture of metals. This is used mainly as an alloy with iron for the desulfurization and deoxidation of steels and more familiarly in cigarette lighter flints. SmCo$_5$ and other alloys of these metals have been used to make extremely strong permanent magnets. LaNi$_5$ has been widely examined as a material for hydrogen storage, with applications in fuel cells, catalytic hydrogenation, and removal of hydrogen from gas mixtures, as it rapidly absorbs hydrogen at room temperature to afford compositions up to LaNi$_5$H$_6$; the hydrogen is given up quickly at 140 °C. The most important application lies in rechargeable batteries for PCs (see Section 3.10).

3.3 Binary Compounds

3.3.1 Trihalides

Most halides are LnX$_3$, but a number of LnX$_2$ are known, as are a handful of tetrafluorides.

Syntheses of the Trihalides

Although the halides can be obtained as hydrates from reaction of the metal oxides or carbonates with aqueous acids, these hydrates are hydrolysed on heating to the oxyhalide and thus the anhydrous halides (which are themselves deliquescent) cannot be made that way.

$$LnX_3 + H_2O \rightarrow LnOX + HX$$

The fluorides are obtained as insoluble hydrates LnF$_3$.0.5H$_2$O by precipitation and the hydrates dehydrated by heating in a current of anhydrous HF gas (or *in vacuo*).

$$LnF_3.0.5H_2O \rightarrow LnF_3 + 0.5\,H_2O$$

Otherwise the anhydrous halides can generally be made by heating the metal with the halogen (except for EuI_3) or gaseous HCl.

$$2\,Ln + 3\,X_2 \rightarrow 2\,LnX_3$$

Another method for the chlorides involves refluxing the hydrated chlorides with thionyl dichloride ($SOCl_2$) for a few hours; an advantage of this method is that the other reaction products are gaseous SO_2 and HCl.

$$LnCl_3.x H_2O + x\,SOCl_2 \rightarrow LnCl_3 + x\,SO_2 + 2x\,HCl$$

A route that works well in practice involves thermal decomposition of ammonium halogenometallates. Adding ammonium chloride to a solution of the metal oxides in hydrochloric acid, followed by evaporation gives halogenometallate salts. These can be dehydrated by heating with excess of ammonium chloride in a stream of gaseous HCl, the resulting anhydrous salt being decomposed by heating *in vacuo* at about 300 °C.

$$Ln_2O_3 + 9\,NH_4Cl + 3\,HCl \rightarrow 2\,(NH_4)_3LnCl_6 + 3\,NH_3 + 3\,H_2O$$

$$2\,(NH_4)_3LnCl_6 \rightarrow 2\,LnCl_3 + 6\,NH_4Cl$$

The anhydrous halides can be purified by sublimation *in vacuo*, but owing to a tendency to react with silica, contact with hot glass, thereby forming the oxyhalide, should be avoided.

Structures of the Trihalides

The lanthanide trihalides demonstrate very clearly the effect of varying the cation and anion radii upon the structure type adopted (Table 3.1).

The early lanthanide fluorides adopt the 'LaF_3' structure (Figure 3.3) based on a metal ion surrounded by a tricapped trigonal prism of fluorides with two additional capping fluorides, giving 11 coordination (9 + 2), whilst the later fluorides have the 'YF_3' structure. This is based on tricapped trigonal prismatic 9 coordination, in which the prism is somewhat distorted. The structure adopted by UCl_3 and several of the lanthanide halides is again a

Table 3.1   Structures of the lanthanide trihalides

	F	Cl	Br	I
La	LaF_3 (11)	UCl_3 (9)	UCl_3 (9)	$PuBr_3$ (8)
Ce	LaF_3 (11)	UCl_3 (9)	UCl_3 (9)	$PuBr_3$ (8)
Pr	LaF_3 (11)	UCl_3 (9)	UCl_3 (9)	$PuBr_3$ (8)
Nd	LaF_3 (11)	UCl_3 (9)	$PuBr_3$ (8)	$PuBr_3$ (8)
Pm	LaF_3 (11)	UCl_3 (9)	$PuBr_3$ (8)	$PuBr_3$ (8)
Sm	YF_3 (9)	UCl_3 (9)	$PuBr_3$ (8)	$FeCl_3$ (6)
Eu	YF_3 (9)	UCl_3 (9)	$PuBr_3$ (8)	
Gd	YF_3 (9)	UCl_3 (9)	$FeCl_3$ (6)	$FeCl_3$ (6)
Tb	YF_3 (9)	$PuBr_3$ (8)	$FeCl_3$ (6)	$FeCl_3$ (6)
Dy	YF_3 (9)	$AlCl_3$ (6)	$FeCl_3$ (6)	$FeCl_3$ (6)
Ho	YF_3 (9)	$AlCl_3$ (6)	$FeCl_3$ (6)	$FeCl_3$ (6)
Er	YF_3 (9)	$AlCl_3$ (6)	$FeCl_3$ (6)	$FeCl_3$ (6)
Tm	YF_3 (9)	$AlCl_3$ (6)	$FeCl_3$ (6)	$FeCl_3$ (6)
Yb	YF_3 (9)	$AlCl_3$ (6)	$FeCl_3$ (6)	$FeCl_3$ (6)
Lu	YF_3 (9)	$AlCl_3$ (6)	$FeCl_3$ (6)	$FeCl_3$ (6)
Y	YF_3 (9)	$AlCl_3$ (6)	$FeCl_3$ (6)	$FeCl_3$ (6)

Values in parentheses indicate the coordination number.

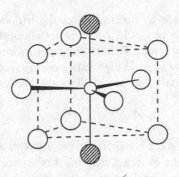

Figure 3.3
The LaF$_3$ structure (reproduced by permission of Macmillan from S.A. Cotton, *Lanthanides and Actinides*, 1991, p. 42).

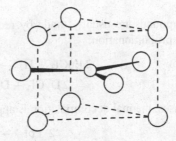

Figure 3.4
The structure of UCl$_3$ and certain LnCl$_3$ [reproduced in modified form, from R.B. King (ed.), Encyclopedia of Inorganic Chemistry, 1st edition, Wiley, Chichester; 1994].

tricapped trigonal prism (Figure 3.4); if one of the face-capping halogens is removed from the 'UCl$_3$' structure, the 8 coordinate PuBr$_3$ structure is generated. Finally, the AlCl$_3$ and BiI$_3$ (also sometimes referred to as 'FeCl$_3$') structures both have octahedral six-coordination.

The trends are similar to those found in halides of the Group I and Group II metals and are explained on an ionic packing model; more anions can be packed round the large lanthanide ions early in the series than around the smaller, later ones; similarly, more of the small fluoride ions can be packed round a given lanthanide ion than is the case with the much larger iodide anion.

Properties of the Trihalides

The trihalides are high-melting solids, with many uses in synthetic chemistry, though their insolubility in some organic solvents means that complexes, such as those with thf (Section 4.3.3), are often preferred.

3.3.2   Tetrahalides

Only the fluorides of Ce, Pr, and Tb exist, the three lanthanides with the most stable (+4) oxidation state. Fluorine is most likely to support a high oxidation state, and even though salts of ions like [CeCl$_6$]$^{2-}$ are known, the binary chloride has not been made. CeF$_4$ can be crystallized from aqueous solution as a monohydrate. Anhydrous LnF$_4$ (Ln = Ce, Pr, Tb) can be made by fluorination of the trifluoride or, in the case of Ce, by fluorination of metallic Ce or $CeCl_3$. All three tetrafluorides have the MF_4 structure with dodecahedral eight coordination. Factors that favour formation of a tetrafluoride include a low value of I_4 for the metal and a high lattice energy (see Section 2.9.1). This is most likely to be found with the smallest halide ion, fluoride. The low bond energy of F_2 is also a supporting factor.

3.3.3 Dihalides

These are most common for metals with a stable (+2) state, such as Eu, Yb, and Sm. As would be expected, they most often occur for the iodide ion, the best reducing agent. A number of dihalides are known for other metals, though some of these do not actually involve the (+2) oxidation state.

Synthetic Routes

These compounds are usually made by reduction using hydrogen (e.g. EuX_2, YbX_2, or SmI_2) or reproportionation.

$$2\,EuCl_3 + H_2 \;\rightarrow\; 2\,EuCl_2 + 2\,HCl$$

$$2\,DyCl_3 + Dy \;\rightarrow\; 3\,DyCl_2$$

In a few cases, thermal decomposition is applicable [LnI_2 (Ln = Sm, Yb); $EuBr_2$].

$$2\,YbI_3 \;\rightarrow\; 2\,YbI_2 + I_2$$

Another method, used especially for diiodides, involves heating the metal with HgX_2.

$$Tm + HgI_2 \;\rightarrow\; TmI_2 + Hg$$

The known dihalides are listed in Table 3.2, together with an indication of the structure and coordination number.

The iodides of Nd, Sm, Eu, Dy, Tm, and Yb are definitely compounds of the +2 ions, with salt-like properties, are insulators, and have magnetic and spectroscopic properties expected

Table 3.2 Structures of the lanthanide dihalides

	F	Cl	Br	I
La				[a]$MoSi_2$ (8)
Ce				[a]$MoSi_2$ (8)
Pr				[a]$MoSi_2$ (8)
Nd		$PbCl_2$ (9)	$PbCl_2$ (9)	$SrBr_2$ (7,8)
Pm				
Sm	CaF_2 (8)	$PbCl_2$ (9)	$PbCl_2$ (9); $SrBr_2$ (7,8)	EuI_2 (7)
Eu	CaF_2 (8)	$PbCl_2$ (9)	$SrBr_2$ (7,8)	EuI_2 (7)
Gd				[a]$MoSi_2$ (8)
Tb				
Dy		$SrBr_2$ (7,8)	SrI_2 (7)	$CdCl_2$ (6)
Ho				
Er				
Tm		SrI_2 (7)	SrI_2 (7)	CdI_2 (6)
Yb	CaF_2 (8)	SrI_2 (7)	SrI_2 (7); $CaCl_2$ (6)	CdI_2 (6)
Lu				

[a] Ln^{III} compounds with the structure $M^{3+}(I^-)_2(e^-)$.

for the M^{2+} ions (thus electronic spectra of EuX_2 resemble those of the isoelectronic Gd^{3+} ions). SmI_2 is proving an important reagent in synthetic organic chemistry (Section 8.3).

Several of the diiodides, however, have a metallic sheen and are very good conductors of electricity, such as LaI_2, CeI_2, PrI_2, and GdI_2. Since they are good conductors in the solid state, the presence of delocalized electrons is indicated. $M^{3+}(I^-)_2(e^-)$ is a likely structure. None of these metals exhibits a stable +2 state in any of its compounds and consideration of factors such as their ionization energies suggest they are unlikely to form stable compounds in this state (see Section 2.9.2).

3.3.4 Oxides

Oxides M_2O_3

As already mentioned, the metals burn easily, rather like Group 2 metals, forming oxides (but with some nitride). The oxides are therefore best made by thermal decomposition of compounds like the nitrate or carbonate.

$$4\,Ln(NO_3)_3 \quad \rightarrow \quad 2\,Ln_2O_3 + 12\,NO_2 + 3\,O_2$$

Most lanthanides form Ln_2O_3, but those metals with accessible +4 and +2 oxidation states can afford other stoichiometries. These may be turned into Ln_2O_3 by synthesis under a reducing atmosphere. CeO_2 can be reduced to Ce_2O_3 using hydrogen. The oxides are somewhat basic and absorb CO_2 from the atmosphere, forming carbonates, and water vapour, forming hydroxides. As expected they dissolve in acid, forming salts, and are convenient starting materials for the synthesis of lanthanide salts, including the hydrated halides.

The sequioxides Ln_2O_3 adopt three structures, depending upon the temperature and upon the lanthanide involved. At room temperature, La_2O_3 to Sm_2O_3 inclusive adopt the A-type structure, which has capped octahedral 7 coordination of the lanthanide. The B-type structure tends to be exhibited by La_2O_3 to Sm_2O_3 at higher temperatures and has three different lanthanide sites, one with distorted 6 coordination and the others with face-capped octahedral 7 coordination. The C-type structure is followed by heavier metals and has 6 coordination, severely distorted from an octahedron.

Oxides MO_2

Cerium, having the stablest (+4) state, is the only metal to form a stoichiometric oxide in this state, CeO_2, which has the fluorite structure. It is white when pure, but even slightly impure specimens tend to be yellow. It can be made by burning cerium or heating salts like cerium(III) nitrate strongly in air. It is basic and dissolves with some difficulty in acid, forming $Ce^{4+}(aq)$, which can be isolated as salts such as the nitrate $Ce(NO_3)_4.5H_2O$. Uses of CeO_2 include self-cleaning ovens and as an oxidation catalyst in catalytic converters.

Praseodymium and terbium form higher oxides, of which a number of phases are known between Ln_2O_3 and LnO_2. Ignition of praseodymium nitrate leads to Pr_2O_3 but further heating in an oxygen atmosphere gives Pr_6O_{11} or even PrO_2; terbium similarly yields Tb_4O_7 and TbO_2.

Oxides MO

Reduction of Eu_2O_3 with Eu above 800 °C (comproportionation) gives EuO.

$$Eu_2O_3 + Eu \rightarrow 3\,EuO$$

This compound and the similar YbO have salt-like (NaCl) structures and are genuine Ln^{II} compounds, being insulators. Similar comproportionation methods using high pressures have been used to obtain SmO and NdO; these are shiny conducting solids, probably containing Ln^{3+} ions.

Oxide Superconductors

Superconductivity is the phenomenon in which a material conducts electricity with virtually zero resistance. For many years until the mid 1980s, the highest temperatures available were around 20 K and required expensive liquid helium (or hydrogen) coolant. If high-temperature superconductors can be made, this has obvious application in area such as power transmission. In 1986, Bednorz and Muller reported that $La_{1.8}Sr_{0.2}CuO_4$ was a superconductor up to 38 K. This sparked intense world-wide activity, and the following year Wu, Chu and others reported that $YBa_2Cu_3O_{7-\delta}$ ($0 \leq \delta \leq 1$) had a T_c (superconducting transition temperature) of 92 K, bringing the phenomenon into the liquid nitrogen range. Steady though less spectacular advances have produced materials like $HgBa_2Ca_2Cu_3O_8$ ($T_c = 133K$).

The structure of $YBa_2Cu_3O_7$ is based on an oxygen-deficient layered perovskite structure and features two types of copper environment (Figure. 3.5), formally containing Cu^{2+} and Cu^{3+}, with both square planar and square pyramidal coordination. Removing the shaded oxygens results in phases down to the semiconductor $YBa_2Cu_3O_6$. The mechanism of superconductivity is still debated, but one theory suggests that it involves an electron passing through the lattice distorting it in such a way that a second electron follows closely in its wake

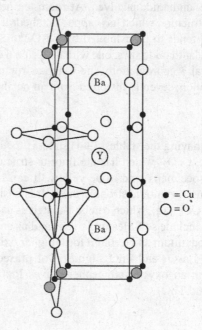

Figure 3.5
The structure of $YBa_2Cu_3O_7$; the shaded atoms are those removed to create oxygen-deficient phases up to $YBa_2Cu_3O_6$ (reproduced by permission of Macmillan from S.A. Cotton, *Lanthanides and Actinides*, 1991, p. 47).

with no hindrance to its passing, the electron pair being known as a 'Cooper pair'. Despite intense activity over the last 15 years no truly commercial material has yet emerged, due to the intrinsically brittle nature of the oxide ceramic materials making fabrication into wires impossible. Current thinking favours making thin films, though with present technology they will only have low current-carrying capacity.

3.4 Borides

A number of stoichiometries obtain, such as LnB_2, LnB_4, LnB_6, LnB_{12}, and LnB_{66}.

The most important are LnB_6. Borides are obtained by heating the elements together at 2000 °C or by heating the lanthanide oxide with boron or born carbide at 1800 °C. They are extremely unreactive towards acids, alkalis, and other chemicals, and are metallic conductors. They contain a continuous three-dimensional framework of $(B_6)^{2-}$ octahedra interspersed with Ln^{3+} ions, indicating an electronic structure (Ln^{3+}) $(B_6)^{2-}$ (e^-) in most cases, though EuB_6 and YbB_6 are insulators, suggesting them to be (Ln^{2+}) $(B_6)^{2-}$. Because of its high thermal stability and melting point (2400 °C), and its metal-like electrical and thermal conductivity, LaB_6 is an important thermionic emitter material for the cathodes of electron guns. Mixed boride materials are also of commercial importance. Nd–Fe–B alloys, with compositions such as $Nd_2Fe_{14}B$, are the strongest permanent magnet materials available today. Other important compounds include lanthanide rhodium boride low-temperature superconductors (such as $ErRh_4B_4$), part of a wider LnM_4B_4 family (M = transition metal).

3.5 Carbides

The lanthanides form carbides with a range of compositions, notably LnC_2, but additionally Ln_2C_3, LnC, Ln_2C, and Ln_3C phases are known (depending upon the lanthanide in question and the conditions of synthesis). LnC_2 adopt the CaC_2 structure, containing isolated C_2^{2-} ions.

3.6 Nitrides

These can be made by direct synthesis from the elements at 1000 °C. They have the NaCl structure and can be hydrolysed to NH_3.

3.7 Hydrides

Ternary hydrides such as $LaNi_5H_6$ have attracted attention as materials for electrodes in fuel cells and for gas storage. Nickel/metal hydride batteries for notebook PCs are a less toxic alternative than Ni/Cd batteries; when charging, hydrogen generated at the negative electrode enters the lattice of the La/Ni alloy (in practice a material such as $La_{0.8}Nd_{0.2}Ni_{2.5}Co_{2.4}Si_{0.1}$ is used to improve corrosion resistance, storage capacity, discharge rate, etc.). The overall cell reaction is:

$$LaNi_5 + 6\,Ni(OH)_2 \rightleftharpoons LaNi_5H_6 + 6\,NiOOH$$

The lanthanides also form simple binary hydrides on combination of the elements at about 300 °C. These compounds have ideal compositions of MH_2 and MH_3, but are frequently non-stoichiometric. Thus lutetium forms phases with ranges $LuH_{1.83}$ to $LuH_{2.23}$ and $LuH_{2.78}$ to $LuH_{3.00}$. Reaction of ytterbium with hydrogen under pressure gives $YbH_{2.67}$. MH_3 are obtained only at higher gas pressures, whilst europium, the lanthanide with the most stable +2 state, forms only EuH_2. The dihydrides are generally good electrical conductors, and thus are thought to be $M^{3+}(H^-)_2(e^-)$, whilst the trihydrides are salt-like nonconductors believed to be $M^{3+}(H^-)_3$. The hydrides are reactive solids, owing to the presence of the easily hydrolysed H^- ion.

3.8 Sulfides

These are quite important compounds. A number of stoichiometries exist, the most important being Ln_2S_3. These can be made by direct synthesis, heating the elements together, or by passing H_2S over heated $LnCl_3$. Eu_2S_3 cannot be prepared by this latter route.

A range of compositions between Ln_2S_3 and Ln_3S_4, the latter formed when Ln_2S_3 lose sulfur on heating, generally occurs. Ln_2S_3 are insulators, genuine Ln(III) compounds, but Ln_3S_4 (having the Th_3P_4 structure) are more complex. Some, like Ce_3S_4, are metallic conductors and are thus $(Ln^{3+})_3(S^{2-})_4(e^-)$; others, like Eu_3S_4 and Sm_3S_4, are semiconductors and may be $(Ln^{2+})(Ln^{3+})_2(S^{2-})_4$. The structures of Ln_2S_3 fall into a pattern. La_2S_3 to Dy_2S_3 adopt the 'Gd_2S_3' structure with 7-coordinate lanthanide ions; Dy_2S_3 to Tm_2S_3 have the 'Ho_2S_3' structure with 6- and 7-coordinate lanthanides, whilst Yb_2S_3 and Lu_2S_3 have the corundum structure with just 6 coordination.

Monsulfides MS are formed by direct combination. They adopt the NaCl structure but have a variety of bonding types. YbS and EuS are genuine Ln^{2+} S^{2-} but CeS exhibits the magnetic properties expected for Ce^{3+} and has a bronze metallic lustre as well, so it thought to be $(Ce^{3+})(S^{2-})(e^-)$. This substance not only has electrical conductivity in the metallic region but also can be machined like a metal too. SmS has some unusual properties; it is usually obtained as a black semiconducting phase, Sm^{2+} S^{2-}, but can reportedly be turned into a golden metallic phase by the action of pressure, polishing or even scratching on single crystals.

Oxysulfides are also rather important. Y_2O_2S is used as a host material for Ln^{3+} ion emitters (e.g. Eu, Tb) in some phosphors, notably those used in TV screens. When mischmetal is used to remove oxygen and sulfur from impure iron and steel, the product is an oxysulfide, which forms an immiscible solid even in contact with molten steel and thus does not contaminate the product.

Question 3.1. Why do Eu and Yb have lower boiling points than the other lanthanides and significantly higher atomic radii?

Answer 3.1 Most lanthanides can be described by an electronic structure (Ln^{3+}) $(e^-)_3$, whereas metals like Eu and Yb can be better described by (Ln^{2+}) $(e^-)_2$. The greater charge on the (+3) ions draws electrons in closer, so it is expected that the metals with this electronic structure will have smaller atoms. Similarly the greater number of electrons in the (Ln^{3+}) $(e^-)_3$ structure leads to stronger metallic bonding than in metals with the structure (Ln^{2+}) $(e^-)_2$ and hence the weaker cohesive force requires less energy to overcome it, resulting in lower enthalpies of vapourization and lower boiling points.

Question 3.2 Suggest why nitrogen is not a suitable inert atmosphere for the following reaction:

$$2\,LnF_3 + 3\,Ca \rightarrow 2\,Ln + 3\,CaF_2$$

Answer 3.2 Like magnesium, which forms Mg_3N_2, hot lanthanides react with nitrogen, forming the nitride LnN (see Section 3.6).

Question 3.3 Produce a balanced equation for the reaction of $LaCl_3$ with silica.
Answer 3.3

$$2\,LaCl_3 + SiO_2 \rightarrow 2\,LaOCl + SiCl_4$$

Question 3.4 Studying the data in Table 3.1 if necessary, explain the patterns in coordination number of the lanthanide trihalides.
Answer 3.4 As the radius of the lanthanide ion decreases on crossing the series from left to right, the coordination number of the metal in a given halide decreases; a similar effect is seen on descending Group 7(17), as the radius of the halide ion increases.

Question 3.5 The melting and boiling points of LaX_3 (X = Cl, Br, I) are 862/1750 °C (X = Cl); 789 and 1580 °C (X = Br); 778 and 1405 °C (X = I). Comment on these values.
Answer 3.5 The melting point is an indication of the magnitude of the lattice energy, whilst the boiling point is a measure of the attraction between the mobile ions. Lattice energy decreases as the size of the anions increase, so it would be least for the iodide. Covalent character in the bonding, which would also decrease the melting point, increases in the same direction.

Question 3.6 Why could CeO_2 be a useful catalyst for oxidations?
Answer 3.6 As with V_2O_5 in the Contact process for making sulfuric acid, the ability of cerium to adopt transition-metal-like behaviour in switching oxidation states means that the cerium oxide can effectively act as an oxygen-storage system.

Question 3.7 LnC_2 are good conductors of electricity and form ethyne, C_2H_2, on hydrolysis (as does CaC_2, traditionally used in 'acetylene' headlamps on early cars and bicycles). Explain these observations.
Answer 3.7 These compounds have the structure $M^{3+}(C_2^{2-})(e^-)$. The delocalized electrons cause the high conductivity. The ethynide ions react with water thus:

$$C_2^{2-}(s) + 2\,H_2O(l) \rightarrow C_2H_2(g) + 2\,OH^-(aq)$$

Question 3.8 Suggest an equation for the reaction of $LaCl_3$ with H_2S.
Answer 3.8

$$2\,LaCl_3 + 3\,H_2S \rightarrow La_2S_3 + 6\,HCl$$

4 Coordination Chemistry of the Lanthanides

By the end of this chapter you should be able to:

- recall which donor atoms and types of ligand form the most stable complexes;
- explain why later lanthanides form complexes with greater stability constants than do the earlier lanthanides;
- explain why polydentate ligands form more stable complexes;
- recall and explain the most common coordination numbers in lanthanide complexes;
- recall the common geometries of complexes;
- work out coordination numbers and suggest geometries of complexes, given a formula;
- explain the choice of particular compounds as Magnetic Resonance Imaging agents.

4.1 Introduction

Forty years ago, very little was known about lanthanide complexes. By analogy with the d-block metals, it was often assumed that lanthanides were generally six coordinate in their complexes. We now know that this is not the case, that lanthanides (and actinides) show a wider variety of coordination number than do the d-block metals, and also understand the reasons for their preferred choice of ligand.

4.2 Stability of Complexes

For the reaction between a metal ion and a ligand

$$M^{n+} + L^{y-} \quad \Leftrightarrow \quad ML^{(n-y)+}$$

a stability constant may be defined approximately

$$\beta_1 = [ML^{(n-y)+}]/[M^{n+}][L^{y-}]$$

Table 4.1 lists $\log[\beta_1]$ values for La^{3+} and Lu^{3+} as well as Sc^{3+} and Y^{3+} with a number of common ligands, whilst Table 4.2 gives the $\log[\beta_1]$ values for all the lanthanide ions complexing with $EDTA^{4-}$, $DTPA^{5-}$, and fluoride.

The values for Lu^{3+} are greater than those for La^{3+} as the smaller Lu^{3+} ion has a greater charge density and stronger electrostatic attraction for a ligand. The stability constants for the smaller halide ligands are greater on account of their higher charge density. Values for multidentate ligands are greater because of entropy factors (see below) and also because

---

Lanthanide and Actinide Chemistry S. Cotton
© 2006 John Wiley & Sons, Ltd.

Table 4.1 Aqueous stability constants ($\log \beta_1$) for lanthanide (3+) and other ions

Ligand	I (mol/dm³)	La³⁺	Lu³⁺	Y³⁺	Sc³⁺	Fe³⁺	Cu²⁺	Ca²⁺	U⁴⁺	UO₂²⁺	Th⁴⁺
F⁻	1.0	2.67	3.61	3.60	6.2ᵃ	5.2	0.9	0.6	7.78	4.54	7.46
Cl⁻	1.0	−0.1	−0.4	−0.1	0	0.63	−0.06	−0.11	0.30	−0.10	0.18
Br⁻	1.0	−0.2		−0.15	−0.07	−0.2	−0.5		0.2	−0.3	−0.13
NO₃⁻	1.0	0.1	−0.2		0.3	−0.5	−0.01	−0.06	0.3	−0.3	0.67ᵃ
OH⁻	0.5	4.7	5.8	5.4	9.0	11.27	6.3ᵃ	1.0	12.2	8.0ᵃ	9.6ᵃ
acac⁻	0.1	4.94	6.15	5.89	8	10	8.16				8
EDTA⁴⁻	0.1	15.46	19.8	18.1	23.1	25.0	18.7	10.6	25.7	7.4	25.3
DTPA⁵⁻	0.1	19.5	22.4	22.05	24.4	28	21.4	10.8			28.8
OAc⁻	0.1	1.82	1.85	1.68		3.38	1.83	0.5ᵃ		2.61	3.89
glycine	0.1	3.1	3.9	3.5		10.0ᵃ	8.12	1.05			

ᵃ = Data for solutions of slightly different ionic strength

Table 4.2 Aqueous stability constants ($\log \beta_1$) for lanthanide (3+) ions with fluoride, EDTA, and DTPA

Ligand	Y³⁺	La³⁺	Ce³⁺	Pr³⁺	Nd³⁺	Pm³⁺	Sm³⁺	Eu³⁺	Gd³⁺	Tb³⁺	Dy³⁺	Ho³⁺	Er³⁺	Tm³⁺	Yb³⁺	Lu³⁺
F⁻	3.60	2.67	2.87	3.01	3.09	3.16	3.12	3.19	3.31	3.42	3.46	3.52	3.54	3.56	3.58	3.61
EDTA⁴⁻	18.08	15.46	15.94	16.36	16.56		17.10	17.32	17.35	17.92	18.28	18.60	18.83	19.30	19.48	19.80
DTPA⁵⁻	22.05	19.48	20.33	21.07	21.60		22.34	22.39	22.46	22.71	22.82	22.78	22.74	22.72	22.62	22.44

once one end of a ligand is attached, there is a higher chance of the other donor atoms attaching themselves (conversely, with a multidentate ligand, if a bond to one donor atom is broken, the others hold).

In a graph of $\log K$ against Z (the atomic number), there is a smooth increase expected as the ionic radius decreases as the greater charge density of the smaller ions leads to more stable complexes. There are inflections, particularly around Gd, possibly due to the change in coordination number of the aqua ion.

For these complexation reactions, ΔH has small values, either exothermic or endothermic; the main driving force for complex formation, particularly where multidentate ligands are involved, is the large positive entropy change. Thus for:

$$[La(OH_2)_9]^{3+}(aq) + EDTA^{4-}(aq) \rightarrow [La(EDTA)(OH_2)_3]^-(aq) + 6\,H_2O(l)$$

$$\Delta G = -87\,\text{kJ}\,\text{mol}^{-1}; \Delta H = -12\,\text{kJ}\,\text{mol}^{-1}; \Delta S = +251\,\text{JK}^{-1}\text{mol}^{-1}$$

In the reaction

$$LnX_3.9H_2O(s) + 3\,L^-(aq) \rightarrow [Ln(L)_3]^{3+}(aq) + 3X^- + 9\,H_2O(l)$$

where L = 2,6-dipicolinate and X is bromate or ethyl sulfate, there is a smooth variation in ΔH across the series, as there is no change in the coordination number of the aqua ion.

In general, lanthanide ions prefer to bind to hard donors such as O and F, rather than to soft bases such as P and S donor ligands. Nitrogen-containing ligands are an apparent anomaly, since they form relatively few complexes; this is at least partly due to the high basicity of such ligands, tending to result in the precipitation of hydroxides, etc. (Lanthanide ammines have been made in solvents like supercritical NH₃.) The higher electronegativity of oxygen and consequent polar nature of ligands like phosphine oxides $R_3P^{\delta+}{=}O^{\delta-}$ is also a factor. Use of nonaqueous solvents of weak donor ability (e.g. MeCN) can lead to the isolation of complexes decomposed by water.

4.3 Complexes

4.3.1 The Aqua Ions

The coordination number of $[Ln(H_2O)_n]^{3+}$ is believed to be 9 for the early lanthanides (La–Eu) and 8 for the later metals (Dy–Lu), with the intermediate metals exhibiting a mixture of species. The nine-coordinate species are assigned tricapped trigonal prismatic structures (Figure 4.1) and the eight-coordinate species square antiprismatic coordination. A considerable amount of spectroscopic data has led to this conclusion; for example, the visible spectrum of Nd^{3+}(aq) and $[Nd(H_2O)_9]^{3+}$ ions in solid neodymium bromate are very similar to each other and quite different to those of Nd^{3+} ions in eight-coordinate environments, correlating with solution X-ray and neutron-diffraction studies, indicating $n \sim 8.9$ in solution. Similarly, neutron-diffraction studies indicate $n \sim 8.5$ for Sm^{3+}(aq) and 7.9 for Dy^{3+}(aq) and Lu^{3+}(aq), whilst solution luminescence studies suggest n values of 9.0, 9.1, 8.3, and 8.4 for Sm^{3+}, Eu^{3+}, Tb^{3+}(aq), and Dy^{3+} respectively.

4.3.2 Hydrated Salts

These are readily prepared by reaction of the lanthanide oxide or carbonate with the acid. Salts of noncoordinating anions most often crystallize as salts $[Ln(OH_2)_9]X_3$ (X e.g. bromate, triflate, ethyl sulfate, tosylate). These contain the $[Ln(OH_2)_9]^{3+}$ ion (Figure 4.1), even for the later lanthanides, where in aqueous solution the eight-coordinate species $[Ln(OH_2)_8]^{3+}$ predominates.

The lanthanide–water distances for the positions capping the prism faces and at the vertices are different; on crossing the series from La to Lu, the Ln–O distance decreases from 2.62 to 2.50 Å for the three face-capping oxygens but change more steeply from 2.52 to 2.29 Å for the six apical oxygens (data for the triflate). In contrast, the perchlorate salts are $[Ln(OH_2)_6](ClO_4)_3$ and eight-coordinate species are found in species like $[Er(OH_2)_8](ClO_4)_3.(dioxane).2H_2O$, $[Eu(OH_2)_8]_2(V_{10}O_{28}).8H_2O$, and as $[Gd(OH_2)_8]^{3+}$ ions encapsulated inside a crown ether ring. Many of these compounds have lattices held together by hydrogen bonds, a factor which clearly affects the solubility and hence the compound isolated.

When anions can coordinate, a wide variety of species obtains. Some examples follow. Among the nitrates, $[Ln(NO_3)_3.(H_2O)_5]$ (Ln = La, Ce) have 11-coordinate lanthanides

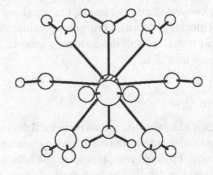

Figure 4.1
The structure of the nonaaqualanthanide ion (reproduced by permission of the American Chemical Society from B.P. Hay, *Inorg. Chem.*, 1991, **30**, 2881).

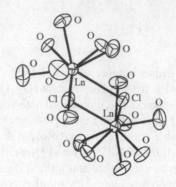

Figure 4.2
Structure of the dimeric cation $[(H_2O)_7LnCl_2Ln(H_2O)_7]^{4+}$ (reproduced by permission of the International Union of Crystallography from A. Habenschuss and F.H. Spedding, *Cryst. Struct. Commun.*, 1982, **7**, 538).

whilst in $[Ln(NO_3)_3.(H_2O)_4]$ (Ln = Pr–Yb, Y) 10 coordination is the rule, and two hydrates containing 9-coordinate $[Lu(NO_3)_3.(H_2O)_3]$ species have been characterized; in general, in lanthanide complexes a coordinated nitrate group is *almost always* bidentate. The chlorides and bromides of La and Ce, $LnX_3 . 7H_2O$, are dimeric $[(H_2O)_7Ln(\mu\text{-}X)_2Ln(OH_2)_7]X_4$ (X = Cl, Br; Figure 4.2), whilst $LnCl_3.6H_2O$ (Ln = Nd–Lu) have antiprismatic $[LnCl_2(OH_2)_6]^+$ ions.

Amongst the hydrated bromides, $LnBr_3.6H_2O$ (Ln = Pr–Dy) are $[LnBr_2(OH_2)_6]Br$, like the heavier rare earth chlorides; and $LnBr_3.8H_2O$ (Ln = Ho–Lu) are $[Ln(OH_2)_8]Br_3$. The structures of the iodides $LnI_3.9H_2O$ and $LnI_3.8H_2O$ also involve pure aqua ions. Several hydrated sulfates exist, where both water and sulfates are bound to the metal, usually with 9 coordination. Thiocyanates exist as $Ln(NCS)_3(H_2O)_6$(Ln = La–Dy) and $Ln(NCS)_3(H_2O)_5$ (Ln = Sm–Eu) molecules.

The acetates, unlike the acetates of transition metals like Cr^{III}, Fe^{III}, and Ru^{III}, do not adopt oxo-centred structures with M_3O cores. Instead, $[Ln(OAc)_3.1.5(H_2O)]$ (Ln = La–Pr) have structures with acetate-bridged chains crosslinked by further acetate bridges; $[Ln(OAc)_3.(H_2O)]$ (Ln = Ce–Pr) have one-dimensional polymeric structures with acetate bridges; and $[Ln(OAc)_3.4H_2O)]$ (Ln = Sm–Lu) are acetate-bridged dimers.

Adding oxalate ions to a solution of lanthanide ions affords quantitative precipitation of the oxalate; this is used in traditional quantitative analysis of lanthanides, as on ignition they are converted into the oxide, the weighing form.

The chloride, bromide, bromate, perchlorate, nitrate, acetate, and iodide salts are all soluble in water, the sulfates sparingly soluble, and the fluorides, carbonates, oxalates, and phosphates insoluble.

4.3.3   Other O-Donors

Complexes of lanthanide salts with neutral donors are generally made by mixing solutions of the ligand and the metal salt dissolved in a nonaqueous solvent such as ethanol or acetonitrile. Complexes of ligands like phosphine and arsine oxides were the first complexes of this type to be studied in detail. The best defined series involve hexamethylphosphoramide [hmpa: $(Me_2N)_3PO$] as a ligand; two series are obtained in which the stoichiometry is maintained across the series, $Ln(hmpa)_6(ClO_4)_3$ and $Ln(hmpa)_3Cl_3$, with structural confirmation for

[Nd(hmpa)$_6$](ClO$_4$)$_3$ and *mer*-Ln(hmpa)$_3$Cl$_3$ (Ln = Pr, Dy, Yb). Apart from the usual synthesis involving the ligand and the appropriate metal salt in ethanol, such complexes can also be made by an unusual 'one-pot' synthesis leading to the isolation of 9-coordinate La(hmpa)$_3$(NO$_3$)$_3$, 7-coordinate La(hmpa)$_4$(NCS)$_3$ and La(hmpa)$_4$Br$_3$.

$$Ln \xrightarrow[\text{toluene, heat}]{\text{NH}_4\text{X/HMPA}} Ln(hmpa)nX_3 \quad (X = NO_3, n = 3; X = Br, NCS, n = 4)$$

Complexes of Ph$_3$PO and Ph$_3$AsO, though well known, have yet to be fully studied. The bulky phenyl groups exert at-a-distance steric effects, yet allow several ligands of this type to coordinate to a metal ion, as in [La(Ph$_3$PO)$_5$Cl] (FeCl$_4$)$_2$ and [Ln(Ph$_3$PO)$_4$Cl$_2$] (CuCl$_3$), both of which have six-coordinate lanthanides. Early lanthanides form Ln(Ph$_3$PO)$_4$(NO$_3$)$_3$ (La–Nd), which have two bidentate and one monodentate nitrates, giving 9-coordinate lanthanides; Lu(Ph$_3$PO)$_4$(NO$_3$)$_3$ is actually [Ln(Ph$_3$PO)$_4$(NO$_3$)$_2$] NO$_3$, with 8 coordination. Nine coordination is also known to occur in Eu(Ph$_3$AsO)$_3$(NO$_3$)$_3$. On the other hand, with a less bulky ligand, lanthanum attains 10 coordination in La(Ph$_2$MePO)$_4$(NO$_3$)$_3$. 'Solvent' is coordinated in Ln(Ph$_3$PO)$_2$(NO$_3$)$_3$(EtOH) (Ln = Ce–Eu), also 9-coordinate. Triflates Ln(Ph$_3$PO)$_4$(OTf)$_3$(OTf = CF$_3$SO$_3$) contain [Ln(Ph$_3$PO)$_4$(OTf)$_2$]$^+$ ions; for La and Nd, one triflate is bidentate and one monodentate, whilst for Er–Lu both triflates are monodentate. A few halide complexes have been characterized structurally, notably for Y, including [Y(Me$_3$PO)$_6$] Br$_3$ and [Y(Ph$_3$PO)$_4$Cl$_2$] Cl, though it would be expected that lanthanide complexes would be analogous.

Complexes of tetrahydrofuran (THF) have been studied in considerable detail in recent years. The anhydrous metal chlorides, widely used in synthesis, are relatively insoluble in organic solvents whilst the ease of replacement of the labile THF molecules in the relatively soluble THF complexes makes them more useful synthetic reagents. Several types of complex have been characterized, including [LaCl$_3$(THF)$_4$] (seven coordinate); LnCl$_3$(THF)$_{3.5}$, actually [LnCl$_2$(THF)$_5$]$^+$ [LnCl$_4$(THF)$_2$]$^-$, containing both seven- and six-coordinate lanthanides (formed by metals such as Sm and Eu); and octahedral [LnCl$_3$(THF)$_3$], only found for Yb and Lu (see Question 4.3, below).

Other complexes to have been studied include Nd(THF)$_4$Br$_3$ and Yb(THF)$_4$(NCS)$_3$, both pentagonal bipyramidal seven-coordinate molecules with two axial halides (generally adopted for LnL$_4$X$_3$ systems). The thiocyanate complexes of the earlier lanthanides, [Ln(THF)$_4$(NCS)$_3$]$_2$, are dimers with two bridging thiocyanates affording eight coordination. Nitrate complexes Pr(THF)$_4$(NO$_3$)$_3$ and Yb(THF)$_3$(NO$_3$)$_3$ have been characterized, as recently have the iodides LnI$_3$(THF)$_4$ [Ln = Pr] and LnI$_3$(THF)$_{3.5}$ [Ln = Nd, Gd, Y], the latter being [LnI$_2$(THF)$_5$][LnI$_4$(THF)$_2$], analogous to some of the chlorides.

Another example of the lanthanide contraction at work, verified by crystallographers, lies in the complexes of the lanthanide nitrates with dimethyl sulfoxide (dmso). The early lanthanides form 10-coordinate Ln(dmso)$_4$(NO$_3$)$_3$ (Ln = La–Sm), whilst the heavier metals form Ln(dmso)$_3$(NO$_3$)$_3$(Ln = Eu–Lu, Y).

4.3.4 Complexes of β-Diketonates

These are an important class of compounds with a general formula Ln(R^1.CO.CH.CO.R^2)$_3$. The acetylacetonates, Ln(acac)$_3$ (R^1 = R^2 = CH$_3$) can readily be made from a lanthanide salt and acetylacetone by adding sodium hydroxide:

$$LnX_3 + 3\,Na(acac) \rightarrow Ln(acac)_3 + 3\,NaX$$

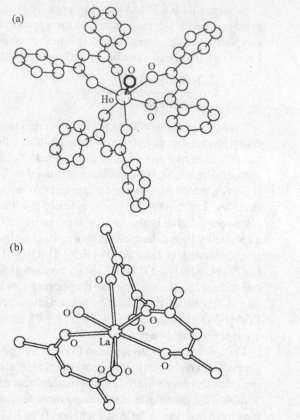

Figure 4.3
(a) The structure of Ho(PhCOCHCOPh)$_3$(H$_2$O) (reproduced by permission of the American Chemical Society from A. Zalkin, D.H. Templeton, and D.G. Karraker, *Inorg. Chem.*, 1969, **8**, 2680). (b) The structure of La(acac)$_3$.(H$_2$O)$_2$ (reproduced by permission of the American Chemical Society from T. Phillips, D.E. Sands, and W.F. Wagner, *Inorg. Chem.*, 1968, **7**, 2299).

They crystallize as hydrates Ln(acac)$_3$.(H$_2$O)$_2$(Ln = La–Ho, Y) and Ln(acac)$_3$.(H$_2$O) (Ln e.g. Yb) though there is a tendency to contain lattice water as well (Figure 4.3 for acac and PhCOCHCOPh complexes). These compounds are difficult to dehydrate, even *in vacuo*; they decompose on heating, and on dehydration at room temperature tend to oligomerize to involatile materials. They form Lewis base adducts like Ln(acac)$_3$.(Ph$_3$PO) and Ln(acac)$_3$.(phen) (7- and 8-coordinate, respectively).

Using bulkier R groups (R^1 = R^2 = CMe$_3$, for example) affords more congested diketonate complexes that are more tractable; Ln(Me$_3$.CO.CH.CO.CMe$_3$)$_3$ [often abbreviated to Ln(dpm)$_3$ or Ln(tmhd)$_3$] are monomers in solution and sublime *in vacuo* at 100–200 °C; in the solid state they are dimers for Ln = La–Dy (CN 7) and monomers for Dy–Lu (CN 6) and have trigonal prismatic coordination. They tend to hydrate readily, forming adducts [e.g. capped trigonal prismatic Ln(dpm)$_3$(H$_2$O)].

Complexes of fluorinated diketones (R^1 = CF$_3$, R^2 = CH$_3$, L = tfac; R^1 = R^2 = CF$_3$, L = hfac; R^1 = cyclo-C$_4$H$_3$S, R^2 = CF$_3$, L = tta; R^1 = CF$_3$CF$_2$CF$_2$, R^2 = CMe$_3$, L = fod) are also important. Again, the initial complexes obtained in synthesis are hydrates that can be dehydrated *in vacuo*. 2-Thenoyltrifluoroacetone complexes Ln(tta)$_3$.2H$_2$O are important in solvent extraction; addition of phosphine oxides gives a synergistic improvement in extraction owing to the formation of phosphine oxide complexes. Phosphine oxide complexes of hfac, $Ln(hfac)_3(Bu_3PO)_2$, can be separated by gas chromatography.

4.3.5 Lewis Base Adducts of β-Diketonate Complexes

Because the diketonates $Ln(R^1.CO.CH.CO.R^2)_3$ are co-ordinatively unsaturated, they complete their coordination sphere by forming adducts with Lewis bases, as already noted. Because of the paramagnetism of nearly all of the lanthanide ions, there are resulting shifts in the resonances in the NMR spectrum of any organic molecule coordinated and this means that these complexes were formerly important as NMR Shift Reagents (see Section 5.5.1). Their stoichiometry and structure was consequently well researched. Both 1:1 and 1:2 adducts are possible, depending on steric factors of both the complex and the ligand. Solubility factors may also mean that the most abundant species in a solution is not that which crystallizes. $Ln(fod)_3$ forms both 1:1 and 1:2 complexes with even quite bulky bases (e.g. 2,4,6-Me_3py). In contrast, the more bulky $Eu(dpm)_3$ forms solid $Eu(dpm)_3L_2$ adducts with both pyridine and $C_3H_7NH_2$, but only $Eu(dpm)_3L$ with the bulkier 2-methylpyridine. Detailed studies of $Ho(dpm)_3$ have shown evidence only for 1:1 complexes with sterically demanding ligands like Ph_3PO, camphor, and borneol.

The structures are known for a number of these compounds; thus, of the eight-coordinate compounds, $Eu(acac)_3(phen)$ is square antiprismatic, $Eu(dpm)_3(py)_2$ and $Nd(tta)_3(Ph_3PO)_2$ are dodecahedral; seven-coordinate $Lu(dpm)_3$(3-picoline) is a face-capped trigonal prism and $Eu(dpm)_3$(quinuclidine) a capped octahedron. Nine-coordination is found in $Eu(dpm)_3$(terpy) and also in $Pr(facam)_3(DMF)_3$(facam = 3-trifluoroacetyl-D-camphorate).

Anionic species of the type $[Ln(R^1.CO.CH.CO.R^2)_4]^-$ are another means of attaining coordinative saturation and are well known, such as the antiprismatic $[Eu(Ph.CO.CH.CO.Ph)_4]^-$ ion.

4.3.6 Nitrate Complexes

These exhibit high coordination numbers. Double nitrates $Mg_3Ln_2(NO_3)_{12}.24H_2O$, which contain $[Ln(NO_3)_6]^{3-}$ ions, are important historically as they were once used for the separation of the lanthanides by fractional crystallization. Use of counter-ions like Ph_4P^+ and Me_4N^+ allow isolation of the 10-coordinate $[Ln(NO_3)_5]^{2-}$ ions.

4.3.7 Crown Ether Complexes

In contrast to the alkali metals, the lanthanides do not form crown ether complexes readily in aqueous solution, due to the considerable hydration energy of the Ln^{3+} ion. These complexes are, however, readily synthesized by operating in non-aqueous solvents. Because many studies have been made with lanthanide nitrate complexes, coordination numbers are often high. Thus 12-coordination is found in $La(NO_3)_3$(18-crown-6) (Figure 4.4), 11 coordination in $La(NO_3)_3$(15-crown-5), and 10 coordination is found in $La(NO_3)_3$(12-crown-4). Other complexes isolated include Nd(18-crown-6)$_{0.75}(NO_3)_3$, which is in fact $[\{Nd$(18-crown-6)$(NO_3)_2\}^+]_3$ $[Nd(NO_3)_6]$. Other lanthanide salts complex with crown ethers; small crowns like 12-crown-4 give 2:1 complexes with lanthanide perchlorates, though the 2:1 complexes are not obtained with lanthanide nitrates where the anion can

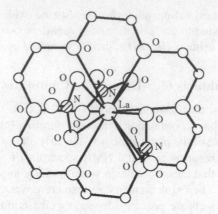

Figure 4.4
The structure of $La(NO_3)_3(18\text{-crown-}6)$ (reproduced by permission of the Royal Society of Chemistry from J.D.J. Backer-Dirks, J.E. Cooke, A.M.R. Galas, J.S. Ghotra, C.J. Gray, F.A. Hart, and M.B. Hursthouse, *J. Chem. Soc., Dalton Trans.*, 1980, 2191).

readily coordinate. Lanthanide chloride complexes often include water in the co-ordination sphere; thus the complex with the formula $ErCl_3(12\text{-crown-}4).5H_2O$ actually contains 9-coordinate $[Er(12\text{-crown-}4)(H_2O)_5]^{3+}$ cations whilst $NdCl_3(18\text{-crown-}6).4H_2O$ is $[Nd(18\text{-crown-}6)Cl_2(H_2O)_2]^+(Cl^-).2H_2O$. On reaction in MeCN/MeOH, 15-crown-5 reacts with neodymium chloride to form complexes with $[Nd(OH_2)_9]^{3+}$ and $[NdCl_2(OH_2)_6]^+$ ions hydrogen-bonded to the crown ether without any direct Nd–crown ether bonds. By carrying out electrocrystallization, the water-free $[Nd(15\text{-crown-}5)Cl_3]$ is obtained. Lanthanides also complex with the noncyclic linear polyethers such as glyme and with other macrocycles such as the calixarenes.

4.3.8   Complexes of EDTA and Related Ligands

These complexes are not only interesting on account of their chemical and structural properties, but they have an important place in the history of lanthanide separation chemistry. Although lanthanide ions do not form strong complexes with N-donor ligands, EDTA and other polyaminopolycarboxylic acids also have a number of carboxylate groups that also coordinate to the Ln^{3+} ion. Early experiments on separating lanthanides were carried out on a tracer scale using citrate complexes, but EDTA and related ligands were found to give lanthanide complexes that afforded better separations on a large scale (Section 1.5); thus log K for the EDTA complexes of Pr^{3+} and Nd^{3+} are 16.36 and 16.56, respectively; the average difference in K between adjacent lanthanides is $10^{0.4}$. EDTA is a hexadentate ligand; since transition metals can also coordinate a water molecule to form complexes like $[Fe(EDTA)(H_2O)]^-$, it is not surprising that lanthanides form complexes with more water molecules bound. The stoichiometry of the complex isolated depends to some extent upon the counter-ion; for example, nine-coordinate $[Ho(EDTA)(H_2O)_3]^-$ ions are found in the crystal when M = K, whilst eight-coordinate $[Ho(EDTA)(H_2O)_2]^-$ ions are found when M = Cs. In general, the early lanthanides form $M[Ln(EDTA)(H_2O)_3]$ and the later ones $[M(EDTA)(H_2O)_2]$. {See Figure 4.5 for the $[La(EDTA)(H_2O)_3]^-$ complex.}

Diethylenetriaminepentaacetic acid (H_5DTPA) is a potentially octadentate ligand (realized in practice) and forms complexes with larger stability constants, of the type $[M(DTPA)(H_2O)]^{2-}$. DTPA complexes have greater stability constants than do EDTA

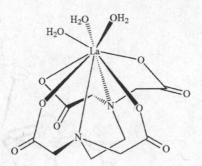

Figure 4.5
The structure of La(EDTA)(H₂O)₃⁻ (reproduced by permission of Macmillan from S.A. Cotton, *Lanthanides and Actinides*, 1991, p. 64).

complexes, since the greater denticity of the ligand makes it harder for the metal ion to be removed. Another factor is that the entropy change on complex formation is likely to be greater, thus making ΔG (and hence K) greater. Compare

$$[La(OH_2)_9]^{3+}(aq) + EDTA^{4-}(aq) \rightarrow [La(EDTA)(OH_2)_3]^-(aq) + 6\,H_2O(l)$$
$$[La(OH_2)_9]^{3+}(aq) + DTPA^{5-}(aq) \rightarrow [La(DTPA)(OH_2)]^{2-}(aq) + 8\,H_2O(l)$$

The synthesis and study of gadolinium complexes as potential magnetic resonance imaging agents is of intense interest at the moment, and is discussed in the section on MRI in Section 5.5.2.

4.3.9 Complexes of N-Donors

Although pyridine and similar bases are well known to form adducts with β-diketonate complexes, relatively few complexes of monodentate N-donors exist. Choice of solvent to restrict the oxophilic tendencies of the lanthanides is important. Thus, ammine complexes, [Yb(NH₃)₈][Cu(S₄)₂].NH₃, [Yb(NH₃)₈][Ag(S₄)₂].2NH₃, and [La(NH₃)₉][Cu(S₄)₂] have been synthesized by reactions in supercritical ammonia. Direct reaction of the lanthanide halides with pyridine gives pyridine complexes of the lanthanides, with the synthesis of [YCl₃py₄] and [LnCl₃py₄].0.5py (Ln = La, Er). These all have pentagonal bipyramidal structures, with two chlorines occupying the axial positions. [MI₃py₄] (M = Ce, Nd) and EuCl₃py₄ have also been reported, whilst square antiprismatic [Sm(meim)₈]I₃ and a few other N-methylimidazole complexes such as [YX₂(N-meim)₅]⁺ X⁻ (X = Cl, Br) have also been made.

In the case of thiocyanate complexes, a wide variety of species of known structure have been made, including (Bu₄N)₃[M(NCS)₆] (M = Y, Pr–Yb; octahedral coordination); (Et₄N)₄ [M(NCS)₇].benzene (M = La, Pr: capped trigonal prismatic); (Me₄N)₄ [M(NCS)₇].benzene (M = Dy, Et, Tb: pentagonal bipyramidal); and (Me₄N)₅ [M(NCS)₈].benzene (M = La–Dy: intermediate between square antiprism and cubic), showing the dependence of coordination number in the solid state upon factors such as the counter-ion.

The best examples of complexes with bidentate N-donors are those with 2,2′-bipyridyl (bipy) and 1,10-phenanthroline (phen). Complexes LnL₂(NO₃)₃ (L = bipy, phen) have been studied in detail and have 10-coordinate structures with all nitrates bidentate

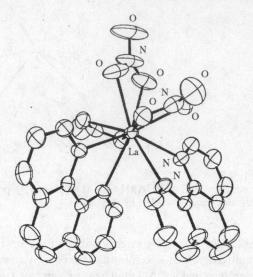

Figure 4.6
The structure of La(phen)$_2$(NO$_3$)$_3$ (reproduced by permission of the American Chemical Society from M. Frechette, I.R. Butler, R. Hynes, and C. Detellier, *Inorg. Chem.*, 1992, **31**, 1650).

(Figure 4.6). Ln(bipy)$_3$(NO$_3$)$_3$ are in fact Ln(bipy)$_2$(NO$_3$)$_3$.bipy (X-ray for Nd). Thiocyanates LnL$_3$(NCS)$_3$ (Ln = Pr; L = phen, bipy) have nine-coordinate monomeric structures, whilst chloride complexes with a range of structures include [Nd(bipy)$_2$Cl$_3$(OH$_2$)].EtOH and [Yb(bipy)$_2$Cl$_3$].

Tridentate N-donor ligands are efficient in separating actinides from lanthanides selectively by solvent extraction, an area of potential great importance in treatment of used nuclear fuel rods. The tridentate ligand 2,2': 6',2" -terpyridyl (terpy) forms a range of complexes. The perchlorate complexes [Ln(terpy)$_3$] (ClO$_4$)$_3$ contain nine-coordinate cations with near-D_3 symmetry, a structure initially deduced from the fluorescence spectrum of the europium compound (Section 5.4) and subsequently confirmed by X-ray diffraction studies (Figure 4.7)

A number of chloride complexes include Pr(terpy)Cl$_3$.8H$_2$O; X-ray diffraction established the presence of a [Pr(terpy)Cl(H$_2$O)$_5$]$^{2+}$ ion. Lanthanide nitrates react with 1 mole of terpy in MeCN to give [La(terpy)(NO$_3$)$_3$(H$_2$O)$_n$] (Ln = La, n = 2; Ln = Ce–Ho, n = 1; Ln = Er–Lu, n = 0) with coordination numbers decreasing from 11 (La) to 9 (Er–Lu); if these are crystallized from water [Ln(terpy)(NO$_3$)$_2$(H$_2$O)$_4$] NO$_3$ (Ln = La–Gd) and [Ln(terpy)(NO$_3$)$_2$(H$_2$O)$_3$]NO$_3$.2H$_2$O (Ln = Tb–Lu) are formed; these have both bidentate and ionic nitrate and nine- and eight-coordinate lanthanides, respectively. Thiocyanate complexes Ln(terpy)$_2$(NCS)$_3$(Ln = Pr, Nd) are also nine coordinate.

4.3.10 Complexes of Porphyrins and Related Systems

One convenient synthesis of these compounds involves heating La(acac)$_3$ with H$_2$TPP (H$_2$TPP = tetraphenylporphyrin) and other porphyrins in high-boiling solvents such as 1,3,5-trichlorobenzene (TCB). Products depend upon conditions and include [Ln(acac)(tpp)] and the double-decker sandwich [Ln$_2$(tpp)$_3$]. Less forcing routes can be used, for example the reaction of [Y{CH(SiMe$_3$)$_2$}$_3$] and yttrium alkoxides with H$_2$OEP

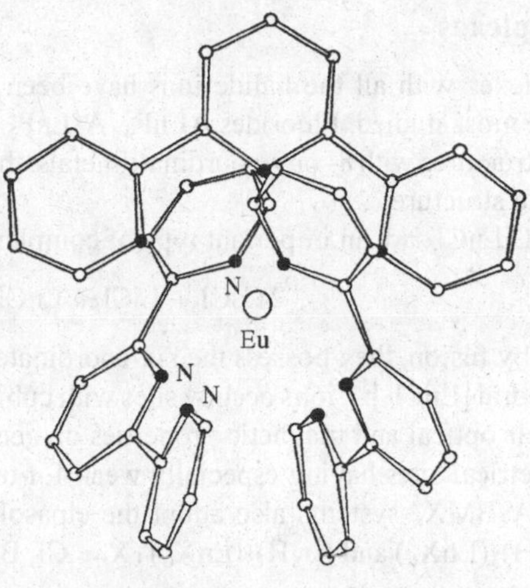

Figure 4.7
The structure of Eu(terpy)$_3^{3+}$ (reproduced by permission of the Royal Society of Chemistry from G.H. Frost, F.A. Hart, C. Heath, and M.B. Hursthouse, *Chem. Commun.*, 1969, 1421).

(octaethylporphyrin) to form compounds like [Y{CH(SiMe$_3$)$_2$}(Oep)]. Bis(porphyrin) complexes and mixed (porphyrin)phthalocyanine complexes are probably complexes of the trivalent lanthanide where one ligand is a dianion and another a one-electron-oxidized π-radical, as in [Ln^{III}(pc)(tpp)] (Ln = La, Pr, Nd, Eu, Gd, Er, Lu, Y).

Complexes, particularly of gadolinium, with a N$_4$-donor macrocycle backbone that also supports carboxylate groups have assumed importance in the context of MRI contrast agents (Sections 5.5.2 and 5.5.3).

Gadolinium texaphyrin (Gd-tex; Figure 4.8) is reported to be an effective radiation sensitizer for tumour cells, whilst the corresponding lutetium compound, which absorbs light in the far-red end of the visible spectrum, is in Phase II trials for photodynamic therapy for brain tumours and breast cancer.

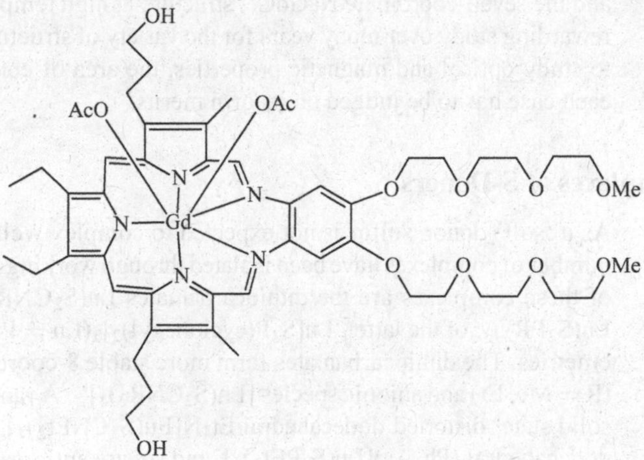

Figure 4.8
Gd-tex.

4.3.11   Halide Complexes

Complexes with all the halide ions have been made, though the fluorides and chlorides are the most studied. Fluorides $ALnF_4$, A_2LnF_5 and A_3LnF_6 exist (A = alkali metal); most have structures with 8- or 9-coordinate metals, though a few A_3LnF_6 have the six-coordinate cryolite structure.

$Cs_2LiLnCl_6$ are an important type of complex.

$$2\,CsCl + LiCl + LnCl_3 \rightarrow Cs_2LiLaCl_6$$

Made by fusion, they possess the six-coordinate elpasolite structure in which the perfectly octahedral $[LnCl_6]^{3-}$ ions occupy sites with cubic symmetry. These have been much studied for their optical and magnetic properties on account of this, the Ln^{3+} ions on their highly symmetrical sites having especially weak f–f transitions in their electronic spectra. Some other A_2BMX_6 systems also adopt the elpasolite structure. Other halide complexes like $(Ph_3PH)_3(LnX_6)$ and $(pyH)_3(LnX_6)$ (X = Cl, Br, I) have been made by solution methods:

$$LnX_3 + 3Ph_3PHX \rightarrow (Ph_3PH)_3LnX_6 \text{ (X = Cl, Br; in EtOH)}$$
$$(Ph_3PH)_3SmBr_6(s) + 6\,HI(l) \rightarrow (Ph_3PH)_3SmI_6(s) + 6\,HBr(l)$$

The iodides are hygroscopic and rather unstable. Other stoichiometries are possible:

$$LiCl + GdCl_3 \rightarrow LiGdCl_4;\, 4\,KI + 2\,Pr + 3\,I_2 \rightarrow 2\,K_2PrI_5$$

including species with more than one lanthanide ion:

$$3\,CsI + 2\,YI_3 \rightarrow Cs_3Y_2I_9$$

Six coordination is the norm for chlorides, bromides, and iodides, but it is not invariable. Formulae are not necessarily a guide to the species present; thus, octahedral $[YbCl_6]^{3-}$ ions are present in M_4YbCl_7 (M = Cs, $MeNH_3$) but Ba_2EuCl_7 contains capped trigonal prismatic $[EuCl_7]^{4-}$ ions. Compounds like $Cs_2Y_2I_9$ attain six coordination through face-sharing of octahedra, Cs_2DyCl_5 has corner-sharing of octahedra, and M_2PrBr_5 (M = K, NH_4) obtain seven coordination though edge-sharing. Structure type can depend on temperature; thus, $NaLnCl_4$ (Ln = Eu–Yb, Y) have the six-coordinate α-$NiWO_4$ structure at low temperature and the seven-coordinate $NaGdCl_4$ structure at high temperatures. To summarize, although rewarding study over many years for the variety of structures adopted and the opportunities to study optical and magnetic properties, the area of complex halides is complicated and each case has to be judged on its own merits.

4.3.12   Complexes of S-Donors

As a 'soft' donor, sulfur is not expected to complex well with lanthanides; nevertheless a number of complexes have been isolated through working in nonpolar solvents. The simplest of these complexes are the dithiocarbamates $Ln(S_2CNR_2)_3$ and related dithiophosphates $Ln(S_2PR_2)_3$; of the latter, $Ln[S_2P(cyclohexyl)_2]_3$(Ln = Pr, Sm) have trigonal prismatic geometries. The dithiocarbamates form more stable 8-coordinate adducts $Ln(S_2CNR_2)_3$.bipy (R = Me, Et) and anionic species $[Ln(S_2CNR_2)_4]^-$. A range of geometries are known in the solid state: distorted dodecahedral $Et_4N[Eu(S_2CNEt_2)_4]$, $Na[Eu(S_2CNEt_2)_4]$, near-perfect dodecahedral $(Ph_4As)[Ln(S_2PEt_2)_4]$, and square antiprismatic $Ph_4P\,[Pr(S_2PMe_2)_4]$.

A number of lanthanide thiolates have been synthesized in recent years and are discussed in Section 4.4.3.

4.4　Alkoxides, Alkylamides and Related Substances

Some of these compounds have been known for over 30 years, but interest in the former in particular has been stimulated recently by their potential use as precursors for deposition of metal oxides using the sol-gel or MOCVD process. These compounds are particularly interesting insofar as many of them involve very bulky ligands and therefore have low coordination numbers (see Section 4.5).

4.4.1　Alkylamides

The volatile air-sensitive silylamides $Ln[N(SiMe_3)_2]_3$ are the best-characterized alkylamides. They are made by salt-elimination reactions in a solvent like THF:

$$LnCl_3 + 3\ LiN(SiMe_3)_2 \rightarrow 3\ LiCl + Ln[N(SiMe_3)_2]_3$$

They are volatile *in vacuo* at 100 °C and dissolve in hydrocarbons; the first three-coordinate lanthanide compounds to be characterized, they have congested three-coordinate geometries (Figure 4.9) due to the bulky $SiMe_3$ groups. They are pyramidal in the solid state with N–Ln–N angles around 114° rather than the 120° expected for a planar structure but they are evidently planar in solution as they have no dipole moment. Theoretical calculations (D.L. Clark *et al.*, 2002; L. Penin *et al.*, 2002) indicate that it is β-Si–C agostic interactions with the lanthanide that are responsible for this small pyramidal distortion.

$Ln[N(SiMe_3)_2]_3$ form stable four-coordinate Ph_3PO and Ph_2CO adducts; when excess of phosphine oxide is used, five-coordinate peroxide-bridged dimers are obtained. $Eu[N(SiMe_3)_2]_3$ forms both 1:1 and 1:2 adducts with Me_3PO. They are capable of binding two very small MeCN groups, the latter reversibly, but no thf adducts have been isolated.

When a deficit of the lithium alkylamide is used, dimeric chloride-bridged silylamides are obtained which are useful synthetic intermediates (Figure 4.10).

Using less bulky alkyl substituents, compounds $[Ln(N(SiHMe_2)_2)_3(thf)_2]$ have been made for La–Lu, Y. These have trigonal bipyramidal structures with axial thf molecules (the scandium analogue is pseudo-tetrahedral $\{Sc[N(SiHMe_2)_2]_3(thf)\}$). Other alkylamides $Ln(NPr^i_2)_3$ (Ln = La, Nd, Yb, Y, Lu) are obtained with the less hindered isopropylamide ligand; these compounds form isolable thf adducts. Simple dimethylamides cannot be obtained, as reaction of $LnCl_3$ with $LiNMe_2$ gives $Ln(NMe_2)_3$ as LiCl adducts.

Use of a highly fluorinated amide has permitted the synthesis of $[Sm[N(SiMe_3)(C_6F_5)]_3]$ in which there are many Sm....F and agostic interactions (Figure 4.11), so that the true coordination number of the metal is really greater than three.

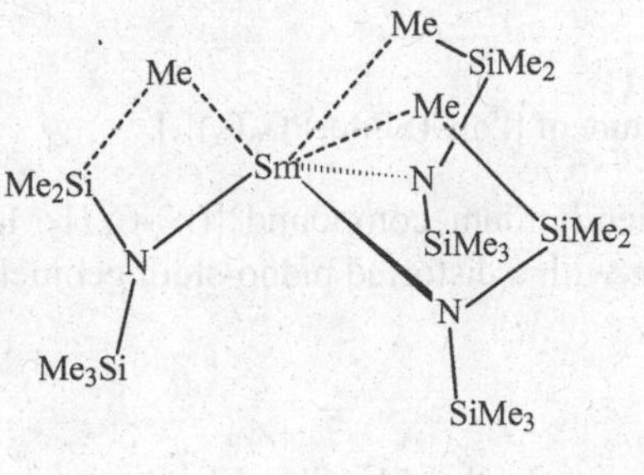

Figure 4.9
The structure of $Ln[N(SiMe_3)_2]_3$.

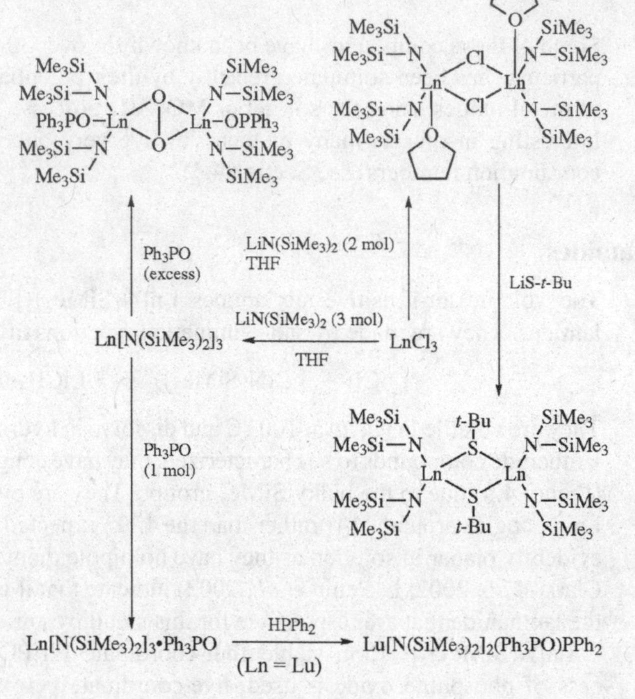

Figure 4.10
Lanthanide bis(trimethylsilyl)amido complexes.

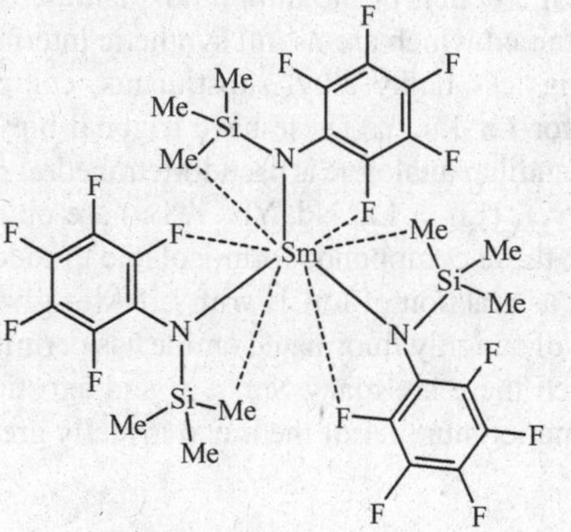

Figure 4.11
The structure of $[Ln[N(SiMe_3)(C_6F_5)]_3]$.

The neodymium compound $[(\eta^6\text{-}C_6H_5Me)Nd\{N(C_6F_5)_2\}_3]$ has a η^6-bonded toluene molecule with a distorted piano-stool geometry.

4.4.2 Alkoxides

The simplest lanthanide alkoxide compounds about which much is known are the iso-propoxides. There is some evidence that under mild conditions a simple $Ln(OPr^i)_3$ can be

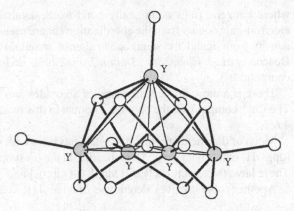

Figure 4.12
The structure of $Y_5O(OPr^i)_{13}$ (reproduced by permission of the American Chemical Society from O. Poncelet, O.W.J. Sartain, L.G. Hubert-Pfalzgraf, K. Folting, and K.G. Caulton, *Inorg. Chem.*, 1989, **28**, 264).

formed, but the product generally obtained from synthesis [which may be described by the synthesizers as '$Ln(OPr^i)_3$'] is an oxo-centred cluster $Ln_5O(OPr^i)_{13}$ (Figure 4.12) (X-ray for Y, Er, Yb, etc.) with a square-pyramidal core.

Syntheses include:

$$NdCl_3(Pr^iOH)_3 + 3\ KOPr^i \rightarrow Nd(OPr^i)_3 + 3\ KCl + 3\ Pr^iOH$$

$$Y + 3\ Pr^iOH \rightarrow Y(OPr^i)_3 + 3/2\ H_2$$

$$Pr[(N(SiMe_3)_2]_3 + 3\ Pr^iOH \rightarrow Pr(OPr^i)_3 + 3\ HN(SiMe_3)_2$$

The reaction conditions for syntheses seem important, as a chlorine-containing cluster $Nd_6(OPr^i)_{17}Cl$ has also been reported, from the reaction of $NdCl_3$ with $NaOPr^i$. Alkoxides of known structure include $Ln(OCH_2Bu^t)_3$, which is a tetramer (Ln = La, Nd), based on a square of lanthanide atoms with bridging alkoxides and no direct metal–metal bonds $[Ln_4(\mu_2\text{-}OR)_8(OR)_4]$ (R = CH_2Bu^t). $La(OBu^t)_3$ was isolated as a trimer containing a triangle of lanthanum atoms (and two coordinated alcohol molecules) $[La_3(\mu_3\text{-}OBu^t)_3(\mu_2\text{-}OBu^t)_3(OBu^t)_4(Bu^tOH)_2]$; $Nd(OCHPr^i)_3$ has been isolated as a dimeric solvate $[Nd_2(\mu_2\text{-}OR)_2(OR)_4L_2]$ (L = py, thf). $Ce(OBu^t)_3$ is believed to be a monomer. It undergoes thermolysis in high yield.

Dimeric structures are also exhibited by $[Ln(OQPh)_3]_2$ (Ln = La, Ce, Y; Q = C, Si), though the bridges are cleaved by Lewis bases, forming adducts such as $Ln(OSiPh)_3(thf)_3$ (*fac*-octahedral; Ln = Y, La, Ce) and $Y(OSiPh)_3(OPBu_3)_2$ (*tbp*).

A range of aryloxides has been made. The diphenylphenoxides $Ln(OR)_3$ (R = 2,6-$Ph_2C_6H_3O$) have three-coordinate pyramidal structures with some additional Ln–ring interactions. When the bulk of the aryloxide is increased by putting two tert-butyl groups into the 2,6-positions of the benzene ring, three-coordinate aryloxides $Ln(O\text{-}2,6Bu^t_2\text{-}4\text{-}RC_6H_2)_3$ can be isolated (R = H, Ln = Y, La, Sm; R = Me, Ln = Y, La, Ce, Pr, Nd, Dy–Er, Yb). Trigonal planar geometry is found in $Y(O\text{-}2,6\text{-}Bu^t_2C_6H_3)_3$ but a pyramidal structure is adopted in $Ce(O\text{-}2,6\text{-}Bu^t_2\text{-}4\text{-}MeC_6H_2)_3$. Synthesis of $Nd(Odpp)_3$ in THF gives $[Nd(Odpp)_3(thf)_2]$, which crystallizes from toluene as $[Nd(Odpp)_3(thf)]$ and sublimes as $[Nd(Odpp)_3]$. The structure of $[Nd(Odpp)_3(thf)_2].2THF$ shows it to be a five-coordinate bis(thf) complex in which the five ligands bind only by oxygen. Removal of the thf ligands affords compounds where benzene rings additionally bond to the lanthanide. The departure of one or two electron-pair donors from the coordination sphere means that the Ln^{3+} ion accepts electron density from ligand π systems and in a sense 'maintains its coordination number'. (see G.B. Deacon *et al., J. Chem. Soc., Dalton Trans.,* 2000, 961 for a fuller account of these unusual compounds.)

There is a more limited chemistry of alkoxides and aryloxides in the $+2$ and $+4$ states. The Ce^{IV} compounds, the only ones known in this oxidation state, are discussed in Section 4.6.2.

Much of the chemistry in the $+2$ state has been with aryloxides; thus 2,6-diphenylphenol (dppOH) reacts with Eu or Yb on heating in the presence of mercury, forming $[Ln(Odpp)_2]$. These have the structures $[Eu_2(Odpp)(\mu\text{-}Odpp)_3]$ and $[Yb_2(Odpp)(\mu\text{-}Odpp)_3]$.

Another route involves alcoholysis of an aryl $(R = 2,6\text{-}Bu^t_2C_6H_3O)$:

$$(C_6F_5)_2Yb \xrightarrow[\text{THF}]{\text{ROH}} Yb(OR)_2(thf)_n \quad (n = 2, 3)$$

$$(C_6F_5)_2Yb \xrightarrow[\text{Et}_2O]{\text{ROH}} Yb(OR)_2(Et_2O)_2$$

In recent years, phosphide complexes have proved accessible:

$$YbI_2(thf)_2 + 2\,KPPh_2 \rightarrow trans\text{-}Yb(PPh_2)_2(thf)_4 + 2\,KI$$

$$Yb(PPh_2)_2(thf)_4 + 4\,MeIm \rightarrow trans\text{-}Yb(PPh_2)_2(meim)_4 + 4\,THF$$

$$(\text{meim} = 1 - \text{methylimidazole})$$

Similar compounds have been synthesized, including $trans\text{-}Sm(Qmes_2)_2(thf)_4(Q = P,$ As; mes = mesityl); $[Nd\{P(SiMe_3)_2\}_3(thf)_2]$ has a trigonal bipyramidal structure.

4.4.3   Thiolates

Lanthanide thiolates are rare, but monomeric species have been obtained using bulky ligands, such as the three-coordinate $Sm(S\text{-}2,4,6\text{-}Bu^t_3C_6H_2)_3$ and the monomeric, *mer*-$[Yb(SPh)_3py_3]$. $Ln(SPh)_3$ reacts with S, forming octanuclear clusters $[Ln_8S_6(SPh)_{12}(thf)_8]$ $(Ln = Ce, Pr, Nd, Sm, Gd, Tb, Dy, Ho, Er)$; analogous pyridine clusters $[Ln_8S_6(SPh)_{12}(py)_8]$ $(Ln = Nd, Sm, Er)$ have also been isolated. Ph_2Se_2 reacts with lanthanide amalgams and pyridine, forming $Ln(SePh)_3(py)_3$ $(Ln = Ho, Tm, Yb)$ which are dimeric with 7-coordinate lanthanides. $[Ln(SePh)_3(thf)_3]$ $(Ln = Tm, Ho, Er)$ have monomeric *fac*-octahedral structures; $[Ln_8E_6(EPh)_{12}L_8]$ clusters $(E = S, Se; Ln = lanthanide; L = Lewis base)$ can be prepared by reduction of Se–C bonds by low-valent Ln or by reaction of $Ln(SePh)_3$. Structures reported include $[Sm_8E_6(SPh)_{12}(thf)_8]$ $(E = S, Se)$ and $[Sm_8Se_6(SePh)_{12}(py)_8]$; they have cubes of lanthanide ions with E^{2-} ions capping the faces and EPh bridging the edges of the cube. Reaction of $Nd(SePh)_3$ with Se to form $[Nd_8Se_6(SePh)_{12}(py)_8]$ shows that this series is not restricted to redox-active lanthanides.

4.5   Coordination Numbers in Lanthanide Complexes

4.5.1   General Principles

Ionic radii of La^{3+} and Lu^{3+} ions in octahedral coordination are 1.032 Å, and 0.861 Å, respectively. The corresponding value for the largest M^{3+} ion of a transition metal is 0.670 Å for Ti^{3+}. Purely on steric grounds, therefore, it would be expected that lanthanide ions could accommodate more than six ligands in their coordination sphere. Coupled with this is the fact that the f-orbitals are 'inner' orbitals, shielded from the effects of the surrounding anions and therefore not able to participate in directional bonding; there are none of the ligand-field effects found in transition metal chemistry with the concomitant preference for octahedral coordination. In summary, the coordination number (CN) adopted by a particular complex is determined by how many ligands can be packed round the central metal ion; coordination numbers between 2 and 12 are known in lanthanide complexes. Geometries are those corresponding to the simple polyhedra that would be predicted from electron-pair repulsion models; tetrahedral (CN 4), trigonal bipyramidal (CN 5), octahedral (CN 6); though with coordination numbers 2 and 3 deviations from simple linear and trigonal planar geometries in the solid state often occur (and are probably due to agostic interactions).

Some 20 years ago, analysis of the coordination numbers for large numbers of coordination compounds of yttrium and the lanthanides indicated that coordination numbers of 8 and 9 are almost equally common, accounting for around 60% of the known structures, and it is unlikely that this distribution has changed significantly. As already pointed out, it is steric factors that determine the coordination number (and geometry) adopted by a lanthanide ion (see e.g. X.-Z. Feng, A.-L. Guo, Y.-T. Xu, X.-F. Li and P.-N. Sun, *Polyhedron*, 1987, **6**, 1041; J. Marçalo and A. Pires de Matos, *Polyhedron*, 1989, **8**, 2431). Saturation in the coordination sphere of the metal can come about in one of two ways.

First-Order Effects

When small ligands like water or chloride bind to a metal, the coordination number is determined by how many ligands can pack round the central metal ion, a number relating to repulsion between the donor atoms directly in contact with the metal, a so-called 'first-order' effect.

Second-Order Effects

Certain 'bulky' ligands have a small donor atom (N, O, C) attached to bulky substituents, examples being certain alkoxides and aryloxides (and related systems), bis(trimethylsilyl)amido [-N(SiMe$_3$)$_2$], and the isolobal alkyl [-CH(SiMe$_3$)$_2$]; and mesityl. These ligands have high second-order steric effects generating crowding round the lanthanide; even though the metal ion is bound to few donor atoms, the bulk of the rest of the ligand shields the metal from other would-be ligands.

Species like $LnCl_6^{3-}$ and $[Ln(H_2O)_9]^{3+}$ are examples where first-order effects determine the coordination number, whilst $Ln[CH(SiMe_3)_2]_3$ and $[Ln\{N(SiMe_3)_2\}_3.(Ph_3PO)]$ are cases where the bulky substituents on the amide and alkyl ligands with high second-order steric effects generate crowding round the lanthanide.

4.5.2 Examples of the Coordination Numbers

Coordination Number 2

Coordination number 2 is represented by the sublimeable two coordinate bent alkyl [Yb$\{$C(SiMe$_3$)$_3\}_2$] (C–Yb–C 137°) and the europium analogue. Although there are only two Yb–C σ bonds, the bending is caused by a number of agostic Yb...H–C interactions (a similar situation applying to some zerovalent Pd and Pt phosphine complexes $M(PR_3)_2$, where R is a bulky group like Bu^t or cyclohexyl).

Coordination Number 3

This is best illustrated by some alkyl and alkylamide LnX_3 systems {X = a bulky group such as $-N(SiMe_3)_2$, $-CH(SiMe_3)_2$, e.g. $Ln[CH(SiMe_3)_2]_3$ (Ln = Y, La, Pr, Nd, Sm, Lu); $Ln[N(SiMe_3)_2]_3$, (Ln = Y, all lanthanides)}. Many transition metals form analogous, planar, three-coordinate alkylamides; the majority of these lanthanide compounds have pyramidal structures in the solid state [but are generally planar in the gas phase and in solution (they have zero dipole moment)]. These distortions appear to involve agostic interactions. Certain alkoxides and aryloxides $Ln(OR)_3$ are formed using bulky ligands (e.g. Ln = Ce; R = 2,6-$Bu^t_2C_6H_3O$).

Coordination Number 4

A few four-coordinate compounds are known, such as $Ln[(N(SiMe_3)_2]_3.(Ph_3PO)$, $[Li(thf)_4]$ $[Ln(NPh_2)_4]$ (Ln = Er, Yb), $[Li(thf)_4]$ $[Lu(2,6-dimethylphenyl)_4]$, and $Li(THF)_4][Ln(CH_2 SiMe_3)_4]$ (Ln = Y, Er, Tb, Yb); there is a tetrahedral geometry as expected. Again, the ligands have high second-order steric effects generating crowding round the lanthanide even though there are only four atoms bound to the lanthanide.

Coordination Number 5

As yet this is a rare coordination number. The three-coordinate silylamides can add two slender ligands to form five-coordinate examples $Ln[N(SiMe_3)_2]_3(NCMe)_2$, whilst there are a few alkyls e.g. $Yb(CH_2Bu^t)_3(thf)_2$. $[Nd\{P(SiMe_3)_2\}_3(thf)_2]$ has also been described. These have trigonal bipyramidal structures, as expected.

Coordination Number 6

Salts $Cs_2LiLnCl_6$ (Ln = La–Lu) have the elpasolite structure with $[LnCl_6]^{3-}$ anions on cubic sites; this geometry could be due to first-order crowding among the six chlorides, but some other salts are known where $LnCl_7$ units occur. Other examples include the reactive alkyls $[Li(L-L)_3]$ $[Ln(CH_3)_6]$ (L-L = $MeOCH_2CH_2OMe$ or $Me_2NCH_2CH_2NMe_2$; Ln = La–Sm, Gd–Lu, Y) and diketonates of bulky ligands such as $[Ln(Bu^tCOCHCOBu^t)_3]$ (Ln = Tb–Lu). Some compounds have geometry distorted to trigonal prismatic because of steric interactions between ligands, such as $[Ln\{S_2P(cyclohexyl)_2\}_3]$. Some thiocyanate complexes $(Bu_4N)_3$ $[M(NCS)_6]$ (M = Y, Pr–Yb) have octahedral coordination (Section 4.3.9) but others like $(Et_4N)_4$ $[M(NCS)_7].benzene$ (M = La, Pr) and $(Me_4N)_5$ $[M(NCS)_8].benzene$ (M = La–Dy) are 7- and 8-coordinate, respectively, a reminder that the coordination number adopted depends upon a subtle balance between factors, including the counter-ion and solvent.

Coordination Number 7

The most common geometries encountered are capped octahedral and capped trigonal prismatic. Many of the known seven-coordinate compounds involve β-diketonate ligands, particularly adducts of the type $Ln(diketonate)_3.L$ (L = Lewis base e.g. H_2O, py). Several thf complexes (Section 4.3.3) of the type $Ln(thf)_4X_3$ (X = Cl, NCS) adopt pentagonal bipyramidal geometries), as does the ytterbium(II) aryl $[Yb(C_6F_5)_2(thf)_5]$. Ba_2EuCl_7 has capped trigonal prismatic $[EuCl_7]^{4-}$ ions. The energy difference between these two geometries is small, witness $(Et_4N)_4 [M(NCS)_7].benzene$ (M = La, Pr: capped trigonal prismatic) and $(Me_4N)_4 [M(NCS)_7].benzene$ (M = Dy, Et, Tb: pentagonal bipyramidal).

Coordination Number 8

Most eight-coordinate lanthanide complexes have dodecahedral or square antiprismatic geometries. The energy difference between these is likely to be small, witness the distorted dodecahedral $Et_4N [Eu(S_2CNEt_2)_4]$, near-perfect dodecahedral $(Ph_4As)[Ln(S_2PEt_2)_4]$, and square antiprismatic $Ph_4P [Pr(S_2PMe_2)_4]$. Similarly, among eight-coordinate thiocyanate complexes, $(Me_4N)_5 [M(NCS)_8].benzene$ (M = La–Dy) are intermediate between square antiprism and cubic; $(Et_4N) [M(NCS)_4(H_2O)_4]$ (M = Nd, Eu) and $(Me_4N)_3 [M(NCS)_6(H_2O)(MeOH)]$ (M = La–Dy, Er) are square antiprismatic; $(Et_4N)_4[M(NCS)_7 (H_2O)]$ (M = La–Nd, Dy, Er) are cubic; and $(Me_4N)_5 [M(NCS)_8].benzene$ (M = La–Dy) are intermediate between square antiprismatic and cubic.

Coordination Number 9

Tricapped trigonal prismatic is the most familiar example of nine-coordinate geometry, adopted for the $[Ln(H_2O)_9]^{3+}$ ions of all lanthanides in a number of crystalline salts. It is also found in a number of chlorides $LnCl_3$ (Ln = La–Gd) and bromides $LnBr_3$ (Ln = La–Pr). This geometry is also sometimes adopted where polydentate ligands are involved, such as the $[Ln(terpy)_3]^{3+}$ ion (terpy = 2,2':6',2"-terpyridyl; see Figure 4.7).

Coordination Numbers 10–12

Sheer congestion of donor atoms around the metal ion and concomitant inter-donor atom repulsions makes these high coordination numbers difficult to attain. They are often associated with multidentate ligands with a small 'bite angle' such as nitrate that take up little space in the coordination sphere, either alone, as in $(Ph_4As)_2[Eu(NO_3)_5]$ or in combination with other ligands, as in $Ln(bipy)_2(NO_3)_3$, $Ln(terpy)(NO_3)_3(H_2O)$ (Ln = Ce–Ho), and crown ether complexes (Section 4.3.7) such as $Ln(12-crown-4)(NO_3)_3$(Ln = Nd–Lu). Other crown ether complexes can have 11 and 12 coordination, e.g. $Eu(15-crown-5)(NO_3)_3$ (Ln = Nd–Lu) and $Ln(18-crown-6)(NO_3)_3$(Ln = La, Nd).

Polyhedra in these high coordination numbers are often necessarily irregular, but when all the ligands are identical, near-icosahedral geometries occur for the 12-coordinate $[Pr(1,8-naphthyridine)_6]^{3+}$ and $[La(NO_3)_6]^{3-}$ ions in crystalline salts. It should also be remembered that the geometries discussed here are found in the solid state, but on dissolution in a solvent, where the influence of counter-ions is lessened, matters may be different (see the aqua ions, Sections 4.3.1 and 4.3.2). In principle, isomers are often possible, but because of the lability of lanthanide complexes they are very rarely observed.

4.5.3 The Lanthanide Contraction and Coordination Numbers

Certain *generalizations* can be made. As noted elsewhere (Section 2.4) the contraction of the $(5s^2 5p^6)$ configuration with increasing nuclear charge, due to penetration inside the f-orbitals, means that the Ln^{3+} (and Ln^{2+}) ions become smaller as the atomic number increases. Because of the decrease in ionic radius as the atomic number of the lanthanide increases, it would be expected that fewer anions could be packed round the central metal ion as the ionic radius decreases; in other words, that the coordination number will decrease with increasing Z. It may similarly be predicted that the coordination number would decrease with increasing size of the anion. This behaviour is clearly observed in the halides and oxides (Chapter 3). To reiterate, in the fluorides LnF_3, the coordination number decreases from 11 in LaF_3 through 9 in SmF_3; similarly in the trichlorides, lanthanum is 9 coordinate in $LaCl_3$ but lutetium is 6 coordinate in $LuCl_3$. The coordination number of lutetium decreases from 9 in LuF_3 to 6 in LuI_3. Similar changes in coordination number are found in many complexes, the best known example being the hydrated $Ln^{3+}(aq)$ ion. Nine-coordinate $[Ln(OH_2)_9]^{3+}$ ions are found in solution for the earlier metals (Ln = La–Sm/Eu) and eight-coordinate $[Ln(OH_2)_8]^{3+}$ ions for later metals (Dy–Lu, Y) with a mixture of species for the intermediate metals.

This decrease in coordination number is not an invariable rule, as examples are known where the coordination number remains constant across the series, as in $Ln(bipy)_2(NO_3)_3$ and $Ln(phen)_2(NO_3)_3$ (Ln = La–Lu) where all are 10 coordinate in the solid state, or $Ln(hmpa)_3Cl_3$, all of which appear to be 6 coordinate. When bidentate ligands are involved, loss of one of them may produce a too dramatic change in CN. The influence of a counter-ion, which will affect the lattice energy and hence the solubility, determining which complex species crystallizes first, can also be important. The best example of this lies in the crystalline hydrated salts of several oxoacids $[Ln(OH_2)_9]X_3$ (Ln = La–Lu, Y; X e.g. BrO_3, $CF_3SO_3^-$, $C_2H_5SO_4$) all of which contain the $[Ln(OH_2)_9]^{3+}$ ion, whereas the coordination number of the $Ln^{3+}(aq)$ ion varies in solution, as noted above. In contrast, the perchlorates $Ln(ClO_4)_3.6H_2O$ contain octahedral $[Ln(OH_2)_6]^{3+}$ ions. Similarly, in the edta complexes $M[Ln(edta)(H_2O)_x]$, nine-coordinate $[Ho(edta)(H_2O)_3]^-$ ions are found in the crystal when M = K whilst eight-coordinate $[Ho(edta)(H_2O)_2]^-$ ions are found when M = Cs.

4.5.4   Formulae and Coordination Numbers

Many transition metals adopt characteristic coordination numbers in their compounds, and consequently it is often possible to deduce whether (and how) a potential ligand will coordinate or not. The lanthanides do not have characteristic coordination numbers, so such assumptions are not possible; ultimately, X-ray diffraction studies are the only sure guide. The presence of coordinated solvent molecules in crystals obtained from polar solvents is generally best deduced thus, hence the assignment of an 8-coordinate structure to $[LaCl_3(phen)_2(H_2O)].MeOH$. Likewise $La(phen)_3Cl_3.9H_2O$ has been shown to be $[La(phen)_2(OH_2)_3]Cl_3.4H_2O.phen$. $[Sm(NO_3)_3(Ph_3PO)_3].2$ acetone and $[Sm(NO_3)_3(Ph_3QO)_2(EtOH)].acetone$ (Q = P, As) have all been shown to have nine-coordinate samarium.

4.6   The Coordination Chemistry of the $+2$ and $+4$ States

These areas have remained relatively undeveloped until recently; with increasing emphasis on the use of nonaqueous solvents, many discoveries in the $(+2)$ state for Eu, Sm, and Yb in particular probably remain to be made.

4.6.1   The (+2) State

Solutions of Eu^{3+}(aq) are reduced by zinc to Eu^{2+}(aq); this can be precipitated as the sulfate, affording a separation from the other Ln^{3+}(aq) ions, which were not reduced under these conditions, used for many years in the purification of the lanthanides. In general the solubilities of the 'inorganic' compounds of the Ln^{2+} ions resemble those of the corresponding compounds of the alkaline earth metals (insoluble sulfate, carbonate, hydroxide, oxalate). Electrolytic reduction is one of a number of methods that can be used to generate solutions of Eu^{2+}, Sm^{2+}, and Yb^{2+} ions; aqueous solutions of Sm^{2+} and Yb^{2+} tend to be short-lived and oxygen-sensitive in particular. These and other divalent lanthanides can be stabilized by the use of nonaqueous solvents such as HMPA and THF, in which they have characteristic colours, different and deeper than those in the isoelectronic Ln^{3+} ions on account of the decreased term separations in the divalent ions (typically, Eu^{2+} pale yellow; Sm^{2+} red; Yb^{2+} green-yellow; Tm^{2+} green; Dy^{2+} brown; Nd^{2+} red).

A significant number of complexes are now known, including $SmCl_2(thf)_5$ and $EuI_2(thf)_5$ (pentagonal bipyramidal); polymeric $[SmCl_2(Bu^tCN)_2]_\infty$ (six coordinate); $LnI_2(N\text{-methylimidazole})_4$ (Ln = Sm, Eu); $EuCl_2(phen)_2$ and $EuCl_2(terpy)$; *trans*-$EuI_2(thf)_4$; $[Yb(hmpa)_4(thf)_2]I_2$, $[Sm(hmpa)_6]I_2$, $[Sm(hmpa)_4I_2]$ and the eight-coordinate $[SmI_2\{O(C_2H_4OMe)_2\}]$ (the latter existing as *cis* and *trans* isomers). Compounds of lanthanides with an accessible Ln^{2+} ion yet previously unknown coordination chemistry are starting to be made, such as $[LnI_2(dme)_3]$ and $[LnI_2(thf)_5]$ (Ln = Nd, Dy), made by reaction of Nd and Dy with I_2 at 1500 °C, followed by dissolution of the product in the appropriate ligand. $[TmI_2(dme)_3]$ has seven-coordinate thulium, with one monodentate dimethoxyethane; the samarium analogue, which contains a larger metal ion, has eight coordination.

The bis(silylamide) complexes $Ln[N(SiMe_3)_2]_2$ (Ln = Eu, Yb) are well characterized; because of the bulky ligands, these tend to exhibit three and four coordination, much as for the corresponding compounds in the +3 oxidation state. Thus $Yb[N(SiMe_3)_2]_2$ is a dimer with two bridging alkylamides, and the $[Eu\{N(SiMe_3)_2\}_3]^-$ ion is trigonal planar (unlike the Eu^{III} analogue). These compounds form adducts with monomeric structures; $M[N(SiMe_3)_2]_2L_2$ (M = Eu, Yb; L, e.g., PBu_3, THF; L_2 = dmpe) are four-coordinate, and $Eu[N(SiMe_3)_2]_2(glyme)_2$ is six-coordinate.

There is a developing chemistry with alkoxide and aryloxide as well as amide ligands, together with organometallics. To give some examples from ytterbium(II) aryloxides, $Yb(Odpp)_2(thf)_3$ (Odpp = 2,6-diphenylphenoxide) is trigonal bipyramidal, whilst with the very bulky 2,6-di-*tert*-butyl-4-methylphenoxy (Odtb) ligand the adduct $Yb(Odtb)_2(thf)_2$ is tetrahedral whilst $Yb(Odtb)_2(thf)_3$ is five-coordinate square pyramidal. Among compounds with the heavier Group VI (16) donor atoms, $Yb(SAr)_2(MeOCH_2CH_2OMe)_2$ and $Ln(Qmes)_2(thf)_n]$ (Ln = Yb, Sm, Eu; Q = Se, Te; mes = 2,4,6-trimethylphenyl) are examples. The most striking organometallics are the two-coordinate alkyls $[Ln[C(SiMe_3)_3]_2]$ (Ln = Yb, Eu), the aryls $LnPh_2(thf)_n$ (Ln = Eu, Yb), with as-yet unknown structures, $Yb(C_6F_5)_2(thf)_4$ of unknown structure, and $Eu(C_6F_5)_2(thf)_5$ having a pentagonal bipyramidal structure, with all thf equatorial. The synthesis has also been reported of tetrahedral $[Eu(dpp)_2(thf)_2]$ (dpp = 2,6-diphenylphenyl). Much of this work is very recent, though cyclopentadienyl-type systems have long been known in the (+2) state. Organometallics are discussed in more detail in Chapter 6.

4.6.2   The (+4) State

Cerium is the only lanthanide with an extensive chemistry in this oxidation state. The cerium (+4) aqua ion, obtained by acid dissolution of CeO_2, is thermodynamically unstable but kinetically stable; the anion sensitivity of the reduction potentials ($E = 1.61$ V in 1M HNO_3; 1.44 V in 1M H_2SO_4; and 1.70 V in 1M $HClO_4$) indicates that complexes are involved rather than just a simple aqua ion. Salts are known, including $CeSO_4.4H_2O$ and $Ce(NO_3)_4.5H_2O$; the latter probably contains 11-coordinate $[Ce(NO_3)_4.(H_2O)_3]$ molecules, like the thorium analogue. Cerium is 10 coordinate in the $[Ce(CO_3)_5]^{6-}$ ion and 12 coordinate in $[Ce(NO_3)_6]^{2-}$, the ion found in the familiar orange-red salt $(NH_4)_2[Ce(NO_3)_6]$, widely used as an oxidizing agent in organic chemistry and in titrimetric analysis.

Cerium(IV) halide complexes include $(NH_4)_4[CeF_8]$ (square antiprismatic) and $(NH_4)_3CeF_7(H_2O)$ (dimeric, with dodecahedral coordination in the $Ce_2F_{12}^{6-}$ ions) which are prepared by crystallization; heating the former decomposes it to $(NH_4)_2CeF_6$, which contains infinite $(CeF_6^{2-})_\infty$ chains. Heating an MCl/CeO_2 mixture in fluorine gives alkali metal salts M_2CeF_6 and M_3CeF_7 (M = Na–Cs); unlike the ammonium salts, these are water-sensitive and cannot be made by wet methods. The ammonium salts may be more stable because hydrogen bonding between the ammonium ions and fluoride ions could contribute to a higher lattice energy.

Although the binary chloride does not exist, a number of hexachloro salts have been made:

$$CeO_2.xH_2O \xrightarrow[\text{diglyme}]{SOCl_2} (H_2 \text{ diglyme}_3)^{++} CeCl_6^{--}$$

$$CeO_2.xH_2O \xrightarrow[\text{EtOH/RCl}]{HCl(g)} R_2CeCl_6 \text{ (R = Ph}_3\text{PH, Et}_4\text{N, Ph}_4\text{As)}$$

These may be converted into the analogous violet, moisture-sensitive salts of the $[CeBr_6]^{2-}$ ion on treatment with gaseous HBr. No salts of the $[CeI_6]^{2-}$ ion have been reported, presumably for redox reasons, though a compound with a $Ce^{(IV)}$–I bond has been made (Figure 4.13).

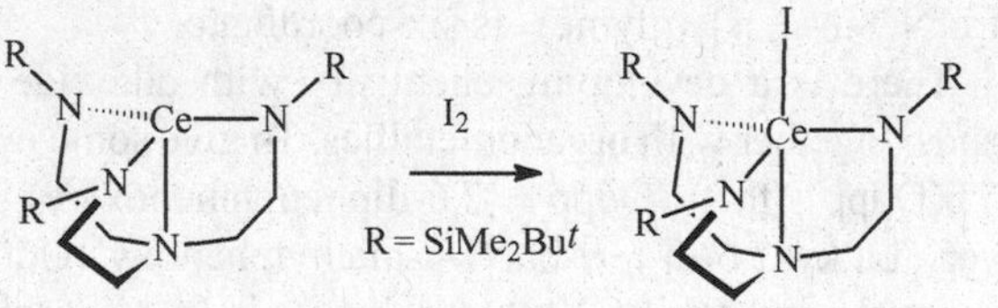

Figure 4.13
A compound with a Ce(lr)-I bond.

The four nitrogen atoms are good donors and reduce the effective charge on the cerium, affecting its reduction potential.

$[Ce\{N(SiMe_3)_2\}_3]$ is an obvious choice of starting material for the synthesis of other Ce^{IV} amides, but oxidation with Cl_2 or Br_2 was not successful; however, use of $TeCl_4$ oxidant led to $[Ce(Cl)\{N(SiMe_3)_2\}_3]$ and similarly Ph_3PBr_2 afforded $[Ce(Br)\{N(SiMe_3)_2\}_3]$ (Figure 4.14). The corresponding iodo compound has not (yet) been synthesized.

A number of neutral complexes of O-donor Lewis bases have been made, such as ten-coordinate *trans*-$Ce(NO_3)_4(Ph_3PO)_2$ and octahedral $CeCl_4L_2$ [L = Ph_3 PO, Ph_3AsO,

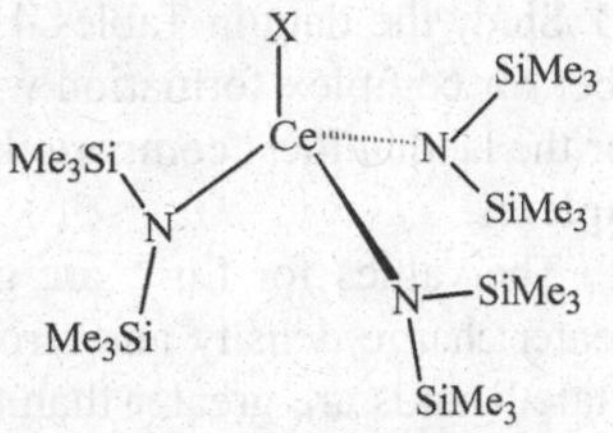

Figure 4.14
X = Br, Cl

$(Me_2N)_3PO$, Bu^t_2SO, $(H_2N)_2CO$]. They are prepared thus:

$$(NH_4)_2Ce(NO_3)_6 \xrightarrow[Me_2CO]{Ph_3PO} Ce(NO_3)_4(Ph_3PO)_2 \quad (trans)$$

$$(NH_4)_2Ce(NO_3)_6 \xrightarrow[Ph_3PO]{HCl(g)} CeCl_4(Ph_3PO)_2 \quad (cis)$$

$$H_2(diglyme)_3CeCl_6 \xrightarrow[EtOAc]{L} CeCl_4L_2 \quad [L = Ph_3AsO; Bu^t_2SO; (Me_2N)_3PO; trans]$$

The $CeCl_4(Ph_3PO)_2$ complex is *cis*, like the uranium analogue, whilst an X-ray diffraction study of $CeCl_4[(Me_2N)_3PO]_2$ (*trans*) showed a close structural correspondence to the corresponding U^{IV} compound.

$Ce(R^1.CO.CH.CO.R^2)_3$ are readily oxidized (O_2) to $Ce(R^1.CO.CH.CO.R^2)_4$, such as $Ce(acac)_4(R^1 = R^2 = Me)$, $Ce(dbm)_4(R^1 = R^2 = Ph)$, and $Ce(tmhd)_4$ $(R^1 = R^2 = Me_3C)$, generally found to have square antiprismatic structures, though $Ce(tmhd)_4$ is closer to dodecahedral. These are volatile dark red solids that are soluble in solvents such as benzene and chloroform; they are volatile, with vapour pressures high enough for Metal Organic Chemical Vapour Deposition (MOCVD) use, whilst they have also been studied as possible alternatives to lead compounds for petrol additives.

Another feature of the chemistry of cerium (IV) is the alkoxides.

$$(pyH)_2CeCl_6 + 4\ ROH + 6\ NH_3 \rightarrow Ce(OR)_4 + 2\ py + 6\ NH_4Cl$$

$$(NH_4)_2Ce(NO_3)_6 + 4\ ROH + 6\ NaOMe \rightarrow Ce(OR)_4 + 2\ NH_3 + 6\ NaNO_3 + 6\ MeOH$$

Most are involatile oligomers, though the isopropoxide is volatile *in vacuo* below 200 °C {its bis(propan-2-ol) adduct $[Ce_2(OPr^i)_8(Pr^iOH)_2]$ is a dimer with 6-coordinate cerium} and it can achieve coordinative saturation in adducts like $[Ce(OR)_4(thf)_2]$ $(R = CMe_3, SiPh_3)$ and $[Ce(OCMe_3)_6]^{2-}$. Heating $[Ce_2(OPr^i)_8(Pr^iOH)_2]$ gives $[Ce_4O(OPr^i)_{14}]$, which is $[Ce_4(\mu_4 -O)(\mu_3-OPr^i)_2(\mu-OPr^i)_8(OPr^i)_4]$, an oxo-centred cluster similar to those found with Ln^{III} alkoxides.

Cerium also forms genuine bis(porphyrin) and bis(phthalocyanine) complexes with the metal in the (+4) oxidation state.

Although other (+4) oxides exist in the form of PrO_2 and TbO_2, no other (+4) aqua ions are known; these oxides oxidize acids on dissolution, forming the M^{3+} (aq) ions. Alkali metal fluoro-complexes Cs_3LnF_7 (Ln = Pr, Nd, Tb, Dy) and M_2LnF_6 (M = Na–Cs, Ln = Tb, Pr) are known for four other lanthanides; as with Ce^{IV}, they have to be made by dry methods, by fluorination of Cs_3LnCl_6 using XeF_2 or MCl/Ln_2O_3 mixtures using F_2.

Question 4.1 Study the data in Tables 4.1 and 4.2 (above). Plot a graph of log β against atomic number for complex formation with EDTA^{4-}. Deduce and explain the patterns in behaviour for the lanthanides; compare the values of the stability constants for EDTA and fluoride complexes.

Answer 4.1 The values for Lu^{3+} are greater than those for La^{3+} as the smaller Lu^{3+} ion has a greater charge density and stronger electrostatic attraction for a ligand. Values for multidentate ligands are greater than those for monodentate ligands, partly because of entropy factors and also because once one end of a ligand is attached, there is a higher chance of the other donor atoms attaching themselves.

Question 4.2 Suggest why the acetates of the lanthanides do not adopt structures with M$_3$O cores as found for many carboxylates of trivalent transition metals (e.g. Cr, Mn, Fe).

Answer 4.2 A Molecular Orbital scheme for species such as [M$_3$O(OAc)$_6$]$^+$ has d–p π-bonding involving the metals and the central oxygen atom. Lanthanide f orbitals are buried too deeply to permit such overlap.

Question 4.3 A summary of structure types in the lanthanide chloride complexes of THF is given in Table 4.3 (data based upon G.B. Deacon, T. Feng, P.C. Junk, B.W. Skelton, A.N. Sobolev and A.H. White, *Aust. J. Chem.*, 1998, **51**, 75).

Use the list of structure types to work out the coordination number of the lanthanide for each compound. Plot a scatter graph of Average coordination number (y axis) against Atomic number (x axis) and comment.

Answer 4.3 There is a decrease in CN as the ionic radius decreases. The isolation of a number of complexes for some of the metal ions may reflect other factors such as the solubility of the complexes and the stoichiometry of the reaction mixtures.

Question 4.4 Why do Ln(acac)$_3$ form Lewis base adducts, when transition metal analogues M(acac)$_3$ (M, e.g., Cr, Fe, Co) do not?

Answer 4.4 The lanthanides are larger than transition metals and therefore Ln(acac)$_3$ are coordinatively unsaturated.

Question 4.5 Why is a nine-coordinate Eu(dpm)$_3$(terpy) known and not (for example) Eu(dpm)$_3$(H$_2$O)$_3$?

Answer 4.5 In terpyridyl, the three donor atoms are part of one molecule and take up less space than three individual molecules, even small ones.

Table 4.3 Lanthanide chloride complexes with THF

| Structure types characterized | | | | | | | | | | | | | |
La	Ce	Pr	Nd	Sm	Eu	Gd	Tb	Dy	Ho	Er	Tm	Yb	Lu
1	2	2	2,3	3	3	3,4	4	4	4	4	4	5,6	6

Description of types:

1. La(thf)$_2$(μ-Cl)$_3$Ln(thf)$_2$(μ-Cl)$_3$La. . . .
2. LaCl(thf)$_2$(μ-Cl)$_2$LnCl(thf)$_2$(μ-Cl)$_2$La . . . .
3. [LnCl$_3$(thf)$_4$].
4. [LnCl$_2$(thf)$_5$]$^+$ [LnCl$_4$(thf)$_2$]$^-$.
5. [Cl$_2$(thf)$_2$Ln(μ-Cl)$_2$Ln(thf)$_2$Cl$_2$].
6. [LnCl$_3$(thf)$_3$].

Question 4.6 Chinese workers have reported compounds with the formula $Ln(acac)_3$ $(Ph_3PO)_3$ (Ln, e.g., Nd, Eu, Ho). Comment on this.

Answer 4.6 Even though acac is the least bulky β-diketonate ligand, it would be surprising if all the Ph_3PO groups were coordinated (recall that Ph_3PO is sterically more demanding than water as a ligand). Ultimately, only X-ray diffraction studies can settle this.

Question 4.7 Why in general does nitrate coordinate to the metal in crown ether complexes but chloride does not?

Answer 4.7 Nitrate has a small bite angle and takes up little more space in the coordination sphere than a chloride ion; forming two Ln–O bonds is energetically more favourable than forming one Ln–Cl bond and can outweigh the considerable $Ln^{3+}-OH_2$ bond energy.

Question 4.8 Lanthanide thiocyanate complexes are N-bonded, not S-bonded. Why is this expected?

Answer 4.8 Ln^{3+} ions are 'hard' acids, which prefer to bond to 'hard' bases. Sulfur is a 'softer' base than nitrogen.

Question 4.9 Explain why europium nitrate forms a 1:1 complex $[Eu(terpy)(NO_3)_3(H_2O)]$ but europium perchlorate forms a 3:1 complex $[Eu(terpy)_3](ClO_4)_3$.

Answer 4.9 Perchlorate is a weakly coordinating anion, so does not compete with terpyridyl; nitrate has reasonably strong coordinating tendencies, usually as a bidentate ligand, and does compete with terpy. Three nitrate groups will occupy six coordinating positions, leaving room for only one terpy.

Question 4.10 Why might $Ce(R^1.CO.CH.CO.R^2)_4$ be expected to be a better MOCVD precursor and petrol additive than $Ce(R^1.CO.CH.CO.R^?)_3$?

Answer 4.10 Ln^{III} β-diketonates are often coordinatively unsaturated (Section 4.3.4) and tend to polymerize. The molecular nature of the Ce^{IV} compounds should confer greater volatility, as well as solubility in nonpolar solvents (e.g. petrol). The Ce^{IV} oxidation state may also assist the oxidation process.

Question 4.11 Suggest coordination numbers for the metal in each of the following:

$$[La(NO_3)_5(OH_2)_2]^{2-}; [Pr(NO_3)_3(thf)_4]; [PrCl_3(15\text{-crown-}5)].$$

Answer 4.11 12, 10, 8 respectively.

Question 4.12 Suggest geometries for (a) $[Eu[(N(SiMe_3)_2)_3]^-$; (b) $[Yb[(N(SiMe_3)_2]_2$ $(Me_2PCH_2CH_2PMe_2)]$; (c) $Ln[(N(SiMe_3)_2]_3(NCMe)_2$; (d) $(Me_4N)_3[Ln(NCS)_7]$; (e) $[Li(tmed)_2]^+ [LuBu^t_4]^-$.

Answer 4.12 (a) Trigonal/triangular; (b) tetrahedral; (c) trigonal bipyramid with axial nitriles; (d) capped octahedron/trigonal prism; (e) tetrahedral.

Question 4.13 Comment on the oxidation state of $[Ce_4O(OPr^i)_{14}]$.

Answer 4.13 Taking O as –2 and OPr^i as –1 leads to an assignment of the oxidation state of Ce as +4, further illustration of the stability of this state, when a redox reaction might have been expected.

Question 4.14 $Ln[N(SiMe_3)_2]_3$ are pentane-soluble and sublime in vacuum at 100 °C, whereas the much lighter $LnCl_3$ have high melting and boiling points and do not dissolve in organic solvents. Explain why.

Answer 4.14 $LnCl_3$ have giant ionic lattice structures, because of the relatively small size of the chloride ion. The nonpolar pentane cannot break up the ionic lattice to dissolve it. In contrast, the bulky silylamide ligands shield the metal ion, preventing oligomerization and also presenting an 'organic exterior' to the solvent pentane. The presence of weak, essentially van der Waals'-type intermolecular forces means that little energy is needed to vapourize the amides and little energy change occurs on dissolution in a nonpolar solvent.

Question 4.15 Three routes for the synthesis of alkoxides involve (i) the reaction of $LnCl_3$ with an alkali metal alkoxide, (ii) reaction of a lanthanide with an alcohol, (iii) the reaction of a silylamide $[Ln\{N(SiMe_3)_2\}_3]$ with an alcohol. What are the advantages and disadvantages of each?

Answer 4.15 (i) The chemicals are readily available. The chloride is fairly insoluble in organic solvents and it is possible that chlorine is retained in the product. (ii) There is no need to make the alkali metal alkoxide, as the alcohol is the starting material. The metal may need careful cleaning and there may be the need for heating and a catalyst. One product is a gas. (iii) The lanthanide silylamide has to be prepared first. Because of the bulky nature of the ligand, it may be inert to substitution, but there are no problems with chloride retention and the reaction should be clean.

Question 4.16 Lanthanide alkoxide clusters do not contain metal–metal bonds. Transition metal carbonyl clusters frequently do. Comment.

Answer 4.16 Transition metal carbonyls have transition metals in low oxidation states where d orbitals can overlap well. f orbital involvement in bonding in lanthanide compounds is minimal and the high oxidation state will contract the orbitals even further.

Question 4.17 Given the Zn^{2+}/Zn reduction potential of -0.76.V, and using values for Ln^{3+}/Ln^{2+} given in Table 2.7, explain why this can be used to separate Eu from other lanthanides that form Ln^{2+} ions.

Answer 4.17 Since $E = -0.76$ V for $Zn^{2+} + 2e^- \rightarrow Zn$, it will reduce Eu^{3+} to Eu^{2+}; since the reduction potential for $Eu^{3+} + e^- \rightarrow Eu^{2+}$ is -0.34 V, the potential for the reaction 2 $Eu^{3+} + Zn \rightarrow 2 Eu^{2+} + Zn^{2+}$ is $+ 0.42$ V, and the reduction is feasible. However, the reduction potentials for the other Ln^{3+} ions are all more negative than that for $Zn^{2+} + 2 e^- \rightarrow Zn$; for the next most easily reduced lanthanide (III) ion, ytterbium, $E = -1.05$ V for $Yb^{3+} + e^- \rightarrow Yb^{2+}$, and thus zinc will not reduce Yb^{3+} to Yb^{2+}.

5 Electronic and Magnetic Properties of the Lanthanides

By the end of this chapter you should be able to:

- know how to use Hund's rules to work out the ground state of lanthanide ions;
- calculate the magnetic moments for these ions, given the appropriate formula;
- understand why spin-only values for the moments are inadequate;
- state why the magnetic moments measured for Sm^{3+} and Eu^{3+} compounds differ from these predictions;
- appreciate why electronic (excitation) spectra show little dependence on compound, for a particular metal ion;
- be aware of the existence of hypersensitive transitions;
- interpret electronic spectra of compounds in terms of energy level diagrams;
- understand the causes of fluorescence in compounds of the lanthanides, particularly compounds of Eu^{3+} and Tb^{3+};
- interpret the fluorescence spectra of Eu^{3+} compounds in terms of the site symmetry;
- understand the principles involved in applications, including Nd^{3+} in lasers, and the use of lanthanides in lighting and TV tubes;
- explain the use of lanthanide complexes as MRI agents and NMR shift reagents.

5.1 Magnetic and Spectroscopic Properties of the Ln^{3+} Ions

These may be accounted for using the Russell–Saunders coupling scheme in which the electron spins are coupled together separately from the coupling of the orbital angular momenta of the electrons, and the orbital moment is unquenched. The ground state for a given lanthanide ion is unaffected by the ligands bound to it – and thus crystal field splittings are weak – because of the shielding of the 4f electrons by the filled 5s and 5p orbitals.

The spins of the individual electrons(s) are coupled together (added vectorially) to give the spin quantum number for the ion (S). The orbital angular momenta (l) of the individual electrons are coupled similarly.

For an f electron, $l = 3$, so that the magnetic quantum number m_l can have any one of the seven integral values between $+3$ and -3. Vectorial addition of the m_l-values for the f electrons for the multi-electron ion affords L, the total orbital angular momentum quantum number:

There is a weaker coupling, spin–orbit coupling, between S and L.

Lanthanide and Actinide Chemistry   S. Cotton
© 2006 John Wiley & Sons, Ltd.

L	0	1	2	3	4	5	6	7
State symbol	S	P	D	F	G	H	I	K

Figure 5.1
State symbols for different values of L.

m_l	3	2	1	0	−1	−2	−3
	↑	↑	↑	↑	↑		

Figure 5.2
Box diagram for Sm^{3+}.

Vector addition of L and S affords the resulting quantum number, J. J can have values of $(L + S), (L + S) − 1; \ldots\ldots (L − S)$. To take an example, if $L = 6$ and $S = 2$, J-values of 8, 7, 6, 5, and 4 are possible.

For any ion, a number of electronic states are possible. The ground state can be determined using Hund's rules (in this order):

1. The spin multiplicity $(2S + 1)$ is as high as possible.
2. If there is more than one term with the same spin multiplicity, the term with the highest L-value is the ground state.
3. For a shell less than half-filled, J for the ground state takes the lowest possible value; for a shell more than half-filled, J for the ground state is the highest possible.

To give an example, working out the term symbol for the ground state of Sm^{3+} (f^5) First complete a 'box diagram', representing orbitals by boxes (7 boxes for 7 f orbitals) and electrons by arrows (Figure 5.2). Put electrons in separate orbitals when the shell is less than half-filled, i.e. choosing the maximum number of unpaired electrons, and choosing to maximize the values of m_l to give the highest L-value (check Hund's rule 2).

So $S = \sum m_s = 5/2$, therefore $2S + 1 = 2(5/2) + 1 = 6$.
$L = \sum m_1 = +3 + 2 + 1 + 0 − 1 = +5$, so it is an H state.
J can have the values of $(L + S); (L + S) − 1; (L + S) − 2; \ldots\ldots.; (L − S)$, so here $J = (5 + 5/2); (5 + 5/2) − 1; (5 + 5/2) − 2 \ldots\ldots (5 − 5/2) = 15/2; 13/2; 11/2; 9/2; 7/2; 5/2$.

Since the shell is less than half-filled, the state with the lowest J-value is the ground state (Hund's third rule), so this is $J = 5/2$.

The term symbol for the ground state of Sm^{3+} is thus $^6H_{5/2}$

5.2 Magnetic Properties of the Ln^{3+} Ions

With the exception of La^{3+} and Lu^{3+} (and of course Y^{3+}) the Ln^{3+} ions all contain unpaired electrons and are paramagnetic. Their magnetic properties are determined entirely by the ground state (with two exceptions we shall encounter), as the excited states are so well separated from the ground state (owing to spin–orbit coupling; see Figure 5.3) and are thus thermally inaccessible.

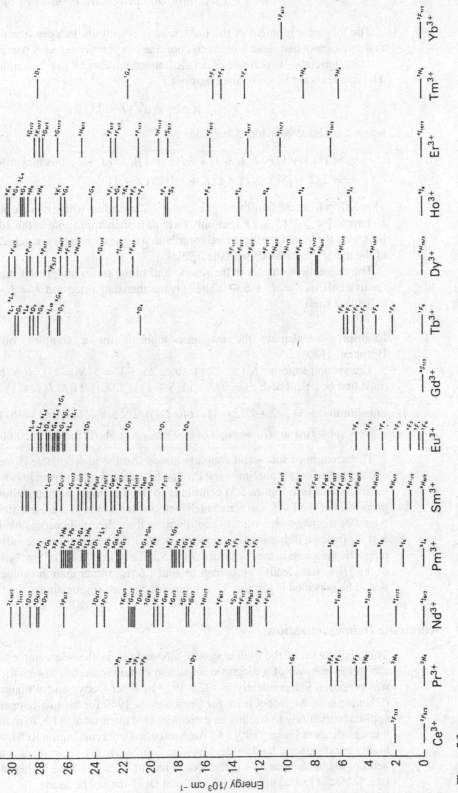

Figure 5.3

Energy levels of the Ln^{3+} ions [from S. Hüfner, in S.P. Sinha (ed.), *Systematics and Properties of the Lanthanides*, Reidel, Dordrecht, 1983; reproduced by permission of the publisher and author].

The magnetic moment of the Ln^{3+} ions is essentially independent of environment, so that one cannot distinguish between coordination geometries as is sometimes possible for transition metals – in the case of octahedral and tetrahedral Co^{2+} complexes, for example. The moments are given by the equation:

$$\mu_{eff} = g_J\sqrt{J(J + 1)}$$

where the Landé g-factor is defined by:

$g_J = [S(S + 1) - L(L + 1) + 3J(J + 1)]/2J(J + 1)$; this may also be written
$g_J = 3/2 + [S(S + 1) - L(L + 1)]/2J(J + 1)$

This differs from the familiar $\mu_{eff} = \sqrt{n(n + 2)}$ formula, where n is the number of unpaired electrons, [or $\sqrt{4S(S + 1)}$, spin-only formula], often applicable to the $3d^n$ transition metal ions, as in the latter case the orbital contribution to the moment is quenched by the interaction of the metal's 3d orbitals with the ligands.

The magnetic moments in the second half of the series are greater than the moments in the first half, as $J = L + S$ for a shell greater than half-filled and $J = L - S$ for a less than half-filled shell.

Example: Calculate the magnetic moment for a complex of Ho^{3+}, such as $Ho(phen)_2(NO_3)_3$.

The ground state is $^5I_8$ $(^{2S+1}L_J)$, since $2S + 1 = 5$, $S = 2$; $L = 6$ (I state); $J = 8$. g_J must first be calculated; $g_J = 3/2 + [S(S + 1) - L(L + 1)]/2J(J + 1)$

Substituting, $g_J = 3/2 + [2(2 + 1) - 6(6 + 1)]/2 \times 8(8 + 1) = 3/2 - 36/144$, so $g_J = 5/4$.

Now substitute in $\mu_{eff} = g_J\sqrt{J(J + 1)}$; $\mu_{eff} = 5/4\sqrt{8(8 + 1)} = 10.60 \mu_B$

The strength of spin–orbit coupling means that the ground state is well separated from excited states, except for Sm^{3+} and Eu^{3+}, where contributions from low-lying paramagnetic excited states (see Figure 5.3) contribute to the magnetic moment. Thus, if the magnetic properties of the Eu^{3+} ion were solely determined by the $^7F_0$ ground state, its compounds would be diamagnetic, whereas contributions from thermally accessible levels such as $^7F_1$ and $^7F_2$ [using a Boltzmann factor of $\exp(-\Delta E/kT)$] lead to the observed room-temperature magnetic moments in the region of 3.5 μ_B. Similarly, in the case of Sm^{3+}, thermal population of the $^6H_{7/2}$ state leads to moments around 1.6 μ_B, rather than the value of 0.845 μ_B that would be expected if just the $^6H_{5/2}$ ground state were responsible.

5.2.1 Adiabatic Demagnetization

This application of the high magnetic moments of lanthanide complexes is a process by which the removal of a magnetic field from certain materials lowers their temperature; it was proposed independently in 1926–1927 by Peter Debye and William Francis Giauque (Giauque won the Nobel prize for Chemistry in 1949 for his low-temperature work). It is applied particularly to cooling an extremely cold material at $\sim$1 K to much lower temperatures, perhaps as low as 0.0015 K. An Adiabatic Demagnetization Refrigerator uses certain highly paramagnetic lanthanide salts [e.g. $Gd_2(SO_4)_3.8H_2O$] immersed in a bath of liquid helium; the lanthanide ions in the second half of the series have higher magnetic moments (see Table 5.1) so compounds of Gd^{3+} and Dy^{3+} tend to be used.

Table 5.1   Magnetic Moments of Ln^{3+} ions at room temperature

	f^n	Ground term	Predicted μ_{eff} (μ_B)	μ_{eff} M(phen)$_2$(NO$_3$)$_3$ (μ_B)
La	0	$^1S_0$	0.00	0
Ce	1	$^2F_{5/2}$	2.54	2.46
Pr	2	$^3H_4$	3.58	3.48
Nd	3	$^4I_{9/2}$	3.68	3.44
Pm	4	$^5I_4$	2.83	
Sm	5	$^6H_{5/2}$	0.85	1.64
Eu	6	$^7F_0$	0.00	3.36
Gd	7	$^8S_{7/2}$	7.94	7.97
Tb	8	$^7F_6$	9.72	9.81
Dy	9	$^6H_{15/2}$	10.63	10.6
Ho	10	$^5I_8$	10.60	10.7
Er	11	$^4I_{15/2}$	9.59	9.46
Tm	12	$^3H_6$	7.57	7.51
Yb	13	$^2F_{7/2}$	4.53	4.47
Lu	14	$^1S_0$	0.00	0

If the paramagnetic substance is placed in a strong external magnetic field, the magnetic dipoles tend to align with the field. This is an exothermic process, as 'aligned' is the lowest energy state, and unaligned dipoles need to lose energy to achieve the lowest state. The heat energy released is removed by the external coolant, and the magnetic material (and its surroundings) cool back to their original temperature (say 1 K). Thermal contact between the material and the liquid helium bath is now broken (a 'heat switch' is turned off) and the magnetic field is turned off. The magnetic dipoles tend to randomize now, absorbing thermal energy from the magnetic material, which cools (along with its immediate surroundings), producing the cooling in the sample.

5.3   Energy Level Diagrams for the Lanthanide Ions, and their Electronic Spectra

Energy level diagrams for all the Ln^{3+} ions are shown in Figure 5.3.

These are based on theoretical predictions, coupled with experimental results. Accurate values have been determined in many experimental situations, such as halide lattices like LnF_3, which closely resemble those of the gaseous ions. In the lanthanide ions, the filled 'outer' 5s and 5p orbitals efficiently shield the 4f electrons from surrounding ligands, with the result that crystal field splittings are of the order of 100 cm^{-1}. The weak crystal field splittings can thus be treated as a perturbation upon the free-ion levels; in contrast, of course, in the 3d metals, CF splittings are large and L–S coupling is weak. A consequence of the lack of CF effects in the lanthanides is that thermal motion of the ligands has very little effect upon them (as is not the case in the 3d situation), so the f–f absorption bands in the spectra are very narrow, almost as narrow as for free (gaseous) ions.

5.3.1   Electronic Spectra

Most lanthanide ions absorb electromagnetic radiation, particularly in the visible region of the spectrum, exciting the ion from its ground state to a higher electronic state, as a consequence of the partly filled 4f subshell. The f–f transitions are excited both by magnetic dipole and electric dipole radiation. Normally the magnetic dipole transitions would not be seen, but in the case of the lanthanides the electric dipole transitions are much weaker than for transition metal complexes and magnetic dipole transitions can often be seen, especially in fluorescence spectra. The magnetic dipole transitions are parity-allowed, whilst the electric dipole transitions are parity-forbidden ('Laporte-forbidden') in the same sense as d–d transitions in transition metal ions. The f–f transitions gain intensity through mixing in higher electronic states (including d states) of opposite parity, either through the (permanent) effects of a low-symmetry ligand field or through asymmetric molecular vibrations that momentarily destroy any centre of symmetry – 'vibronic coupling' – though the effect is still weaker than in transition metal complexes.

Not all the lanthanide ions give rise to f–f transitions, including obviously the f^0 and f^{14} species, La^{3+} and Lu^{3+}. Likewise there are no f–f transitions for the $f^1(Ce^{3+})$ and f^{13} (Yb^{3+}) ions, as with only a single L-value there is no upper 4f state. Transitions between $^2F_{5/2}$ and $^2F_{7/2}$ are seen in the case of Ce^{3+} as a rather broad band in the infrared region around 2000 cm^{-1}. Ce^{3+} and Yb^{3+} do, however, give rise to broad $4f^n \rightarrow 4f^{n-1}5d^1$ transitions (as indeed do many lanthanides). Even an ion like Eu^{3+}, which has several absorptions in the visible region of the spectrum, has only weak absorptions, so many of its compounds appear colourless; the only three tripositive ions whose compounds are invariably coloured are Pr^{3+}, Nd^{3+}, and Er^{3+}.

Colours of the Ln^{3+} ions in aqueous solution are listed in Table 5.2. A typical spectrum is that of Pr^{3+}, shown in Figure 5.4; note the sharp absorption bands, with extinction coefficients below unity.

The electronic spectra of lanthanide compounds resemble those of the free ions, in contrast to the norm in transition metal chemistry; the crystal-field splittings can be treated as a perturbation on the unsplit $^{2S+1}L_J$ levels. Complexes thus have much the same colour as the corresponding aqua ions. Thus one cannot distinguish between coordination geometries as with octahedrally and tetrahedrally coordinated Co^{2+}, for example (let alone note profound colour differences). As already noted, a further consequence of the weak CF splittings is the sharpness of the f–f transitions. The spectrum of the complex $Eu(NO_3)_3$(15-crown-5) shown in Figure 5.5 is another example of this.

The previous discussion has centred upon the Ln^{3+} ions. Most Ln^{2+} ions are not stable in solution, but have been prepared artificially in lattices by doping CaF_2 with the Ln^{3+}

Table 5.2   Colours of Ln^{3+} ions in aqueous solution

	f^n	Colour
La^{3+}	f^0	Colourless
Ce^{3+}	f^1	Colourless
Pr^{3+}	f^2	Green
Nd^{3+}	f^3	Violet
Pm^{3+}	f^4	Pink
Sm^{3+}	f^5	Pale yellow
Eu^{3+}	f^6	Colourless
Gd^{3+}	f^7	Colourless
Tb^{3+}	f^8	V.pale pink
Dy^{3+}	f^9	Pale yellow
Ho^{3+}	f^{10}	Yellow
Er^{3+}	f^{11}	Rose
Tm^{3+}	f^{12}	Pale green
Yb^{3+}	f^{13}	Colourless
Lu^{3+}	f^{14}	Colourless

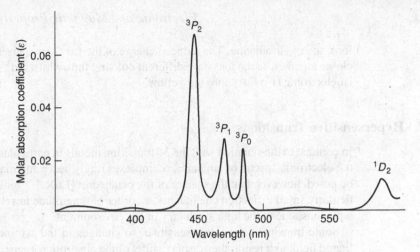

Figure 5.4
Electronic absorption bands in the spectrum of $PrCl_3$ (aq) reproduced with permission from S.A. Cotton, Lanthanides and Actinides, Macmillan (1991) p. 30.

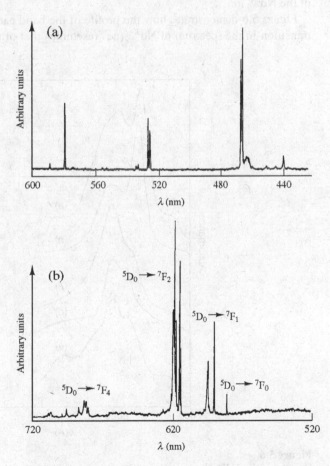

Figure 5.5
(a) Excitation spectrum at 77 K of $Eu(NO_3)_3 \cdot 15$-crown-5 ($\lambda_{anal} = 618$ nm). (b) Fluorescence spectrum at 77 K ($\lambda_{exc} = 397.7$ nm) (reproduced by permission of the American Chemical Society from J.-C.G. Bunzli, B. Klein, G. Chapuis, and K.J. Schenk, *Inorg. Chem.*, 1982, **21**, 808).

ions, then γ-irradiating. The reduced charge of the Ln^{2+} ion causes the energy levels to be closer together, so the ions have different colours; thus whilst Gd^{3+} ions are colourless, the isoelectronic (f^7) Eu^{2+} ions are yellow.

5.3.2 Hypersensitive Transitions

In contrast to the situation with the 3d transition metals in particular, the 4f–4f transitions in the electronic spectra of lanthanide complexes rarely serve any diagnostic purpose. It may be noted, however, that the spectra of the octahedral $[LnX_6]^{3-}$ ions (X = Cl, Br) have particularly small extinction coefficients, an order of magnitude lower than the corresponding aqua ions, due to the high symmetry of the environment.

Some transitions are 'hypersensitive' to changes in the symmetry and strength of the ligand field; as a result, they display shifts of the absorption bands, usually to longer wavelength, as well as band splitting and intensity variation. It is most marked for Ho^{3+}, Er^{3+} and particularly for the $^4I_{9/2} \rightarrow \, ^2H_{9/2}$, $^4F_{5/2}$ and $^4I_{9/2}, \rightarrow \, ^4G_{5/2}$, $^4G_{7/2}$ transitions in the case of the Nd^{3+} ion.

Figure 5.6 demonstrates how the profile of the band caused by the $^4I_{9/2} \rightarrow \, ^2H_{9/2}$, $^4F_{5/2}$ transition in the spectrum of Nd^{3+} (aq) resembles that of the tricapped trigonal prismatic

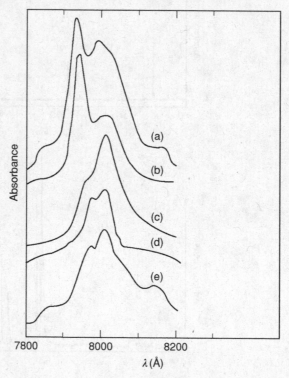

Figure 5.6
Spectra of the Nd^{3+} $^4I_{9/2} \rightarrow \, ^2H_{9/2}$, $^4F_{5/2}$ transitions: (a) solid $Nd(BrO_3)_3 \cdot 9H_2O$; (b) $5.35 \times 10^{-2}M$ Nd^{3+} (aq); (c) $5.35 \times 10^{-2}M$ Nd^{3+} in 11.4M HCl; (d) solid $NdCl_3 \cdot 6H_2O$; (e) Solid $Nd_2(SO_4)_3 \cdot 8H_2O$ (reproduced by permission of the American Chemical Society from D.G. Karraker, *Inorg. Chem.*, 1968, **7**, 473).

$[Nd(H_2O)_9]^{3+}$ ions in $Nd(BrO_3)_3.9H_2O$, but is significantly different from that of the 8-coordinate Nd^{3+} in the hydrated sulfate and chloride salts. This was important evidence in favour of the aqua ion being 9 coordinate. The spectrum of Nd^{3+} in concentrated HCl is significantly different, and suggests that an 8-coordinate species is present here.

5.4 Luminescence Spectra

Many lanthanide ions exhibit luminescence, emitting radiation from an excited electronic state, the emitted light having sharp lines characteristic of f–f transitions of a Ln^{3+} ion. As will be seen later, this can be enhanced considerably by attaching a suitable organic ligand (e.g. a β-diketonate, phenanthroline, crown ether, etc.) to the lanthanide.

The process occurs as summarized in Figure 5.7.

The mechanism of luminescence of a lanthanide complex is as follows. An electron is promoted to an excited singlet state in a ligand upon absorption of a quantum of energy (e.g. from ultraviolet light). This photon drops back to the lowest state of the excited singlet, from where it can return to the ground state directly (ligand fluorescence) or follow a non-radiative path to a triplet state of the ligand. Thence it may either return to the ground state (phosphorescence) or alternatively undergo non-radiative intersystem crossing, this time to a nearby excited state of a Ln^{3+} ion, whence it can return to the ground state either by non-radiative emission or by metal-ion fluorescence involving an f–f transition.

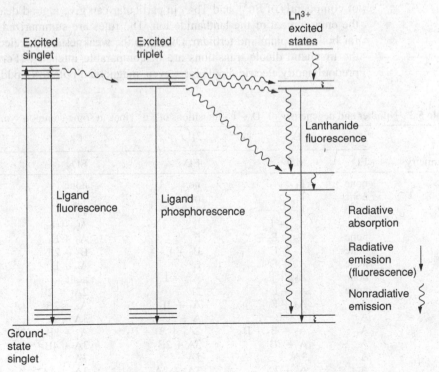

Figure 5.7
Luminescence in lanthanide complexes.

Certain Ln^{3+} ions have excited states lying slightly lower in energy than the triplet states of typical ligands, and exhibit strong metal-ion fluorescence, most markedly for Eu^{3+} and Tb^{3+}, and possibly for others such as Sm^{3+} and Dy^{3+}. Among the other Ln^{3+} ions, La^{3+} and Lu^{3+} have no f^n excited state; Gd^{3+} has all its excited states above the ligand triplet states while the others have a large number of excited states promoting energy loss by a non-radiative route. Tb^{3+} and Eu^{3+} are thus the two most useful ions for these studies. They luminesce with green and red colours, respectively. For Tb^{3+} the main emissions responsible are $^5D_4 \rightarrow {}^7F_n (n = 6{-}0)$ with $^5D_4 \rightarrow {}^7F_5$ the strongest, whilst for Eu^{3+}, $^5D_0 \rightarrow {}^7F_n$ are seen ($n = 4{-}0$) with the transitions to $^7F_0$, $^7F_1$, and $^7F_2$ the most useful. The process is important in lighting applications, and also in both qualitative and quantitative analysis, and in luminescent imaging.

Although the energy level separations are little affected by ligand, nevertheless they are not the same in all complexes. Careful study of the $^7F_0 \rightarrow {}^5D_0$ separation in a number of Eu^{3+} complexes has shown its shift towards lower energies to depend upon the donor atoms involved, the actual values of the separation varying between $17\,232$ cm^{-1} in $[Eu(dpa)_3]^{3-}$ (dpa = dipicolinate) and $17\,280$ cm^{-1} in $[Eu(H_2O)_9]^{3+}$ in some 30 complexes studied, reflecting changes in covalency in the bond.

The ligand field in a complex ion removes the degeneracy of a given $^{2S+1}L_J$ term partly or completely, the extent of this and the resultant splitting of the emission line depending upon the symmetry of the ligand field (Table 5.3). This means that, in many cases, the individual transitions in the luminescence spectra consist of more than one line (subject to resolution).

Study of the intensity and splitting pattern of certain transitions in the fluorescence spectra of compounds of Eu^{3+} and Tb^{3+} in particular can give a good deal of information about the environment of the lanthanide ion. The rules are summarized in Tables 5.4 and 5.5 for both europium and terbium. Owing to the weakness of the electric dipole transitions, the magnetic dipole transitions are of comparable intensity. (For Eu^{3+}, $^5D_0 \rightarrow {}^7F_x$ are predominantly electric dipole for x even, magnetic dipole for x odd).

Table 5.3 Number and degeneracy of $^5D_0$–^7F_J transitions of Eu^{3+} ions in some common symmetries

	$^7F_0$	$^7F_1$	$^7F_2$	$^7F_3$	$^7F_4$
Symmetry	ED	MD	ED	ED	ED
I_h	none	T_{1g}	none	none	none
O_h	none	T_{1g}	none	T_{1g}	none
T_d	none	T_1	T_2	T_1	T_2
D_{4h}	none	$A_{2g} + E_g$	E_g	$A_{2g} + E_g$	none
D_{4d}	none	$A_2 + E_3$	E_1	$A_2 + E_3$	$B_2 + E_1$
D_{2d}	none	$A_2 + E$	$B_2 + E$	$B_2 + 2E$	$B_2 + 2E$
D_{3h}	none	$A_2' + E''$	E'	$A_2' + E''$	$A_2'' + 2E'$
D_{3d}	none	$A_{2g} + E_g$	$A_{1g} + E_g$	none	none
D_3	none	$A_2 + E$	$2E$	$2A_2 + 2E$	$A_2 + 3E$
C_{3v}	A_1	$A_2 + E$	$A_1 + 2E$	$A_1 + 2E$	$2A_1 + 3E$
C_3	A	$A + E$	$A + 2E$	$3A + 2E$	$3A + 3E$
C_{2v}	A_1	$A_2 + B_1 + B_2$	$2A_1 + B_1 + B_2$	$A_1 + 2B_1 + 2B_2$	$3A_1 + 2B_1 + 2B_2$
C_2	A	$A + 2B$	$3A + 2B$	$3A + 4B$	$5A + 4B$
C_i	A	$3A$	$5A$	$7A$	$9A$
C_s	A'	$A' + 2A''$	$3A' + 2A''$	$3A' + 4A''$	$5A' + 4A''$

Table 5.4 Features of $^5D_0 \rightarrow {}^7F_J$ luminescent transitions for Eu^{3+}

J	Main character	Region (nm)	Intensity	Comments
0	ED	577–581	V.weak	Absent in high symmetry – 'forbidden'
1	MD	585–600	Strong	Intensity largely independent of environment
2	ED	610–625	V.weak to v.strong	Absent if ion on inversion centre; hypersensitive
3	ED	640–655	V.weak	'Forbidden'
4	ED	680–710	Medium to strong	Environment-sensitive

Table 5.5 Features of $^5D_4 \rightarrow {}^7F_J$ luminescent transitions for Tb^{3+}

J	Region (nm)	Intensity	Comments
6	480–505	Medium to strong	Environment-sensitive
5	535–555	Strong to v.strong	Good probe
4	580–600	Medium to strong	Environment-sensitive
3	615–625	Medium	
2	640–655	Weak	Environment-sensitive

As the luminescence technique is one that can be applied particularly to lanthanide complexes we will discuss it in some depth; and since interpretation is simplest in the case of Eu^{3+}, so discussion will centre on that, particularly since it is most involved in applications. In the case of the europium(III) ion, the $^7F_0$ ground state is unsplit, so that transitions to it give straightforward information about the excited state. Because the $^5D_0$ state is also unsplit, if more than one component is seen for this transition, it shows more than one europium site. This can be seen in the luminescence spectrum of the crown-ether complex $[Eu(NO_3)_3(15\text{-}crown\text{-}5)]$ shown in Figure 5.5, which gives some useful information. This compound contains a pentadentate macrocycle and three bidentate nitrates, leading to 11-coordinate europium. A single, very sharp line is seen for the $^5D_0 \rightarrow {}^7F_0$ transition, indicating only one europium site. Secondly, the fact that this electric-dipole transition is seen indicates that the site symmetry is not especially high (i.e. it is not on an inversion centre). The $^5D_0 \rightarrow {}^7F_1$ transition (magnetic-dipole allowed and relatively insensitive to environment) appears as a doublet and singlet, a splitting consistent with the approximately pentagonal pseudosymmetry seen in its crystal structure (in a lower-symmetry environment, the doublet would be further split). The $^5D_0 \rightarrow {}^7F_2$ transition is electric dipole in origin; again, it is absent if the ion is on an inversion centre, and its intensity is very sensitive to environment, making it a good 'probe', so its strength here indicates a relatively low symmetry of the site.

An important example historically of the application of luminescence spectroscopy (1969) is $[Eu(terpy)_3](ClO_4)_3$, whose spectrum (Figure 5.8) supported the assigned geometry before crystallographic details were available (crystallographic structural determination was a *much* slower business at that time). The $^5D_0 \rightarrow {}^7F_0$ transition is absent, whilst the $^5D_0 \rightarrow {}^7F_1$ transition consists of a singlet and a very slightly split (0.4 nm) doublet. The $^5D_0 \rightarrow {}^7F_2$ transition is two split doublets, whilst the $^5D_0 \rightarrow {}^7F_4$ transition is composed of three slightly split doublets and a singlet. If the symmetry were perfectly D_3, none of the doublets would be split (check Table 5.3), so these splittings can be explained by a very slight descent of symmetry below D_3.

The effect of site symmetry upon luminescence is not confined to structural determination, it also affects lighting applications. Compare the emissions of Eu^{3+} in the two oxides $NaInO_2$ and $NaGdO_2$ (Figure 5.9); both have structures based on NaCl, but in

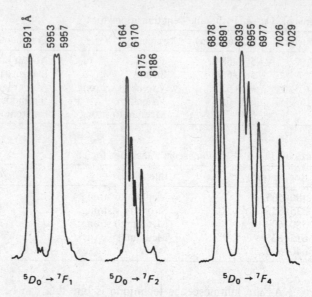

Figure 5.8
Fluorescence spectrum of $[Eu(terpy)_3]^{3+}$ ions in $[Eu(terpy)_3]\,(ClO_4)_3$, reproduced from D.A. Durham, G.H. Frost, and F.A. Hart, *J. Inorg. Nucl. Chem.*, 1969, **31**, 833 by permission of Pergamon Press PLC.

the former the europium ions are on centrosymmetric sites, so the $^5D_0 \rightarrow {}^7F_2$ transition is absent, and the emission is predominantly around 590 nm from the $^5D_0 \rightarrow {}^7F_1$ transition. In the case of Eu^{3+} in $NaGdO_2$, the europium site lacks an inversion centre, so in this case the emission is largely from the strongly radiating (electric dipole allowed) $^5D_0 \rightarrow {}^7F_2$ transition at 610 nm, the difference in wavelength producing a significant shift in colour (see section 5.4.4 for more on this).

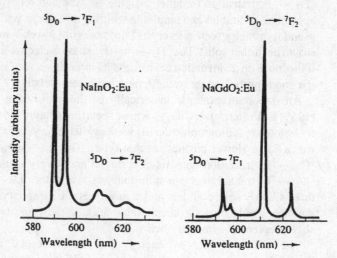

Figure 5.9
Luminescence from Eu^{3+} substituted in $NaInO_2$ and $NaGdO_2$, showing the effect of site symmetry upon emission (adapted from on B. Blasse and A. Bril, *J. Chem. Phys.*, 1966, **45**, 3327).

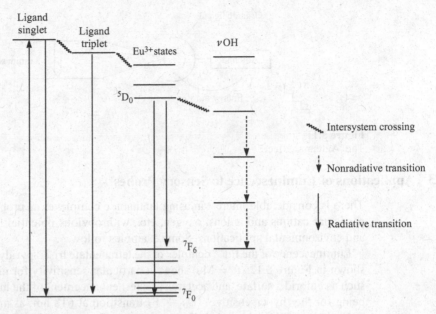

Figure 5.10
Quenching of luminescence of Eu^{3+}.

5.4.1 Quenching

One problem with luminescence in aqueous solution is that another pathway is available for deactivation of the excited state of the lanthanide, in the form of vibrational energy transfer to water molecules in particular. Figure 5.10 shows this process [in addition to showing the processes involved in Eu^{III} luminescence (only a selection of the radiative transitions are shown]. This 'quenching' of luminescence can be minimized (apart from working in the solid state, or using non-aqueous solvents) by using multidentate ligands which exclude waters from the coordination sphere of the metal, and also by using ligands which tend to encapsulate the lanthanide ion.

5.4.2 Antenna Effects

Luminescence from lanthanides is inherently weak (Laporte-forbidden transitions). Deactivation by vibrational transfer can be reduced, as discussed in the last section; another way that gives considerably enhancement is to use a chromophore as a ligand, which absorbs a suitable wavelength of radiation strongly (acts as an 'antenna') which it can then transfer to the lanthanide and excite it to the emissive state, using excitation in the region 330–430 nm (Figure 5.11). The most likely acceptor levels for Eu^{3+} and Tb^{3+} are 17 200 and 20 400 cm^{-1} respectively, so the triplet level in the acceptor ligand needs to be above 22 000 cm^{-1}, otherwise competing thermally activated back-energy-transfer occurs. Some of the most suitable ligands that have been utilized in the past 20 years include Lehn's tris(bipyridyl) cryptand; calixarenes; and substituted macrocycles containing phenanthridines. The acceptor chromophore does not need to be directly bound to the lanthanide, but should be close for best results. Some lanthanides emit in the near-IR, notably Yb^{3+}, Nd^{3+}, and Er^{3+}; these can be excited directly with an optical parametric oscillator.

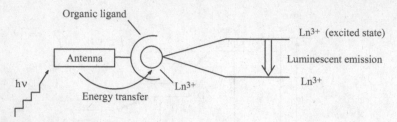

Figure 5.11
The 'Antenna' effect.

5.4.3 Applications of Luminescence to Sensory Probes

There is considerable interest in using lanthanide complexes as probes for the presence of particular cations and anions, oxygen, etc., with obvious potential in biological, clinical, and environmental applications. Some examples follow.

Luminescence of the Eu^{3+} complex of the tetradentate tris(2-pyridylmethyl)amine ligand shown in Figure 5.12 (R = Me) shows a particular sensitivity for nitrate (over other ions such as chloride, sulfate, and acetate), greatest enhancement of the luminescence spectrum being for the 'hypersensitive' $^5D_0 \rightarrow {}^7F_2$ transition at 618 nm, as might be expected (see T. Yamada, S. Shinoda, and H. Tsukube, *Chem. Commun.*, 2002, 218).

The terbium complex, in contrast, exhibits greatest sensitivity for chloride. Using the achiral ligand (R = H) exhibits similar selectivity for these anions, but with rather less sensitivity.

A terbium complex (Figure 5.13) is a molecular logic gate corresponding to a two-input INHIBIT function; the output (a terbium emission line) is observed only when the 'inputs', the presence of proteins and the absence of oxygen, are both satisfied.

A similar complex (Figure 5.14, M = Tb) exhibits a luminescence enhancement by a factor of 125, comparing the emissive neutral complex with its protonated form (pH 3), protonation (of the phenanthridyl nitrogen) suppressing electron transfer. Protonation reduces the energy of the ligand triplet state (the triplet is 1500 cm^{-1} higher than the terbium $^5D_4$ state in the neutral complex, compared with only 800 cm^{-1} higher in the protonated form), favouring deactivation of the terbium complex by back-energy-transfer to the triplet. Conversely, ligand-based fluorescence increases appreciably on protonation. The triplet state of the phenanthridyl moiety is quenched by O_2, so removing oxygen from a solution of the complex steadily increases the terbium lifetime and emission intensity. This complex thus acts as an oxygen sensor, as does the N-methylated analogue (independent of pH in the range 2–9). As strong terbium luminescence only occurs in the absence of both acid and

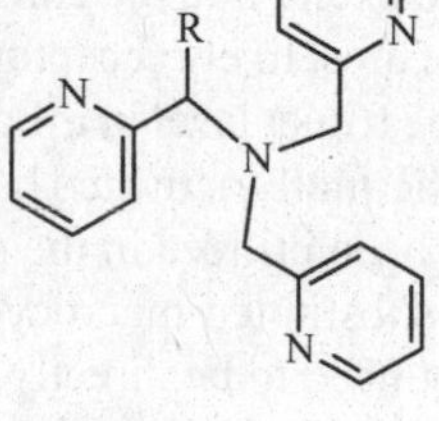

Figure 5.12
A tris(2-pyridylmethyl)amine ligand.

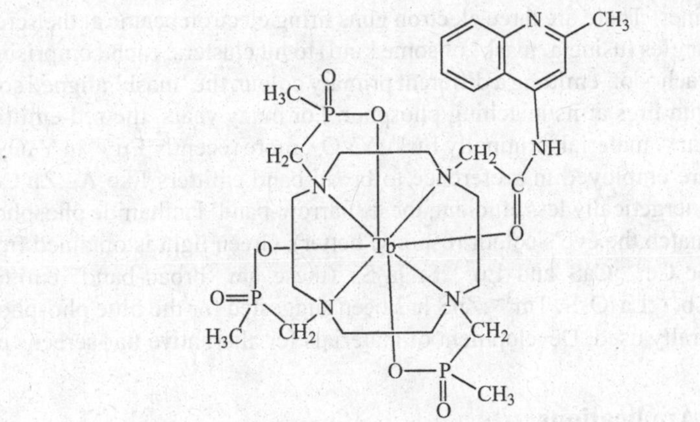

Figure 5.13
See T. Gunnlaugsson, D.A. MacDonail, and D. Parker, *Chem. Commun.*, 2000, 93.

O_2, this complex can be considered as a molecular NOR gate, with H^+ and O_2 as inputs and the Tb luminescence as the output.

The europium complex of this ligand (Figure 5.14, M = Eu) exhibits considerable Eu^{3+} luminescence enhancement on acidification, comparing the neutral complex with the protonated form at pH 1.5. The N-methylated analogue exhibits europium luminescence that is selectively quenched by chloride ions in the presence of phosphate, citrate, lactate, and bicarbonate, all potentially present in a cellular situation, showing potential as a luminescent sensor in bioassays. Such complexes have further been incorporated into sol–gel thin films and glasses that show good stability to leaching and photodegradation.

5.4.4 Fluorescence and TV

Colour televisions and similar displays are the largest commercial market for lanthanide phosphors, with over 100 million tubes manufactured a year. About 2 g of phosphor is used in each tube. The lanthanide involvement in a traditional colour TV tube works along these

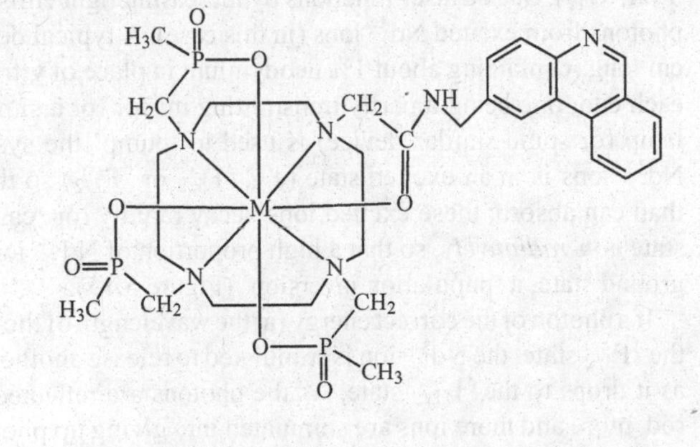

Figure 5.14
(see D. Parker and J.A.G. Williams, *Chem. Commun.*, 1998, 245; D. Parker, K. Senanaayake, and J.A.G. Williams, *Chem. Commun.*, 1997, 1777).

lines: There are three electron guns firing electron beams at the screen from subtly different angles (using a 'mask' of some kind) to hit clusters, each comprising three phosphor 'dots', each 'dot' emitting a different primary colour, the 'mask' aligned so the appropriate electron gun fires at its matching phosphor. For many years, the red-emitting phosphor has used a Eu^{3+} material, originally Eu^{3+}:YVO_4, more recently Eu^{3+} in Y_2O_2S or Eu^{3+}:Y_2O_3. These are employed in preference to broad-band emitters like Ag:Zn,CdS as although they are energetically less efficient, these 'narrow-band' lanthanide phosphors are brighter (and also match the eye's colour response better). Green light is obtained from either Cu,Al:Zn,CdS or Ce^{3+}:CaS and Eu^{2+}:$SrGa_2S_4$ (these are 'broad-band' emitters) or the narrow-line Tb^{3+}:La_2O_2S. Tm^{3+}:ZnS has been suggested for the blue phosphor but Ag,Al:ZnS is generally used. Development of materials for alternative flat-screens proceeds apace.

5.4.5 Lighting Applications

There is intense interest in using rare earths in lighting applications. The so-called tricolour lamps use three narrow-band lanthanide light-emitting materials with maxima around 450, 540, and 610 nm; such as Eu^{2+}:$BaMgAl_{10}O_{17}$; (Ce,Gd,Tb)MgB_5O_{10}; and Eu:Y_2O_3, respectively. This gives a good colour rendering at high efficiency. Research continues in developing materials for better projection materials, flat displays, and plasma displays. Eu^{2+}-based materials can be used for blue phosphors but tend to suffer from degradation under vacuum UV excitation, whilst a problem with the widely used Eu^{3+}:(Y,Gd)BO_3 'red' phosphor is that, owing to the high symmetry of the Eu^{3+} site, the $^5D_0 \rightarrow {}^7F_1$ emission is very pronounced and the colour is orange-biased (see Section 5.4); however, if the symmetry is too low, there is too much far-red and IR radiation from the $^5D_0 \rightarrow {}^7F_{4,6}$ transitions! Thus new materials are continually being synthesized and tested.

One unusual application is a UV emitter containing Eu^{2+}:SrB_4O_7; insects can 'see' the UV and be attracted to it, so it has been used in insect traps.

5.4.6 Lasers

(Light Amplification by Stimulated Emission of Radiation) Various lanthanide ions can be used in lasers, different ions operating at different frequencies. The most popular, however, is the neodymium laser, most usually using Nd^{3+} ions in yttrium aluminium garnet (YAG; $Y_3Al_5O_{12}$). Such a laser functions by increasing light emission by stimulating the release of photons from excited Nd^{3+} ions (in this case). A typical device consists of a YAG rod a few cm long (containing about 1% neodymium in place of yttrium) that is fitted with a mirror at each end, one being a partly transmitting mirror (or a similar device). A tungsten-halogen lamp (or some similar device) is used to 'pump' the system to ensure that an excess of Nd^{3+} ions is in an excited state (e.g. $^4F_{5/2}$ or $^4F_{7/2}$) so that more ions can emit electrons than can absorb; these excited ions decay rapidly (or 'cascade') to the long-lifetime $^4F_{3/2}$ state *non-radiatively*, so that a high proportion of Nd^{3+} ions are in this state rather than the ground state, a 'population inversion' (Figure 5.15).

If a photon of the correct energy (at the wavelength of the laser transition) hits a Nd^{3+} ion in the $^4F_{3/2}$ state, the Nd^{3+} ion is stimulated to release another photon of the same wavelength, as it drops to the $^4I_{11/2}$ state. As the photons are reflected backwards and forwards in the rod, more and more ions are stimulated into giving up photons (thus depopulating the $^4F_{3/2}$ state) and eventually the build up of photons is so great that they emerge from the rod as an intense beam of coherent monochromatic light (wavelength 1.06 μm, in the near-IR). The $^4I_{11/2}$ state is an excited level of the ground state, which is not thermally populated and

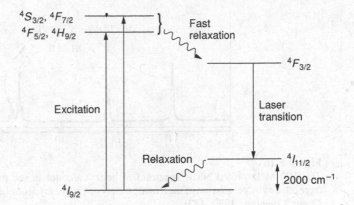

Figure 5.15
A 'four-level' Nd^{3+} laser. Reproduced by permission of Macmillan from S.A. Cotton, *Lanthanides and Actinides*, Macmillan, 1991, p. 32.

so undergoes rapid relaxation to the ground state, maintaining the 'population inversion' (whereupon the laser process can recommence). Neodymium is thus said to act as a 'four-level' laser.

5.4.7 Euro Banknotes

These exhibit green, blue, and red luminescent bands under UV irradiation, as a security measure. The red bands are doubtless due to some Eu^{3+} complex, probably with a β-diketonate or some similar ligand. As we have seen, there are Eu^{2+} complexes that could cause the green and blue luminescence. Researchers at the University of Twente in the Netherlands suggest that a likely candidate for the source of the green colour is $SrGa_2S_4:Eu^{2+}$, and that the blue colour may be caused by $(BaO)_x.6Al_2O_3:Eu^{3+}$. It's quite appropriate that Euro notes contain europium, really.

5.5 NMR Applications

5.5.1 β-Diketonates as NMR Shift Reagents

Paramagnetic lanthanide β-diketonate complexes $Ln(R^1COCHCOR^2)_3$ produce shifts in the NMR spectra of Lewis base molecules capable of forming adducts with them and are thus often referred to as Lanthanide Shift Reagents (LSRs) though all paramagnetic lanthanide complexes can exhibit shifted resonances. Molecules were chosen, such as $Eu(dpm)_3$ ($R^1 = R^2 = Me_3C$), which were quite soluble in non - polar solvents. The magnitude of the proton shifts depends upon the distance of the proton from the site of coordination to the lanthanide ion.

Immense activity in this area in the early 1970s resulted as the use of LSRs enabled simplification of the spectra of organic molecules without the use of high-frequency spectrometers. The spreading-out of the spectrum and differential nature of the shifts removed degeneracies and overlaps, whilst study of the shifts, particularly when more than one LSR was used, permitted spatial assignment of the protons (or other resonant nucleus) with concomitant structural information about the organic molecule.

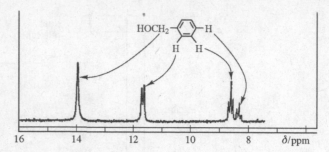

Figure 5.16
The 100 MHz proton NMR spectrum of benzyl alcohol in the presence of Eu(dpm)$_3$ (0.39 mol)
(reproduced by permission of the Royal Society of Chemistry from J.K.M. Sanders and D.H. Williams,
Chem. Commun., 1970, 422).

One example of such an application is shown in Figure 5.16. At 100 MHz, the aromatic
region of the ^{1}H NMR spectrum of $C_6H_5CH_2OH$ is a singlet, but on adding Eu(dpm)$_3$, which
coordinated to the OH group, the aromatic protons are shifted by amounts that depend on
their distance from the Eu and are susceptible to first-order analysis.

Nowadays this technique is less generally used owing to the spread of high-frequency
spectrometers, but chiral reagents like a lanthanide camphorato complex find applications
[Figure 5.17 shows Eu(tfc)$_3$]. for example, when such a chiral shift reagent binds to a racemic
mixture, two diastereoisomeric forms of the complex are formed, each with different peaks
in the NMR spectrum; each signal can be integrated and used to calculate the enantiomeric
excess, a quick way of estimating the yield of each isomer.

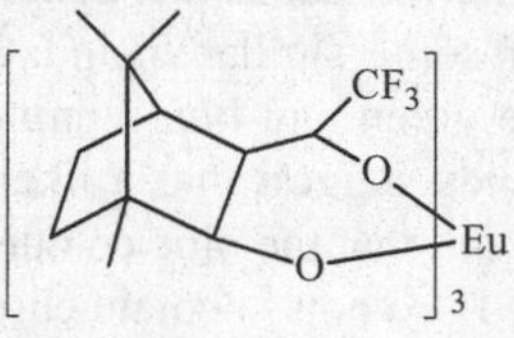

Figure 5.17
A chiral shift reagent.

Figure 5.18 shows the ^{1}H NMR spectrum of a mixture of 1-phenylethylamine and Yb(tfc)$_3$
in CDCl$_3$. The (S)-(+) and (R)-(−) isomers here give clearly distinguished peaks (the amine
resonances are well downfield and not displayed).

5.5.2 Magnetic Resonance Imaging (MRI)

The use of gadolinium(III) complexes to assist diagnosis in this expanding area of medicine
is now most important.

In the MRI experiment, a human body is placed on a horizontal table that is slid into the
centre of the magnetic field of the MRI scanner (essentially a pulsed FT NMR spectrometer).
MRI relies on detecting the NMR signals from hydrogen atoms in water molecules (which
make up ∼60% of the human body) and distinguishes between water molecules in healthy
and diseased tissue (since water molecules in cancerous tissue have much longer relaxation
times). MRI works well in soft tissue and has been used particularly to identify lesions in
the brain and spinal cord. When brain activity occurs, blood flow to that region increases,
which causes a contrast change between the active and inactive regions.

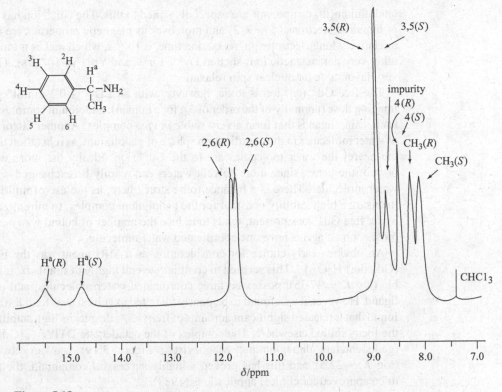

Figure 5.18
The ^{1}H NMR spectrum of a mixture of 1-phenylethylamine and Yb(TFC)$_3$ in CDCl$_3$ (reproduced with permission of the Editor from T. Viswanathan and A. Toland, *J. Chem. Educ.*, 1995, **72**, 945).

In order to make diagnoses about different parts of the body, it is necessary to obtain a two-dimensional image of the signals. This is done by placing the patient in a magnetic field gradient (using gradient coils to create different fields at many points in a piece of tissue) so that otherwise identical water molecules yield resonances as slightly different frequencies at each point, depending upon their position in the field, with a computer processing the data to give a digitized image of a spatially organized signal. The signal intensity depends upon the relaxation times of the protons, so, in order to enhance the contrast to differentiate between healthy and diseased tissue, paramagnetic contrast agents are used.

5.5.3 What Makes a Good MRI Agent?

The choice is dictated by a combination of several factors:

1. High magnetic moment;
2. Long electron-spin relaxation time;
3. Osmolarity similar to serum;
4. Low toxicity;
5. High solubility in water;
6. Targeting tissue;
7. Coordinated water molecules;
8. Large molecule with long rotational correlation times.

Gadolinium(III) compounds are especially suited to this. The Gd^{3+} ion has a large number of unpaired electrons ($S = 7/2$) and moreover its magnetic properties are isotropic. It has a relatively long electron-spin relaxation time, $\sim 10^{-9}$ s, which makes it more suitable than other very paramagnetic ions such as Dy^{3+}, Eu^{3+}, and Yb^{3+} ($\sim 10^{-13}$ s). These factors are very favourable for nuclear spin relaxation.

The free Gd^{3+}(aq) ion is toxic, however, with an $LD_{50} \sim 0.1$ mmol/kg, less than the imaging dose (normally of the order of 5 g for a human); gadolinum complexes are therefore used, using ligands that form a very stable *in vivo* complex. Another factor is the presence of water molecules in the coordination sphere of gadolinium, as relaxation times are shorter the nearer the water molecules are to the Gd^{3+} ion. Ideally the more water molecules present the better, since more solvent waters can readily be exchanged with coordinated water molecules. There is a balance to be struck here, as the use of multidentate ligands to ensure a high stability constant for the gadolinium complex, to minimize the amount of toxic, free Gd^{3+} ions present, tends to reduce the number of bound waters, and in practice most contrast agents have one coordinated water molecule.

An obvious early choice for consideration as a MRI agent was the EDTA complex, $[Gd(edta)(H_2O)_3]^-$. This seemed to combine several highly desirable factors, as it was stable (log $K = 17.35$); possessed three coordinated water molecules; and utilized a cheap ligand. However, it was found to be poorly tolerated in animal tests (and it was subsequently found that it released significant amounts of free Gd^{3+}, despite its high stability constant), so the focus shifted elsewhere. The complex of the octadentate $DTPA^{5-}$, $[Gd(dtpa)(H_2O)]^{2-}$ (gadopentetate dimeglumine; Magnevist®; Figure 5.19) was an obvious extension (log $K \sim 22.5$), and this has proved a highly successful compound, the first compound to be approved for clinical applications (1988).

Another compound to be widely used is $[Gd(dota)(H_2O)]^-$ (gadoterate meglumine; Dotarem®; Figure 5.20), using a cyclic ligand in an attempt to form more stable complexes (log $K \sim 24$–25). Another example is $[Gd (do3a(H_2O)_2]$ (Figure 5.21).

Apart from charged species, neutral complexes are also used. One example of this is $[Gd\text{-}hp\text{-}do3a]$; another is $[Gd(dtpa\text{-}bma)(H_2O)]$ (gadodiamide; Omniscan®; Figure 5.22). In these complexes, the loss of charge is achieved by removing a carboxylate group, or by

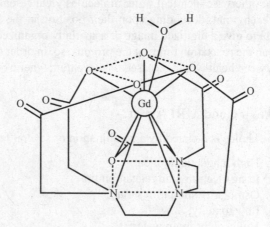

Figure 5.19
The structure of $Gd(dtpa)(H_2O)^{2-}$ (reproduced by permission of Macmillan from S.A. Cotton, *Lanthanides and Actinides*, 1991, p. 64).

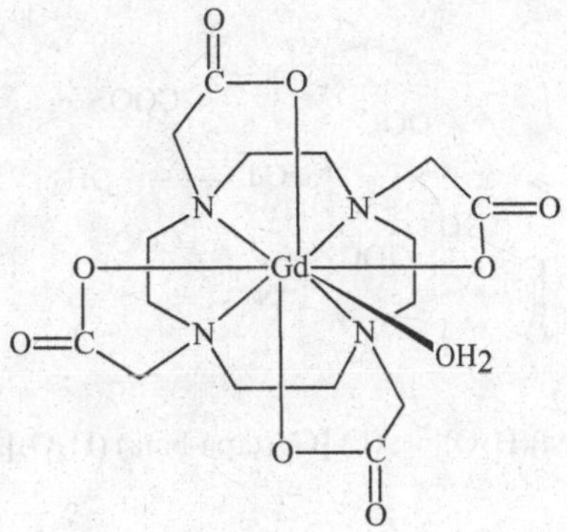

Figure 5.20
[Gd(dota)(H$_2$O)]$^-$.

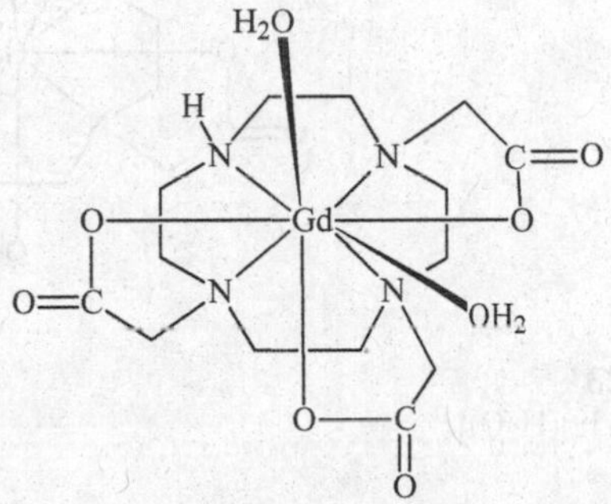

Figure 5.21
[Gd(do3a)(H$_2$O)$_2$].

converting two carboxylates into amides, respectively. Both charged and neutral complexes have advantages. Neutral complexes have less osmotic effect, so are more likely to have osmolalities similar to blood, whilst charged species are more hydrophilic, and therefore are more soluble.

Replacement of an NH group in [Gd-(do3a)(H$_2$O)$_2$] by sidechains containing OH groups was designed to enhance solubility in the agents Gd-hp-do3a (ProHance®; Figure 5.23) and Gd-do3a-butrol (Gadovist®; Figure 5.24).

Another gadolinium MRI agent, of a different type, is Gadolite ® (approved for use in the USA and UK in 1996). This is Gd^{3+}-exchanged zeolite NaY that has been shown to be an effective contrast agent for the gastrointestinal tract, defining it from adjacent tissues. With standard MRI imaging techniques, bowel fluid in the gastrointestinal tract can mimic diseases such as tumours. Gadolite is stable at pH 2.5–5 for several hours, and although gadolinium gets leached at lower pH, the oral toxicity of gadolinium is low.

5.5.4 Texaphyrins

A new type of complex with a different spectroscopic application is provided by the texaphyrins, compounds of 'extended' porphyrins where the ring contains five donor nitrogens. These have attracted considerable interest because of their possible medicinal applications.

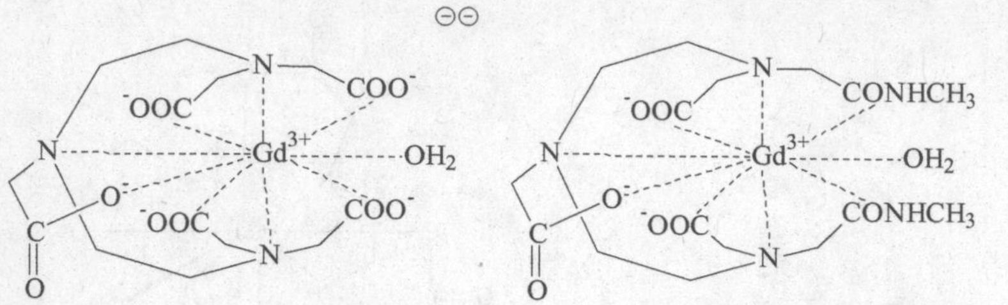

Figure 5.22
(a) $[Gd(dtpa)(H_2O)]^{2-}$, (b) $[Gd(dtpa-bma)(H_2O)]$.

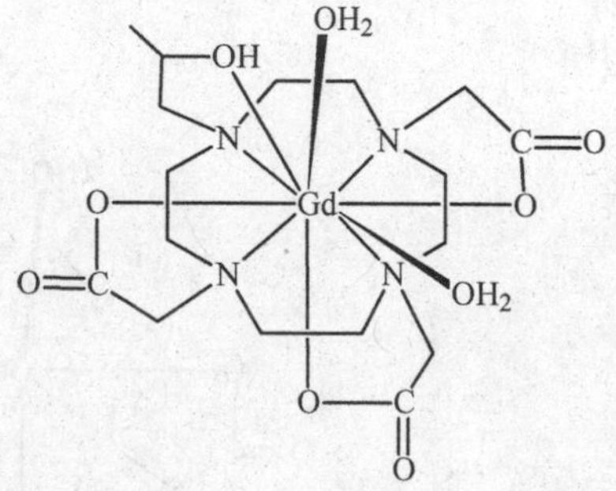

Figure 5.23
$[Gd(hp-do3a)(H_2O)]$.

Two texaphyrin complexes are undergoing clinical trials; a gadolinium compound (Gd-tex; XCYTRIN®; Figure 5.25) is an effective radiation sensitizer for tumour cells. It assists the production of reactive oxygen-containing species, whilst the presence of the Gd^{3+} ion means that the cancerous lesions to which it localizes can be studied by MRI. It is being investigated in connection with pancreatic tumours and brain cancers. One form of the lutetium analogue (LUTRIN®) is being developed for photodynamic therapy for breast cancer, and another (ANTRIN®) is being developed for photoangioplasty, where it has potential for treating arteriosclerosis by removal of atherosclerotic plaque.

5.6 Electron Paramagnetic Resonance Spectroscopy

Although most lanthanide ions are paramagnetic, because of rapid relaxation effects, spectra can be obtained only at low temperatures (often 4.2 K) in most cases. From the point of view of the chemist, EPR spectra are readily obtained (at room temperature) only from the f^7 Gd^{3+}, with its $^8S_{7/2}$ ground state. The sublevels of this state are degenerate in the absence of a crystal field (in a free Gd^{3+} ion), but are split into four Kramers' doublets, with M_J-values of $\pm 1/2$, $\pm 3/2$, $\pm 5/2$ and $\pm 7/2$. The application of a magnetic field removes the degeneracy of each doublet, and transitions can occur on irradiation with microwave radiation, subject to the usual selection rule of $\Delta M_J = \pm 1$.

In the absence of a zero-field splitting (z.f.s.), all transitions occur at the same field (corresponding to a g-value of 2.00), but as the z.f.s. increases, transitions occur at higher

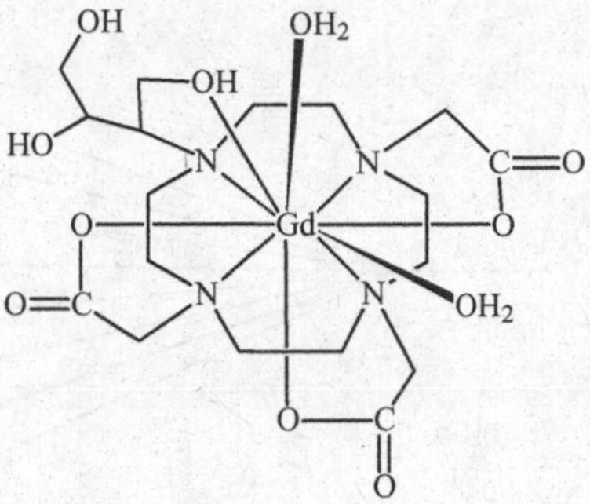

Figure 5.24
[Gd(do3a-butrol)(H$_2$O)].

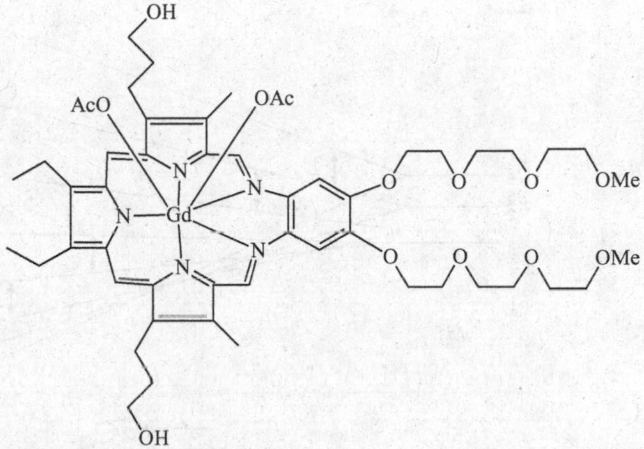

Figure 5.25
Gd-tex.

and lower fields (corresponding to 'effective' g-values above and below 2) (Figure 5.26). The situation is not dissimilar to the high-spin $Cr^{3+}(d^3)$ and Fe^{3+} and Mn^{2+} (d^5) ions. In the case of the three-coordinate silylamide $[Gd[N(SiMe_3)_2)_3]$, where there is a very strong axial distortion, signals are observed with the effective g-values $g\perp = 8$, $g_{\parallel} = 2$.

5.7 Lanthanides as Probes in Biological Systems

The lanthanides form a series of ions of closely related size and bonding characteristics and in many respects resemble Ca^{2+}, for which they often substitute isomorphously in biological systems. Since different Ln^{3+} ions can be probed with particular spectroscopic techniques (e.g. Eu^{3+} and Tb^{3+}, fluorescence; Gd^{3+}, EPR; Nd^{3+}, electronic spectra), in favourable circumstances it should be possible to obtain information about the binding site of spectroscopically inactive Ca^{2+} in several ways. Systems studied include the calcium-binding sites in calmodulin, trypsin, parvalbumin, and the Satellite tobacco necrosis virus.

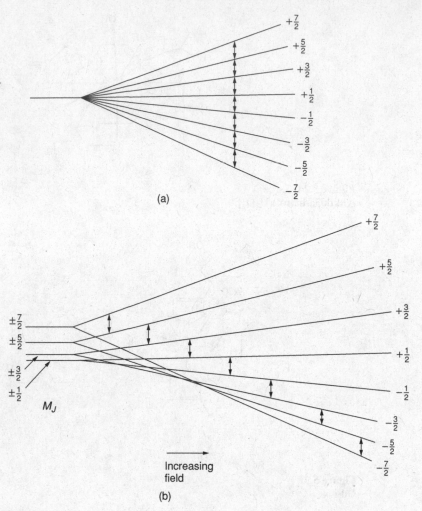

Figure 5.26
Showing the effect of zero-field splitting on the EPR transitions in Gd^{3+}.

Lanthanide chelates have been used as labels of antigens and antibodies in fluoroimmunological analysis, and in areas such as the determination of steroids in animals, of hormones and viral antigens in humans, and of herbicides in water.

Question 5.1 Work out the ground states for Pm^{3+} (f^4), Eu^{3+} (f^6), Er^{3+}(f^{11}), Pr^{4+}(f^1) and Eu^{2+} (f^7).

Answer 5.1 Pm^{3+} ($^5I_4$); Eu^{3+} ($^7F_0$); Er^{3+}($^4I_{15/2}$); Pr^{4+}($^2F_{5/2}$); Eu^{2+} ($^8S_{7/2}$).

Question 5.2 Calculate the magnetic moment for Pr^{3+} ions and Dy^{3+} ions.

Question 5.3 Explain why f–f transitions in the electronic spectra of lanthanide complexes are weaker than d–d transitions in the corresponding spectra of transition metal complexes.

Answer Forming a Ln^{3+} ion from a lanthanide atom by removing electrons stabilizes the orbitals in the order 4f > 5d > 6s. The 6s and 5d orbitals are emptied and the ions have the electron configurations $[Xe] 4f^n$. The 4f orbitals have become core-like, not involved in bonding, with negligible crystal-field effects. Both d–d and f–f transitions are parity-forbidden. In the case of d–d spectra, vibronic coupling lowers the symmetry round the metal ion, so d/p mixing can occur and the d–d transitions are to some extent 'allowed'. Because the 4f orbitals are so contracted compared with d orbitals, f/p mixing and vibronic coupling are much more limited, thus the f–f transitions are less 'allowed' and so extinction coefficients of the absorption bands are smaller.

Question 5.4 Why is the $Ce^{3+}(4f^1)$ ion colourless whereas Ti^{3+} solutions ($3d^1$) are purple?

Question 5.5 Explain why the electronic spectra of Pr^{3+} ($4f^2$) consist of a number of weak sharp lines, whereas those in the spectrum of V^{3+} ($3d^2$) are stronger and broader and less numerous.

Question 5.6 Why do La^{3+}, Lu^{3+}, or Gd^{3+} compounds not show luminescence properties?

Question 5.7 Comment on the emission spectrum of $[Eu(Me_3CCOCHCOCMe_3)_3(2,9-dimethyl-1,10-phenanthroline)]$ at 77 K (Figure 5.27). It shows the $^5D_0 \rightarrow {}^7F(J = 0–2)$ transitions. Consult Section 5.4 if required.

Comment: The $^5D_0 \rightarrow {}^7F_0$ transition consists of 2 lines (the spacing is 1.17 nm, 35 cm^{-1}), suggesting that there are two different Eu^{3+} sites. There are 6 bands in the region of the $^5D_0 \rightarrow {}^7F_1$ transition between 584.6 and 598.8 nm, and at least 7 bands in the region of the $^5D_0 \rightarrow {}^7F_2$ transition between 610.0 and 616.6 nm.

Consulting Table 5.13 indicates that a maximum of 3 bands is predicted for the former and 5 bands for the latter, again suggesting that there are two distinctly different europium sites. In fact, the crystal structure of this compound shows that there are two distinctly different molecules present in the unit cell, both with square antiprismatic coordination of europium, with occupancy factors of 59% and 41% (see R.C. Holz and L.C. Thompson, *Inorg. Chem.*, 1993, **32**, 5251; another example of this behaviour can be found in D.F. Moser, L.C. Thompson, and V.G. Young, *J. Alloys Compd.*, 2000, **303–304**, 121.)

Question 5.8 The emission of the macrocyclic complex shown in Figure 5.28 (M = Eu) has been studied as a function of pH and metal ion concentration.
A. Emission was weak over the pH range 12–8.5, but on reducing the pH below 8.5 emission was greatly enhanced until it was almost constant between pH 6.5 and 5. On further reduction to pH 3, emission was gradually quenched. These changes were completely reversible. Comment upon this.
B. The effect of adding different metals ions on the emission was investigated at pH 7.4. Groups I and II ions, as well as Zn^{2+}, had no effect. Addition of Cu^{2+} ions shifted the absorption spectrum of the phenanthroline ligand slightly from 255 to 280 nm and caused a substantial quenching of the europium luminescence. This was reversed, with recovery of the europium emission, on adding EDTA solution. Comment on this.
Answer A. Excitation of the phananthroline 'antennas' sensitizes the Eu^{3+} excited state. Under alkaline conditions, this does not emit, possibly because of reduction to the

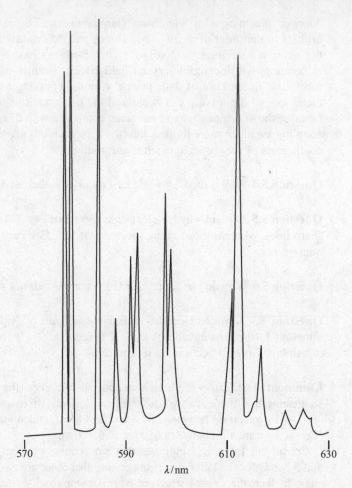

Figure 5.27
The emission spectrum of [Eu(Me$_3$CCOCHCOCMe$_3$)$_3$(2,9-dimethyl-1,10-phenanthroline)] at 77 K, showing the $^5D_0 \rightarrow F_J$ ($J = 0$–2) transitions (reproduced with permission of the American Chemical Society from R.C. Holz and L.C. Thompson, *Inorg. Chem.*, 1993, **32**, 5251).

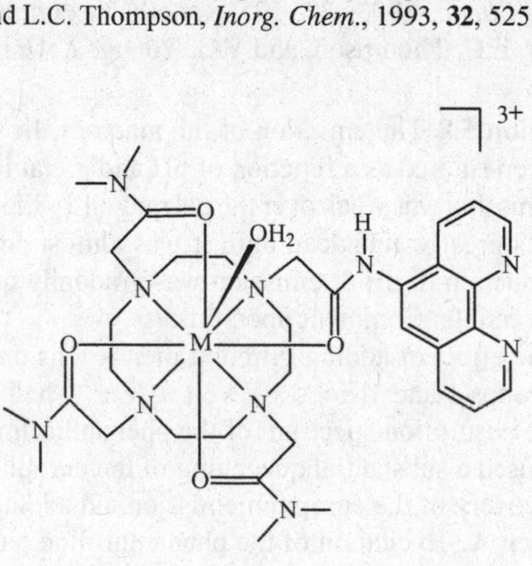

Figure 5.28

non-emitting Eu^{2+} state or because deprotonation of the amide group affects its ability to populate the excited state of Eu^{3+}. In acidic solution, basic nitrogen atoms in the phenanthroline group can be protonated, altering the electronic structure of the ligand and thereby affecting the ability of the ligand to populate the $^5D_0$ state of europium.

Answer B. Cu^{2+} ions have a strong affinity for N-donor ligands like ammonia, so are expected to bind well to phenanthroline (causing the shift in its absorption spectrum). The phenanthroline group acts as an antenna and coordination to copper obviously affects this. EDTA, however, is a hexadentate ligand that binds copper ions more strongly, detaching the copper from the phenanthroline group and restoring its antenna ability.

Refs: T. Gunnlaugsson, J.P. Leonard, K. Sénéchal, and A.J. Harte, *J. Am. Chem. Soc.*, 2003, **125**, 12062; T. Gunnlaugsson, J.P. Leonard, K. Sénéchal, and A.J. Harte, *Chem. Commun.*, 2004, 782.

Question 5.9 List and explain the desireable qualities of a good MRI contrast agent.

Question 5.10 From your knowledge of the chemistries of calcium and the lanthanides, suggest reasons why lanthanides are good at substituting at calcium-binding sites in, e.g., proteins.

Answer 5.10 They have similar ionic radii; similar high coordination numbers; both are 'hard' Lewis acids that prefer O-donor ligands.

6 Organometallic Chemistry of the Lanthanides

By the end of this chapter you should be able to:

- recall that these compounds are very air- and water-sensitive;
- recall that the bonding has a significant polar character;
- recall that some of these compounds have small molecular structures with appropriate volatility and solubility properties;
- recall appropriate bonding modes for the organic groups;
- recall that these compounds may exhibit low coordination number;
- suggest suitable synthetic routes;
- suggest structures for suitable examples;
- recall that these compounds may be useful synthons (starting materials for synthesis);
- recall that the 18-electron rule is not a predictive tool here.

6.1 Introduction

Compounds of these metals involving either σ- or π-bonds to carbon are generally much more reactive to both air and water than those of the d-block metals. Thus there is no lanthanide equivalent of ferrocene, an unreactive air- and heat-stable compound. They are often thermally stable to 100 °C or more, but are usually decomposed immediately by air (and are not infrequently pyrophoric). Within these limitations, lanthanide organometallic compounds have their own special features, often linked with the large size of these metals.

6.2 The +3 Oxidation State

As usual with lanthanides, this is the most important oxidation state, but there are compounds in the 0, +2 and +4 oxidation states that are discussed later.

Simple homoleptic σ-bonded alkyls and aryls have been difficult to characterize. They are most readily achieved by using bulky ligands or completing the coordination sphere with neutral ligands or by forming 'ate' complexes.

6.2.1 Alkyls

Reaction of 3 moles of methyllithium with $LnCl_3$ (Ln = Y, La) in THF yielded what are probably $Ln(CH_3)_3$ contaminated with LiCl and whose IR spectra showed the presence of

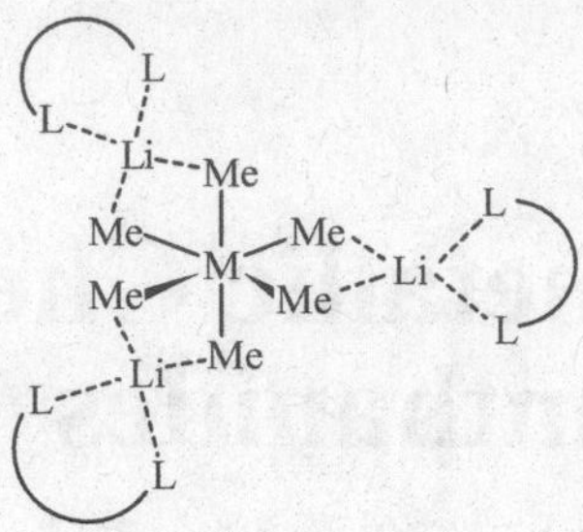

Figure 6.1
The structure of $[Li(L-L)]_3$ $[M(CH_3)_6]$.

THF. If an excess of CH_3Li is added to MCl_3 in THF, in the presence of a chelating ligand (L-L), either tetramethylethylenediamine (tmed or tmeda) or 1,2-dimethoxyethane (dme) complexes $[Li(tmed)]_3$ $[M(CH_3)_6]$ or $[Li(dme)]_3$ $[M(CH_3)_6]$ are obtained for Sc, Y, and all lanthanides except Eu (see Figure 6.1).

$$MCl_3 + 6\ MeLi + 3\ L-L \rightarrow [Li(L-L)]_3\ [M(CH_3)_6] + 3\ LiCl$$

A number of anionic complexes with the bulky *tert*-butyl ligand have been made, containing tetrahedral $[LnBu_4^t]^-$ ions {Ln = Er, Tb, Lu, confirmed (X-ray) for $[Li(tmed)_2]^+$ $[LuBu_4^t]^-$}.

$$LnCl_3 + 4\ Bu^tLi + n\ THF \rightarrow [Li(THF)_4][Ln(Bu^t)_4] + 3\ LiCl$$

$$Ln(OBu^t)_3 + 4\ Bu^tLi + 2\ TMED \rightarrow [Li(tmed)_2][Ln(Bu^t)_4] + 3\ Bu^tOLi$$

Moving to the CH_2SiMe_3 ligand, the neutral alkyl $[Sm(CH_2SiMe_3)_3(THF)_3]$ has been synthesized and shown to have a *fac*-octahedral structure, whilst $[Ln(CH_2SiMe_3)_3(THF)_2]$ (Ln = Er, Yb, Lu) are five coordinate. It seems that these alkyls are extremely unstable for lanthanides larger than Sm. Anionic species are obtained by reaction with excess of $LiCH_2SiMe_3$ in the presence of THF or TMED to solvate the lithium ion; compounds $[Li(thf)_4][Ln(CH_2SiMe_3)_4]$ (Ln = Y, Er, Tb, Yb) and $[Li(tmed)_2][Ln(CH_2SiMe_3)_4]$ (Ln = Y, Er, Yb, Lu).

In order to make simple alkyls of the even bulkier $-CH(SiMe_3)_2$ ligand, an indirect approach is needed. Direct reaction of $LnCl_3$ with $LiCH(SiMe_3)_2$ in ether solvents gives products like $[(Et_2O)_3Li(\mu\text{-}Cl)Y\{CH(SiMe_3)_2\}_3]$; this can be avoided by using a chloride-free starting material, by using a solvent like pentane that does not coordinate to an alkali metal, and by using a a bulky aryloxide ligand that does not bridge between a lanthanide and lithium. Thus, starting with $[Ln\{OC_6H_3Bu_2^t\text{-}2,6\}_3]$, alkyls $[Ln\{CH(SiMe_3)_2\}_3]$ have been reported (Ln = Y, La, Pr, Nd, Sm, Er, and Lu at present). The general synthetic route is:

$$[Ln\{OC_6H_3Bu_2^t\text{-}2,6\}_3] + 3\ LiCH(SiMe_3)_2 \rightarrow [Ln\{CH(SiMe_3)_2\}_3] + 3\ LiOC_6H_3Bu_2^t\text{-}2,6$$

These compounds have three-coordinate triangular pyramidal structures, similar to those found in the silylamides $[Ln\{N(SiMe_3)_2\}_3]$.

6.2.2 Aryls

Extended reaction at room temperature between powdered Ln (Ln = Ho, Er, Tm, Lu) and Ph_2Hg in the presence of catalytic amounts of LnI_3 affords the σ-aryls *fac*-$LnPh_3(thf)_3$. These have molecular structures with octahedral coordination of the lanthanides.

$$2\ Ln + 3\ Ph_2Hg + 6\ THF \rightarrow 2\ LnPh_3(thf)_3 + 3\ Hg\ (Ln = Ho, Er, Tm, Lu)$$

[Carrying out the reaction using Eu and Yb yields EuPh$_2$(THF)$_2$ and YbPh$_2$(THF)$_2$, see Section 6.5.1.]

Using the bulker 2,6-dimethylphenyl ligand results in [Li(THF)$_4$]$^+$ [Ln(2,6-Me$_2$C$_6$H$_3$)$_4$]$^-$ for the two smallest lanthanides, ytterbium (lemon) and lutetium (colourless).

$$LuCl_3 + 4\,C_6H_3Me_2Li + 4THF \rightarrow [Li(THF)_4][Lu(C_6H_3Me_2)_4] + 3LiCl$$

6.3 Cyclopentadienyls

6.3.1 Compounds of the Unsubstituted Cyclopentadienyl Ligand (C$_5$H$_5$ = Cp; C$_5$Me$_5$ = Cp*)

Cyclopentadienyl (and substituted variants thereof) has been the most versatile ligand used in organolanthanide chemistry. Unlike the situation with the d-block metals, where a maximum of two pentahapto- (η^5-)cyclopentadienyls can coordinate, up to three η^5-cyclopentadienyls can be found for the lanthanides, in keeping with the higher coordination numbers found for the f-block elements. In terms of the space occupied, a η^5-cyclopentadienyl takes up three sites in the coordination sphere. Three types of compound can be obtained, depending upon the stoichiometry of the reaction mixture:

$$LnCl_3 + NaCp \rightarrow LnCpCl_2(THF)_3 + NaCl \text{ (in THF; Sm–Lu)}$$

$$LnCl_3 + 2\,NaCp \rightarrow LnCp_2Cl + 2\,NaCl \text{ (in THF; Gd–Lu)}$$

$$LnCl_3 + 3\,NaCp \rightarrow LnCp_3 + 3\,NaCl \text{ (in THF; all Ln)}$$

The mono(cyclopentadienyl) complexes all have pseudo-octahedral structures (Figure 6.2). Another route for their synthesis is

$$LnCl_3(THF)_3 + \tfrac{1}{2}\,HgCp_2 \rightarrow CpLnCl_2(THF)_3 + \tfrac{1}{2}\,HgCl_2$$

Cases are known of other anionic ligands present, and the triflate CpLu(O$_3$SCF$_3$)$_2$(thf)$_3$ can be used as a synthon for a dialkyl:

$$CpLu(O_3SCF_3)_2(THF)_3 + 2\,LiCH_2SiMe_3 \rightarrow CpLu(CH_2SiMe_3)_2(THF)_3 + 2\,LiO_3SCF_3$$

The structures of the two other families of compounds are more complicated. After synthesis, LnCp$_2$Cl are isolated by removal of the THF solvent, followed by vacuum sublimation. LnCp$_2$Cl tend to be associated, most usually as symmetric dimers [Cp$_2$Ln(μ-Cl)$_2$LnCp$_2$], but DyCp$_2$Cl has a polymeric chain structure and GdCp$_2$Cl is a tetramer. These compounds can be isolated only for Gd–Lu and Y, the smaller lanthanides, for the unsubstituted cyclopentadienyl ring, but, with the bulkier Cp* ligands, [Cp*$_2$Ln(μ-Cl)$_2$LnCp*$_2$] are obtainable for metals like La as well. A few other systems like LnCp$_2$X (X = Br, I) have also been made.

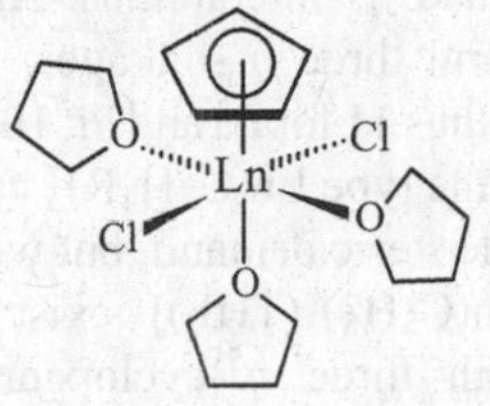

Figure 6.2
Structure of CpLnCl$_2$(THF)$_3$.

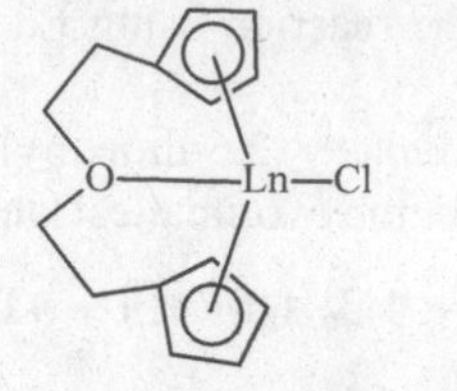

Figure 6.3
Structure of [{O(CH$_2$CH$_2$ C$_5$H$_4$)$_2$}LnCl].

Adducts [Cp*$_2$LnCl(thf)] have been isolated from THF solutions for most lanthanides and a number of other [Cp$_2$LnCl(thf)] have been made. (The THF can be replaced by other Lewis bases such as MeCN and CyNC (Cy = cyclohexyl), whilst the chloride ligand may be substituted too.) Under some circumstances, the product of synthesis is a species like the remarkable blue, pentane-soluble 'ate' complex, [Cp*$_2$Nd(μ-Cl)$_2$Li(thf)$_2$]. Another way of obtaining bis(cyclopentadienyl) complexes is to link the two rings. If the link is functionalized with a donor atom, coordinative saturation can be achieved without solvent coordination (Figure 6.3) and without the formation of an 'ate' complex, which might otherwise be obtained (Figure 6.4).

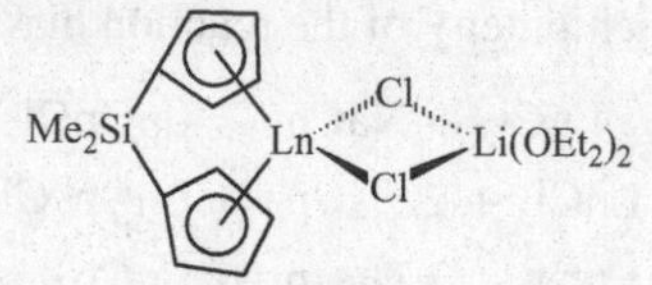

Figure 6.4
Structure of [{Me$_2$Si(C$_5$H$_4$)$_2$} In (μ-Cl)$_2$ Li/OEt$_2$)$_2$].

Cases are known where these ligands bridge two metals, instead of chelating (Figure 6.5).

Tris(cyclopentadienyl) compounds Ln(C$_5$H$_5$)$_3$ are isolated from synthesis in THF as adducts [Ln(C$_5$H$_5$)$_3$(THF)] (see below, this Section, and Figure 6.7), which undergo vacuum sublimation to afford pure Ln(C$_5$H$_5$)$_3$. These compounds are thermally stable but air- and moisture-sensitive, and are labile enough to react rapidly with FeCl$_2$, forming ferrocene. Their coordination spheres are unsaturated, to judge by their readiness to form adducts with Lewis bases like THF, RNC, RCN, and pyridine, and also by their solid-state structures.

The structures of Ln(C$_5$H$_5$)$_3$ are shown in Figure 6.6. The coordination number of the metal increases with increasing size of the lanthanide. Assuming that an η^5-C$_5$H$_5$ group occupies three coordination sites, then only ytterbium forms a simple molecular '9 coordinate' [Yb(η^5-C$_5$H$_5$)$_3$] . The compounds of slightly larger metals like Y and Er associate through weak Van der Waals' forces, affording a compound where the coordination number is slightly greater than 9. Early lanthanides form compounds where the metals additionally form some η^1 and η^2 interactions. At the other end of the series, lutetium is evidently too small to form three η^5-linkages. Coordination numbers of the lanthanides in these compounds are thus 11 for La and Pr, 10 for Nd, 9 (+) for Y, Er and Tm, 9 for Yb, and 8 for Lu. Compounds of the type Ln(C$_5$H$_4$R)$_3$ can be made for substituted cyclopentadienyl ligands when R has little steric demand, but with bulky ligands this is frequently not possible.

Adducts [Ln(C$_5$H$_5$)$_3$(THF)] exist for all lanthanides. They have ten-coordinate lanthanides with three η^5-cyclopentadienyl rings and a monodentate THF, having pseudo-tetrahedral geometry round the lanthanide (Figure 6.7). They are one of those rare cases when the same coordination number is maintained across the lanthanide series.

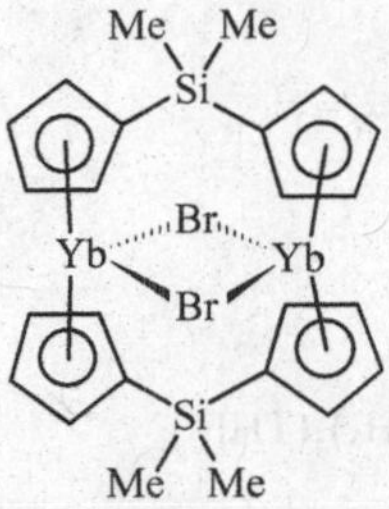

Figure 6.5
Structure of [{He$_2$Si(C$_5$H$_4$)$_2$}$_2$ Yb$_2$ Br$_2$].

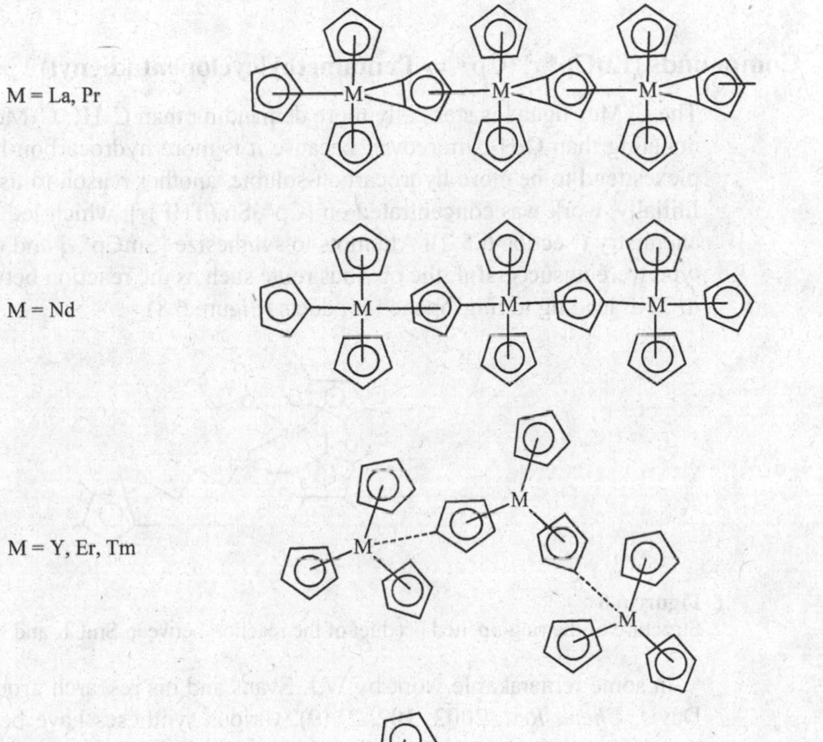

M = La, Pr

M = Nd

M = Y, Er, Tm

M = Yb

M = Lu

Figure 6.6
Structures of Ln(C$_5$H$_5$)$_3$.

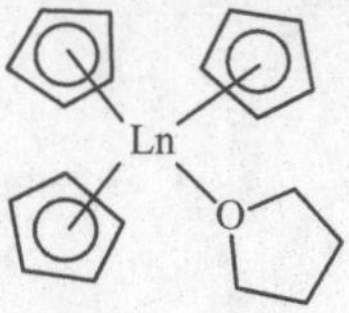

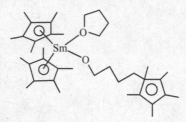

Figure 6.7
Structure of [Ln(C₅H₅)₃(THF)].

Other adducts whose structure has been confirmed by X-ray diffraction include [La(C₅H₅)₃(NCMe)₂]; [Ln(C₅H₅)₃(NCEt)] (Ln = La, Pr, Yb); [Pr(C₅H₅)₃(CNCy)]; and [Ln(C₅H₅)₃(py)] (Ln = Nd, Sm).

6.3.2   Compounds [LnCp*₃] (Cp* = Pentamethylcyclopentadienyl)

The C₅Me₅ ligand is sterically more demanding than C₅H₅; C₅Me₅ also is more electron-donating than C₅H₅; moreover, because it is more hydrocarbon-like, C₅Me₅ (Cp*) complexes tend to be more hydrocarbon-soluble, another reason to use it for solution studies. Initially, work was concentrated on [Cp*₂Sm(THF)₂], which led to extremely interesting chemistry (Section 6.5.2). Attempts to synthesize [SmCp*₃] and other compounds of this type were unsuccessful, the obvious route such as the reaction between SmCl₃ and NaCp* in THF leading to ring-opened products (Figure 6.8).

Figure 6.8
Structure of the ring-opened product of the reaction between SmCl₃ and NaCp* in THF.

In some remarkable work by W.J. Evans and his research group (W.J. Evans and B.L. Davis, *Chem. Rev.*, 2002, **102**, 2119), various syntheses have been developed which do not involve the use of THF, such as the reaction between tetramethylfulvene and a dimeric hydride, and the reaction of KCp* with a cationic bis(pentamethylcyclopentadienyl) species (Figure 6.9).

The latter route has been applied to making all these compounds for La–Sm and Gd, as well as some even more congested molecules such as [La(C₅Me₄R)₃] (R = Et, SiMe₃, Prⁱ). The difficulty in isolating the strongly oxophilic [LnCp*₃] systems is such that synthesis conditions require not only the absence of Lewis bases such as THF, RNC, and RCN, all possibly donors, but in some cases required both the use of silylated glassware and the absence of even traces of THF vapour from gloveboxes used.

When [SmCp*₃] was eventually isolated, it was found to be [Sm(η⁵-C₅Me₅)₃] (Figure 6.10) and also that three of these bulky ligands fit round samarium with an increase in bond length from the usual 2.75 Å to 2.82 Å.

Apart from the novel ring-opening reaction with THF, [SmCp*₃] inserts RNC and CO and is an ethene polymerization catalyst. Another reaction uncharacteristic of [LnCp₃] systems

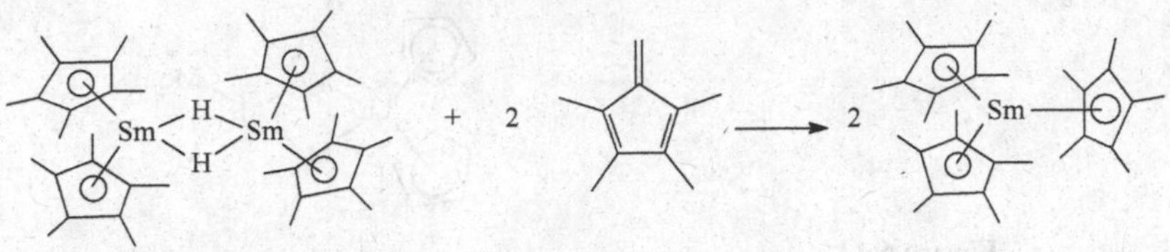

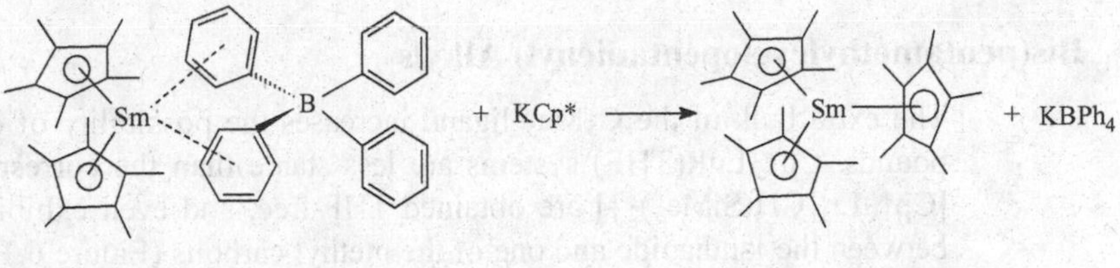

Figure 6.9
Routes for the synthesis of [SmCp*$_3$].

is its ready hydrogenolysis:

$$2\,[\text{SmCp*}_3] + 2\,\text{H}_2 \rightarrow [\text{Cp*}_2\text{Sm}(\mu\text{-H})_2\text{SmCp*}_2] + 2\,\text{Cp*H}$$

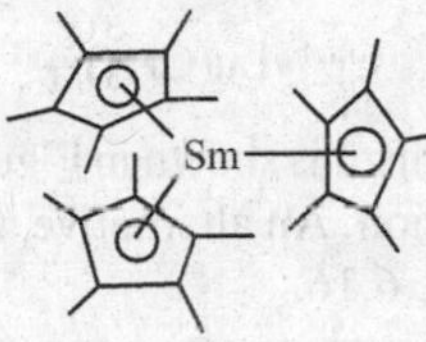

Figure 6.10
Structure of [SmCp*$_3$].

6.3.3 Bis(cyclopentadienyl) Alkyls and Aryls LnCp$_2$R

These compounds can be synthesized for the later lanthanides in particular. Reaction of LnCp$_2$Cl with LiAlMe$_4$ gives a μ-dimethyl-bridged species (Figure 6.11). Similar, but less stable, ethyls can also be made.

$$\text{LnCp}_2\text{Cl} + \text{LiAlMe}_4 \rightarrow \text{Cp}_2\text{Ln}(\mu\text{-Me}_2)\text{AlMe}_2 + \text{LiCl}$$

These heterometallic dimers undergo cleavage of the bridge with pyridine (Figure 6.12) to give [LnCp$_2$R]$_2$.

$$2\,\text{Cp}_2\text{Ln}(\mu-\text{Me}_2)\text{AlMe}_2 + 2\,\text{py} \rightarrow \text{Cp}_2\text{Ln}(\mu-\text{Me}_2)\text{LnCp}_2 + 2\,\text{AlMe}_3.\text{py}$$

Similarly, reactions of LiR with LnCp$_2$Cl {which, strictly, should be written to give the product [Cp$_2$Ln(μ-Cl)$_2$LnCp$_2$]} resulted in the synthesis of LnCp$_2$R (R, e.g., Bu^n, Bu^t, CH$_2$SiMe$_3$, p-C$_6$H$_4$Me, etc.) for later lanthanides:

$$\text{LnCp}_2\text{Cl} + \text{LiR} \rightarrow \text{LnCp}_2\text{R} + \text{LiCl}$$

When this reaction is carried out in THF, a solvent that can coordinate well, the product is LnCp$_2$R(THF).

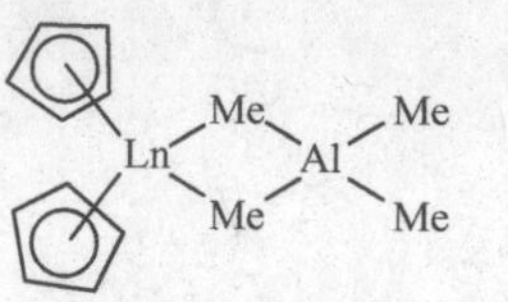

Figure 6.11
Structure of $[Cp_2Ln(\mu\text{-}Me_2)\,AlMe_2]$.

6.3.4 Bis(pentamethylcyclopentadienyl) Alkyls

The extra bulk of the C_5Me_5 ligand increases the possibility of crowding in these compounds. $Cp*_2LnR(THF)$ systems are less stable than the corresponding $Cp_2LnR(THF)$. $[Cp*_2Ln\{CH(SiMe_3)_2\}]$ are obtained THF-free, and even exhibit an agostic interaction between the lanthanide and one of the methyl carbons (Figure 6.13). The agostic Ln . . . C distance also decreases with increasing atomic number (size). The interaction is too weak to be seen on the NMR timescale.

$Cp*_2Lu(CH_3)$ is a remarkable compound. In the solid state, it is an unsymmetrical dimer (Figure 6.14), but in solution this is in equilibrium with the monomer. Monomeric $Cp*_2Lu(CH_3)$ undergoes some remarkable reactions with inert substances, demonstrating a very high Lewis acidity. Thus it activates CH_4, a reaction that can be followed using isotopically labelled methane.

$$Cp*_2Lu(CH_3) + {}^{13}CH_4 \rightleftharpoons Cp*_2Lu({}^{13}CH_3) + CH_4$$

A possible mechanism is shown in Figure 6.15. The mechanism shown above is bimolecular, involving association. An alternative, that does not involve initial coordination of methane, is shown in Figure 6.16.

$Cp*_2Lu(CH_3)$ also activates benzene, and readily undergoes hydrogenolysis:

$$Cp*_2Lu(CH_3) + C_6H_6 \rightarrow Cp*_2Lu(C_6H_5) + CH_4$$
$$2\,Cp*_2Lu(CH_3) + 2\,H_2 \rightarrow [Cp*_2Lu(H)]_2 + 2\,CH_4$$

It undergoes ready insertion of alkenes into the Lu–C bond and is a very active catalyst for the polymerization of ethene.

Some compounds are known with just one Cp-type ligand (Figure 6.17).

$$[La\{CH(SiMe_3)_2\}_3] + Cp*H \rightarrow [Cp*La\{CH(SiMe_3)_2\}_2] + CH_2(SiMe_3)_2$$

It features agostic interactions with two methyl groups; if the lanthanum atom were just bound to an η^5-cyclopentadienyl and also formed two an σ-bonds to alkyl groups, it would be regarded as five coordinate and thus coordinatively unsaturated and electron-deficient. The agostic interactions relieve this.

Another mono(pentamethylcyclopentadienyl) alkyl is $[Li(tmed)_2]\,[Cp*LuMe_3]$.

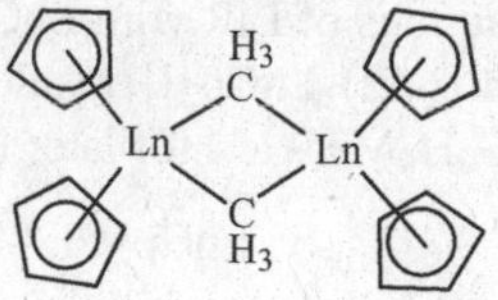

Figure 6.12
Structure of $[Cp_2Ln(\mu\text{-}Me_2)\,LnCp_2]$.

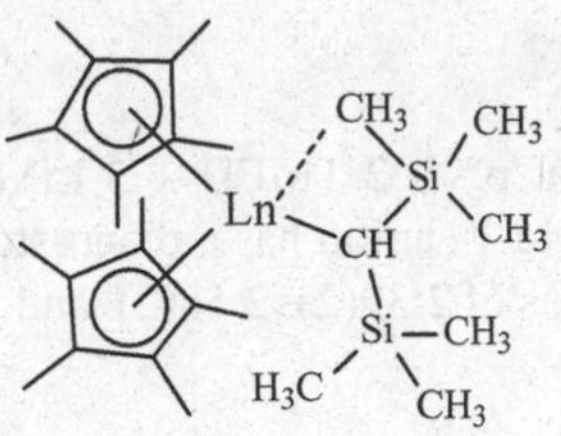

Figure 6.13
Structure of [Cp*$_2$Ln{CH(SiMe$_3$)$_2$}].

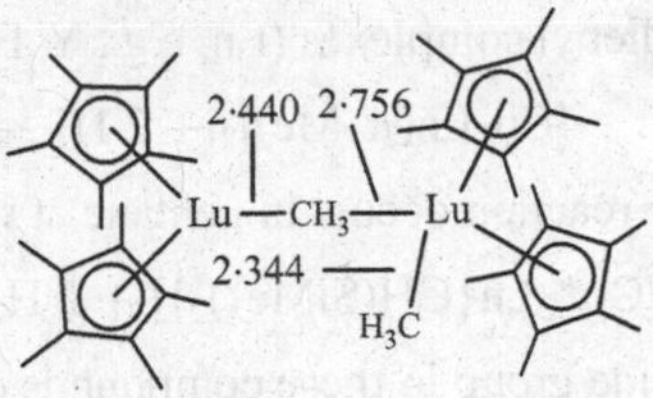

Figure 6.14
Structure of [Cp*$_2$Lu(μ-Me)LuCp*$_2$Me].

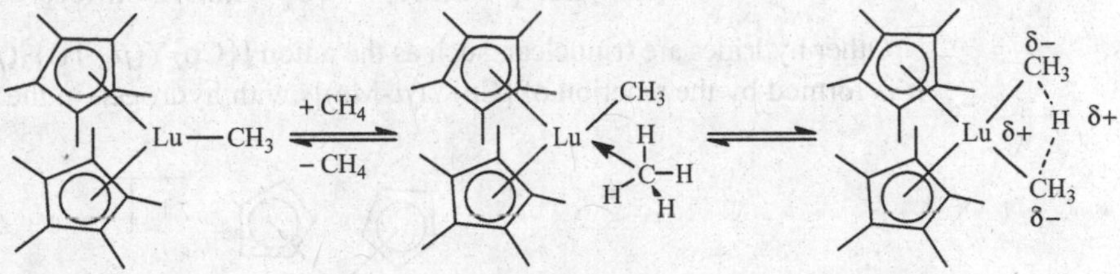

Figure 6.15
Reaction of [Cp*$_2$LuMe] with CH$_4$.

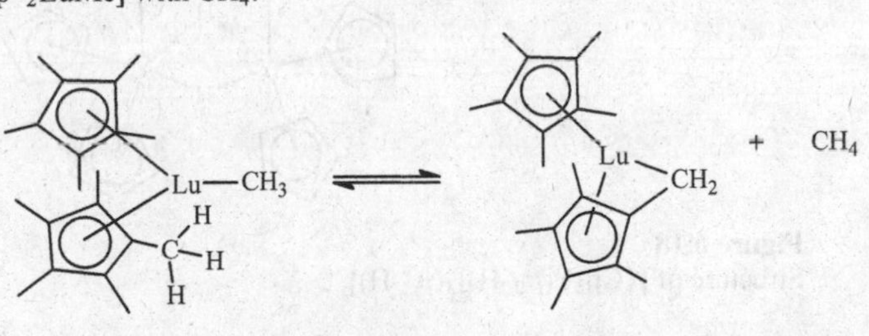

Figure 6.16
Intramolecular elimination of methane from [Cp*$_2$LuMe].

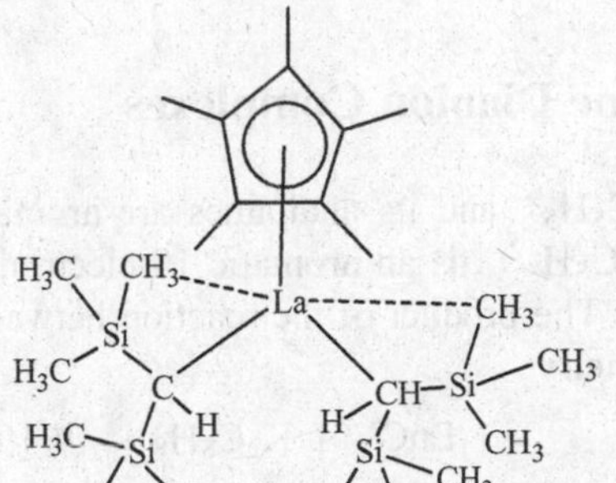

Figure 6.17
Structure of [Cp*La{CH(SiMe$_3$)$_2$}$_2$].

$LuCl_3 + NaCp^* + 2\ TMED + 3\ LiMe \rightarrow [Li(tmed)_2]\ [Cp^*LuMe_3] + NaCl + 2\ LiCl$. The $[Cp^*LuMe_3]^-$ anion has a piano-stool structure with some asymmetry in the reported Lu–C distances of 2.39(2), 2.56(2), and 2.59(2) Å.

6.3.5 Hydride Complexes

These can be made by various routes, including hydrogenolysis of Ln–C bonds; thermolysis of alkyls; and substitution of halide. Thus hydrogenolysis of the dimeric methyl-bridge cyclopentadienyl complexes (Ln, e.g., Y, Er, Yb, Lu) in THF affords $[Cp_2Ln(thf)(\mu\text{-}H)]_2$:

$$[Cp_2Ln(\mu\text{-}Me)]_2 + 2\ H_2 \rightarrow Cp_2Ln(THF)(\mu\text{-}H)]_2 + 2\ CH_4$$

A similar reaction occurs in pentane at room temperature (Ln, e.g., Y, La–Nd, Sm, Lu):

$$2\ [Cp^*_2Ln\{CH(SiMe_3)_2\}] + 2\ H_2 \rightarrow [Cp^*_2Ln(\mu\text{-}H)]_2 + CH_2(SiMe_3)_2$$

The hydride group in these compounds can be very reactive:

$$[Cp^*_2LnH] + C_6H_6 \rightarrow [Cp^*_2Ln(C_6H_5)] + H_2$$
$$[Cp^*_2LuH] + SiMe_4 \rightarrow [Cp^*_2LnCH_2SiMe_3] + H_2$$

Other hydrides are trinuclear, such as the anion $[\{Cp_2Y(\mu_2\text{-}H)\}_3(\mu_3\text{-}H)]^-$ (Figure 6.18) It is formed by the reaction of $[Cp_2Y(\mu\text{-}Me)]_2$ with hydrogen in the presence of LiCl.

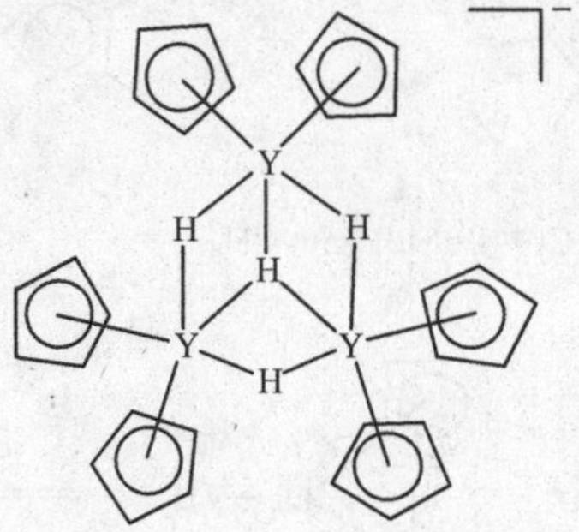

Figure 6.18
Structure of $[\{Cp_2Y(\mu_2\text{-}H)\}_3(\mu_3\text{-}H)]^-$.

Reduction of $[Cp_2LuCl(THF)]$ with Na/THF gives a similar trinuclear species, $[Na(THF)_6]\ [\{Cp_2Lu(\mu_2\text{-}H)\}_3(\mu_3\text{-}H)]$.

6.4 Cyclooctatetraene Dianion Complexes

Just as $C_5H_5^-$ and its analogues are aromatic 6-electron ligands, the cyclooctatetraene dianion, $C_8H_8^{2-}$, is an aromatic 10-electron ligand. Several types of complex have been obtained. The product of the reaction between $LnCl_3$ and $K_2C_8H_8$ in THF depends upon stoichiometry:

$$LnCl_3 + K_2C_8H_8 \rightarrow \tfrac{1}{2}\ [(C_8H_8)Ln(THF)Cl]_2 + 2\ KCl$$
$$LnCl_3 + 2\ K_2C_8H_8 \rightarrow K[Ln(C_8H_8)_2] + 3\ KCl$$

Both compounds have been obtained for most lanthanides. The anion in $K(dme)_2$ $[Ln(C_8H_8)_2]$ has a 'uranocene'-like sandwich structure (Section 13.7), whilst $[(C_8H_8)Ce(THF)_2Cl]_2$ has a dimeric structure with bridging chlorides (Figure 6.19). All

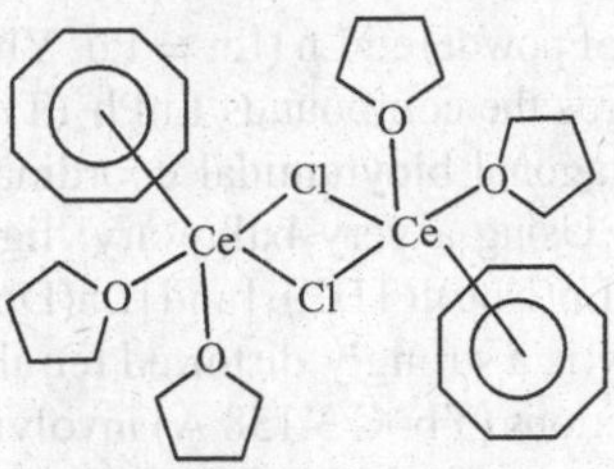

Figure 6.19
Structure of dimeric [(C$_8$H$_8$)Ce(THF)$_2$Cl]$_2$.

these compounds seem essentially 'ionic', reacting with UCl$_4$ to form [U(C$_8$H$_8$)$_2$] quantitatively.

Co-condensation of the metal vapour and cyclooctatraene at 77 K gives Ln$_2$(C$_8$H$_8$)$_3$ (Ln, e.g., Ce, Nd, Er); crystallization from THF gives [(C$_8$H$_8$)Ln(THF)$_2$][Ln(C$_8$H$_8$)$_2$]. Mixed-ring complexes [(C$_8$H$_8$)Ln(Cp)(THF)$_n$] can be made by routes such as equimolar amounts of [(C$_8$H$_8$)Ln(THF)Cl] and NaCp or [CpLnCl$_2$(THF)$_3$] with K$_2$C$_8$H$_8$, though 'one-pot' routes work equally as well. Unsolvated complexes would be coordinatively unsaturated, hence the coordinated THF, as in [(C$_8$H$_8$)Pr(C$_5$H$_5$)(THF)$_2$] and [(C$_8$H$_8$)Y(C$_5$H$_4$Me)(THF)], but use of bulky substituted cyclopentadienyl gives unsolvated compounds like [(C$_8$H$_8$)Lu{C$_5$(CH$_2$Ph)$_5$}] and [(C$_8$H$_8$)LuCp*] (Figure 6.20).

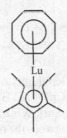

Figure 6.20
Structure of [(C$_8$H$_8$)LuCp*].

6.5   The +2 State

6.5.1   Alkyls and Aryls

A number of simple alkyls and aryls have been made. The use of a bulky silicon-substituted *tert*-butyl ligand permitted the isolation of simple monomeric (bent) alkyls [Ln{C(SiMe$_3$)$_3$}$_2$] (Ln = Eu, Yb) which are easily sublimeable *in vacuo*. Unlike lanthanide cyclopentadienyls, it is very rare for simple σ-bonded organolanthanides to be stable enough to be sublimeable *in vacuo*. Evidently the very bulky organic ligands enclose the metal ions very well, so that the intermolecular forces are essentially weak Van der Waals' interactions between the organic groups, ensuring the molecules are volatile at low temperatures. Other such compounds include [Yb{C(SiMe$_3$)$_2$(SiMe$_2$X)}] (X = CH=CH$_2$; CH$_2$CH$_2$OEt) and Grignard analogues [Yb{C(SiMe$_3$)$_2$(SiMe$_2$X)}I.OEt$_2$] (X = Me, CH=CH$_2$, Ph, OMe), synthesized from RI and Yb. The alkylytterbium iodides have iodo-bridged dimeric structures, containing four-coordinate ytterbium when X = Me, but five coordinate for X = OMe, due to chelation.

Reaction of powdered Ln (Ln = Eu, Yb) and Ph_2Hg in the presence of catalytic amounts of LnI_3 affords the compounds $LnPh_2(THF)_2$. The pentafluorophenyl $[Eu(C_6F_5)_2(THF)_5]$ has the pentagonal bipyramidal coordination geometry familiar for simple coordination compounds. Using a very bulky aryl ligand, Dpp (Dpp = $2,6\text{-}Ph_2C_6H_3$), results in the isolation of $[Ln(Dpp)I(THF)_3]$ and $[Ln(Dpp)_2(THF)_2]$ (Ln = Eu, Yb). In $[Yb(Dpp)_2(THF)_2]$ the geometry is a strongly distorted tetrahedron; there are additionally two weak $\eta^1 - \pi$-arene interactions (Yb–C 3.138 Å) involving α-carbons of the terphenyl groups. Structures are not yet known for $LnPh_2(THF)_2$ (Ln = Eu, Yb); a coordination number of four seems low, as these are not bulky ligands. Possibly they may not have monomeric structures, or there are agostic interactions involving Ln....H–C bonding.

6.5.2 Cyclopentadienyls

Simple cyclopentadienyls $[MCp_2]$ (M = Eu, Yb) can be made by reaction of cyclopentadiene with solutions of the metal in liquid ammonia. The initial products are ammines $[MCp_2(NH_3)_x]$, which can be desolvated by heating *in vacuo*. $[SmCp_2]$ is also known. These compounds have been little studied on account of their low solubilities; using instead the pentamethylcyclopentadienyl group confers greater solubility in aromatic hydrocarbons like toluene.

Synthetic routes include:

$$SmI_2(THF)_2 + 2\,KCp^* \rightarrow [Cp^*_2Sm(THF)_2] + 2KI$$
$$EuCl_3 + 3\,NaCp^* \rightarrow [Cp^*_2Eu(THF)(Et_2O)] + 3\,NaCl + \tfrac{1}{2}\,Cp^*\text{–}Cp^*$$

These compounds often desolvate readily, thus recrystalllization of these two compounds from toluene gives $[Cp^*_2Ln(THF)]$ (Ln = Sm, Eu). Gentle heating of $[Cp^*_2Sm(THF)]$ *in vacuo* forms $[Cp^*_2Sm]$. Whilst it is to be expected that $[Cp^*_2Ln(THF)_n]$ will have bent structures, the angle between the planes of the Cp* rings in crystals of $[Cp^*_2Ln]$ is 130–140°. It has been suggested that a bent molecule is polarized and can give stronger dipole–dipole forces attracting it to neighbouring molecules in the solid state. However, this is not the whole story, as electron diffraction results indicate an angle of 160° in the gas phase, and another possibility is that intramolecular nonbonded interactions are influential.

Besides forming adducts with Lewis bases (THF, Et_2O, Ph_3PO, etc.), $[Cp^*_2Sm]$ exhibits some remarkable reactions with dinitrogen (reversible) and carbon monoxide (irreversible) (Figure 6.21).

6.5.3 Other Compounds

Simple 'cyclooctatetraenediyls' $[M(C_8H_8)]$ (M = Eu, Yb) can be made by reaction of cyclooctatetraene with solutions of the metal in liquid ammonia. Another route to $[M(C_8H_8)]$ (Sm, Yb) is the reaction of the metals with cyclooctatetraene using iodine as catalyst. These compounds are not monomers, but the Lewis base adduct $[M(C_8H_8)(py)_3]$ has a 'piano-stool' structure.

Reaction of cyclooctatetraene with solutions of ytterbium and 2 moles of potassium in liquid ammonia gives $K_2[M(C_8H_8)_2]$; the solvate $[K(dme)]_2[Yb(C_8H_8)_2]$ (DME = 1,2-dimethoxyethane) has a sandwich uranocene-type structure for the anion.

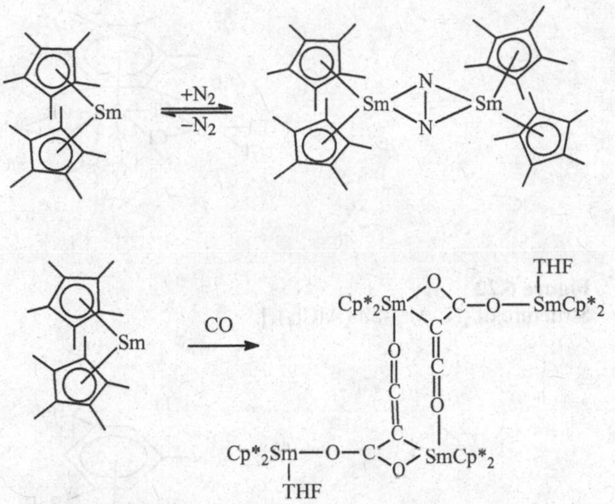

Figure 6.21
Products of the reaction of [SmCp*$_2$] with N$_2$ and CO.

6.6 The +4 State

Few compounds are known in this state, confined to Ce^{IV}; even so, the oxidizing power of Ce^{IV}, together with the tendency for species like LiR or NaR to be reducing, makes these rare. Alkoxides [Cp$_3$Ce(OR)] (R = Pr^i, Bu^t) and [Cp$_2$Ce(OBu^t)$_2$] have been synthesized. One such route is:

$$[Ce(OPr^i)_4] + 3\ CpSnMe_3 \rightarrow [Cp_3Ce(OPr^i)] + 3\ SnMe_3(OPr^i)$$

This uses an organotin compound as it is a weaker reducing agent than the alkali metal compounds usually used, and so less likely to result in reduction to Ce^{III}. The other compounds with claims to be in this oxidation state are [Ce(C$_8$H$_8$)$_2$] and analogues. [Ce(C$_8$H$_8$)$_2$] can be prepared by a 'witches' brew' approach from a mixture of [Ce(OPr^i)$_4$(Pr^{i}OH)], AlEt$_3$, and cyclooctatetraene at 140 °C, or by AgI oxidation of K [Ce(C$_8$H$_8$)$_2$]; the latter reaction can be reversed using metallic potassium as the reducing agent. Spectroscopic evidence support the view that this compound is [Ce^{3+}\{(C$_8$H$_8$)$_2$\}$^{3-}$] rather than [Ce^{4+}(C$_8$H$_8$$^{2-}$)$_2$]. The related [Ce\{1,3,6-(Me$_3$Si)$_3$C$_8$H$_5$\}$_2$], more stable and more hydrocarbon-soluble, has also been synthesized.

6.7 Metal–Arene Complexes

A few of these have been made in which lanthanides are refluxed with arenes under Friedel–Crafts conditions, such as [(C$_6$Me$_6$)Sm(AlCl$_4$)$_3$] (Figure 6.22). These appear to have the lanthanide in 'normal' oxidation states; the arene group is bound η^6. Interactions between a lanthanide and an arene ring are also well-authenticated in what would otherwise be coordinatively unsaturated aryloxides, like [Yb(OC$_6$H$_3$Ph$_2$)$_3$].

A different type of compound (Figure 6.23) has been made by co-condensation of vapourized lanthanide atoms with 1,3,5-tri-*t*-butylbenzene at 77 K. These compounds are thermally stable for many lanthanides at room temperature and above, and are strongly coloured,

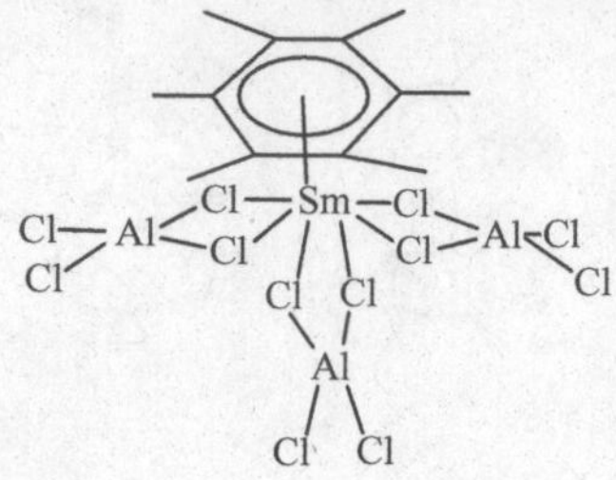

Figure 6.22
Structure of [(C$_6$Me$_6$)Sm(AlCl$_4$)$_3$].

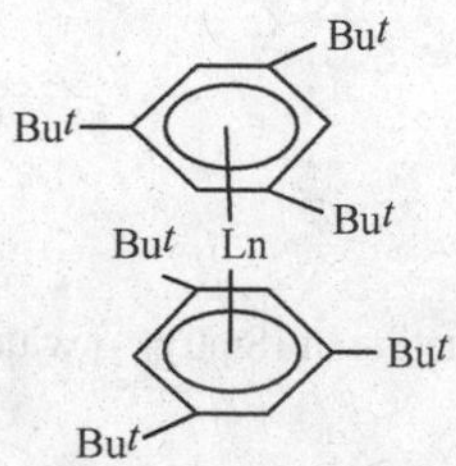

Figure 6.23
Structure of lanthanide bis(1,3,5-tri-*t*-benzene) compound.

pentane-soluble solids (the yttrium and gadolinium compounds, for example, can be sublimed at 100 °C/10^{-4} mbar with only partial decomposition).

The compounds of Nd, Gd–Er, Lu, and Y are stable to around 100 °C; the La, Pr, and Sm compounds decompose at 0, 40 and −30 °C, respectively. The compounds of Ce, Eu, Tm, and Yb could not be isolated. This cannot be explained simply by steric effects or redox behaviour of the metals; (i) compounds can be isolated with quite large early metals (Nd, Gd) but not with small ones like Yb; (ii) the compounds formally involve metals in the (0) state, so metals like Eu and Yb would have been expected to be successful in supporting low oxidation states. Steric factors may have their place, as the compounds of the earlier lanthanides (La–Pr) are less stable than might be expected, possibly because the ligands may not be bulky enough to stabilize the lanthanide atom against decomposition pathways.

This pattern is irregular but has been correlated with the f^{n}s^2 → f^{n-1}d^{1}s^2 promotion energies, as it is believed that an accessible d^{1}s^2 configuration is necessary for stable bonding. A bonding model has been developed similar to that used for compounds like dibenzenechromium, requiring delocalization of metal e$_{2g}$ electrons onto the aromatic rings. Bond dissociation energies have been measured in the range of 200–300 kJ/mole, and are quite comparable with known bis(arene) complexes of transition metals.

6.8 Carbonyls

One area which demonstrates the contrast between the lanthanides and the d-block transition metals is the absence of stable carbonyls. Co-condensation of the vapours of several metals (Pr, Nd, Eu, Gd, Ho, Yb) with CO in argon matrices at 4 K leads to compounds identified as [Ln(CO)$_x$] ($x = 1-6$) from their IR spectra. They decompose on warming above 40 K.

Other compounds containing carbonyl groups are known, involving a transition metal. Most of these feature isocarbonyl linkages. Thus $[Cp*_2Yb]$ reacts with $Co_2(CO)_8$ in THF solution to form $[Cp*_2Yb(THF)\text{-}O\text{-}C\text{-}Co(CO)_3]$. A few compounds actually contain metal–metal bonds involving a lanthanide, thus:

$$Cp_2LuCl + NaRuCp(CO)_2 \rightarrow NaCl + Cp_2Lu(THF)RuCp(CO)_2$$

The Lu–Ru bond length is 2.966 Å.

Question 6.1 Suggest why $[Li(L\text{-}L)]_3$ $[Eu(CH_3)_6]$ cannot be isolated.
Answer 6.1 Europium is the lanthanide with the most stable divalent state, so reduction to a Eu^{II} compound is taking place.

Question 6.2 Predict the coordination geometry of $[Ln(CH_2SiMe_3)_3(THF)_2]$.
Answer 6.2 Trigonal bipyramidal, as with main group AX_5 species. This will be more stable (repulsions minimized) than a square pyramidal structure.

Question 6.3 $[Yb(CH_2SiMe_3)_3(THF)_2]$ can be made by reaction of ytterbium chips with ICH_2SiMe_3 in THF. Would this route be suitable for $[Eu(CH_2SiMe_3)_3(THF)_2]$?
Answer 6.3 Because of the high stability of the Eu^{II} state, it is possible that oxidation might only occur as far as a europium(II) species like $[Eu(CH_2SiMe_3)_2(THF)_2]$. The reaction does not seem to have been attempted yet.

Question 6.4 Ether solutions of $[Lu\{CH(SiMe_3)_2\}_3]$ react with KX (X = Cl, Br), forming $[(Et_2O)K(\mu\text{-}Cl)Lu\{CH(SiMe_3)_2\}_3]$. This reaction is not shown by the lanthanum analogue. Comment on this.
Answer 6.4 Lutetium is smaller than lanthanum, so would have a greater charge density and thus be a better electrophile, but this is still a surprising reaction. Coordination of the chloride to potassium and lutetium evidently compensates for the loss of lattice energy.

Question 6.5 Chemically $[Ln\{CH(SiMe_3)_2\}_3]$ behave as Lewis acids; besides reacting with water, like lanthanide organometallics in general, they also are attacked by other nucleophiles such as amines and phenols, forming the silylamides and aryloxides, respectively. Write equations for the reactions of $[Sm\{CH(SiMe_3)_2\}_3]$ with $HN(SiMe_3)_2$ and $HOC_6H_2Bu^t_2\text{-}2,6\text{-}Me\text{-}4$.
Answer 6.5 $[Sm\{CH(SiMe_3)_2\}_3] + 3\ HOC_6H_2Bu^t_2\text{-}2,6\text{-}Me\text{-}4 \rightarrow [Sm\{OC_6H_2Bu^t_2\text{-}2,6\text{-}Me\text{-}4\}_3] + 3\ CH_2(SiMe_3)_2$

$[Sm\{CH(SiMe_3)_2\}_3] + 3\ HN(SiMe_3)_2 \rightarrow [Sm\{N(SiMe_3)_2\}_3] + 3\ CH_2(SiMe_3)_2$

Question 6.6 Comment on the formulae of the alkyls formed by the lanthanides with the CH_3, CH_2SiMe_3, CMe_3 and $CH(SiH_3)_2$ ligands (Section 6.2.1).
Answer 6.6 The coordination number decreases from 6 in $[LnMe_6]^{3-}$ through 5 and 4 in $[Ln(CH_2SiMe_3)_3(thf)_2]$, $[LnBu^t_4]^-$ and $[Ln(CH_2SiMe_3)_4]^-$ to 3 in $[Ln\{(CH(SiH_3)_2)_3\}]$, simply in order of the increasing bulk of the ligands, as hydrogens are replaced in the methyl group by $SiMe_3$ groups.

Question 6.7 $[Cp_2LnCl]$ are dimers in benzene solution but monomers in THF. Why?
Answer 6.7 THF is a good Lewis base and cleaves the bridges, forming monomeric complexes $[Cp_2LnCl(THF)]$. Benzene is too weak a donor to coordinate to the lanthanide.

Question 6.8 Predict the structure of the complex formed between pyrazine (1,4-diazine) and $[Yb(C_5H_5)_3]$.
Answer 6.8 $[\{(C_5H_5)_3Yb\}(pyrazine)\{Yb(C_5H_5)_3\}]$ (see Figure 6.24).

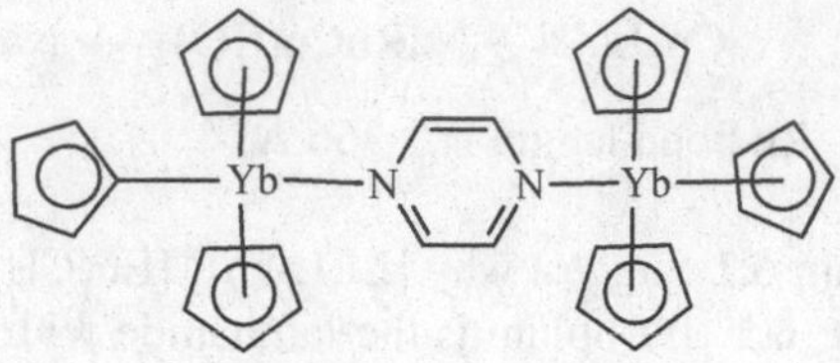

Figure 6.24
$[(C_5H_5)_3Yb(\mu\text{-pyrazine})Yb(C_5H_5)_3]$.

$LnCp_2R$ (R e.g. Bu^n Bu^t, CH_2SiMe_3, $p\text{-}C_6H_4Me$, etc.) can be made for later lanthanides:

$$LnCp_2Cl + LiR \rightarrow LnCp_2R + LiCl$$

When this reaction is carried out in THF, the product is $LnCp_2R(thf)$ (Figure 6.15).
Question 6.9 Structures of a number of lutetium alkyls $LuCp_2R(thf)$ have been examined by crystallographers. Some structural data are listed in Table 6.1; comment on there.

Table 6.1   Bond lengths in some Cp_2Lu complexes

Compound	Lu–C (Å)	Lu–O (Å)
$Cp_2Lu(Bu^t)(thf)$	2.47(2)	2.31(2)
$Cp_2Lu(CH_2SiMe_3)(thf)$	2.376(17)	2.288(10)
$Cp_2Lu(p\text{-}MeC_6H_4)(thf)$	2.345(39)	2.265(28)
$Cp_2Lu(Cl)(thf)$		2.27(1)

Answer 6.9   Both the Lu–C σ-bond and Lu–O bond lengths in $LuCp_2Bu^t(thf)$ look unusually long in comparison with the other compounds, suggesting that there is steric crowding here.

Question 6.10   The Lu–C σ-bond distance in $[Cp_2Lu(CH_2)_3NMe_2]$ (Figure 6.25) is 2.22 (1) Å. Compare it with the compounds in the preceding question and comment on this.
Answer 6.10   The Lu–C distance in $[Cp_2Lu(CH_2)_3NMe_2]$ is shorter than any of the others (it is the shortest Lu–C σ-bond reported) and must reflect an uncrowded environment around lutetium due to the presence of a chelate ring in place of two monodentate ligands.

Question 6.11   Read Section 6.3.4 and predict the products of reaction of $Cp*_2Ln(CH_3)$ on gentle warming with $SiMe_4$, another 'inert' molecule.
Answer 6.11 $[Cp*_2Ln(CH_2SiMe_3)_3] + CH_4$

Question 6.12   Predict the products of reaction of $[Cp*_2LnH]$ with Et_2O.
Answer 6.12 $[Cp*_2Ln(OEt)]$ and EtOH.

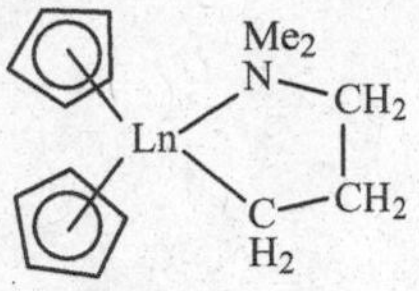

Figure 6.25
Structure of [(Cp$_2$Lu(CH$_2$)$_3$NMe$_2$].

Question 6.13 Why does using the pentamethylcyclopentadienyl group make these compounds more soluble?
Answer 6.13 The Cp* group on account of its bulk shields the metal more. It makes coordinative saturation more likely, so that there is less tendency to oligomerize, and the molecule presents a more 'organic' exterior to the solvent.

Question 6.14 In many of its reactions [Cp*$_2$Sm] is oxidized to Sm^{III} compounds. Suggest products of such reactions between THF solutions of [Cp*$_2$Sm] and (i) HgPh$_2$, (ii) Ag$^+$ BPh$_4^-$, (iii) C$_5$H$_6$.
Answer 6.14 (i) [Cp*$_2$SmPh(THF)]; (ii) [Cp*$_2$Sm(THF)$_2$]$^+$ BPh$_4^-$; (iii) [Cp*$_2$Sm(Cp)].

Question 6.15 Why do stable binary carbonyls of these metals not exist? (Consider what d-block transition metals need to do to form stable carbonyls that the lanthanides cannot do.)
Answer 6.15 Transition metals need both vacant orbitals to accept electron pairs from the σ-donor (ligand) and also filled d orbitals for π back-donation (back-bonding) to the ligand (note that neither the very early nor the very late transition metals form stable binary carbonyls). CO is a rather weak σ-donor, and lanthanides tend to complex only with good σ-donors. For π-bonding to occur, the metal must possess suitable orbitals to π-bond to the ligands, and the 'inner' 4f orbitals are unsuited to this, unlike the d$_{xy}$, d$_{xz}$, and d$_{yz}$ orbitals of transition metals.

7 The Misfits: Scandium, Yttrium, and Promethium

By the end of this chapter you should be able to:

- recognize that yttrium resembles the later lanthanides in its chemistry;
- recall that scandium differs from both the transition metals and the lanthanides in its chemistry;
- appreciate that promethium is a typical lanthanide in the types of compounds it forms, despite its radioactivity and the difficulties posed to its study.

7.1 Introduction

Scandium and yttrium are elements in Group IIIA (3) of the Periodic Table, usually placed above La (or Lu). Their treatment is frequently grouped with the lanthanides in textbooks (often for reasons of convenience). Promethium is, however, a radioactive lanthanide not encountered in the vast majority of laboratories.

Both scandium and yttrium are electropositive metals with similar reduction potentials to the lanthanides (E° $Sc^{3+}/Sc = -2.03$ V; E° $Y^{3+}/Y = -2.37$ V; compare values of -2.37 V and -2.30 V for La and Lu, respectively). The ionic radii of Sc^{3+} and Y^{3+} are 0.745 Å and 0.900 Å, respectively (in six coordination). The former is much smaller than any Ln^{3+} ion but yttrium is very similar to Ho^{3+} (radius 0.901 Å); purely on size grounds, it would be predicted that yttrium would resemble the later lanthanides but that scandium would exhibit considerable differences, and this expectation is largely borne out in practice. Table 7.1 lists stability constants for typical complexes of Sc^{3+} and Y^{3+}, together with values for La^{3+} and Lu^{3+}.

7.2 Scandium

When Mendeleev produced his original Periodic Table in 1869, he left a space for a metallic element of atomic mass 44 preceding yttrium. The first fairly pure scandium compounds were isolated by Cleve in 1879, but it was not until 1937 that the element itself was isolated. Although a relatively abundant element, it is fairly evenly distributed in the earth's crust and has no important ores, though it is the main component of the rare ore thortveitite ($Sc_2Si_2O_7$), thus being relatively expensive. In fact, it is mainly obtained as a by-product from uranium extraction.

Lanthanide and Actinide Chemistry   S. Cotton
© 2006 John Wiley & Sons, Ltd.

Table 7.1 Aqueous stability constants (log β_1) for complexes of Sc, Y, La and Lu

Ligand	I (mol/dm^3)	Sc^{3+}	Y^{3+}	La^{3+}	Lu^{3+}
F$^-$	1.0	6.2^a	3.60	2.67	3.61
Cl$^-$	1.0	0	−0.1	−0.1	−0.4
Br$^-$	1.0	−0.1	−0.15	−0.2	
NO$_3{}^-$	1.0	0.3		0.1	−0.2
OH$^-$	0.5	9.3*	5.4	4.7	5.8
acac$^-$	0.1	8.00	5.89	4.94	6.15
EDTA^{4-}	0.1	23.1	18.1	15.46	19.8
DTPA^{5-}	0.1	24.4	22.05	19.5	22.4
OAc$^-$	0.1		1.68	1.82	1.85

^{a} = Data for solutions of slightly different ionic strength.

Scandium is a soft and silvery metal; it is electropositive, tarnishing rapidly in air and reacting with water. It is generally obtained by chemical reduction, for example the reaction of the trifluoride with calcium. It has only one natural isotope, ^{45}Sc.

7.2.1 Binary Compounds of Scandium

Scandium oxide, Sc_2O_3, is a white solid (mp 3100 °C; Mn_2O_3 structure) made by burning the metal or by heating compounds such as the hydroxide or nitrate. It has a bcc structure with 6-coordinate scandium and has amphoteric properties, dissolving in alkali as $[Sc(OH)_6]^{3-}$ ions. All four scandium(III) halides are known, all but the fluoride (WO_3 structure) having the $FeCl_3$ structure. All are white solids. They can be obtained directly from the elements and in some cases by dehydration of the hydrated salts as well as by thermal decomposition of $(NH_4)_3ScX_6$ (X = Cl, Br), a method also used for the lanthanides. Other significant binary compounds include the fcc dihydride ScH_2; unlike most of the lanthanides, scandium does not form a trihydride.

Scandium forms a number of well-defined lower halides, especially chlorides. These are synthesized by high-temperature reaction of Sc with $ScCl_3$. They include ScCl, Sc_2Cl_3 ('mouse fur'), Sc_5Cl_8, Sc_7Cl_{10}, and Sc_7Cl_{12}. They typically have chain structures with metal–metal bonding and octahedral clusters of scandium atoms. ScCl is isostructural with ZrBr, Sc_5Cl_8 contains infinite chains of Sc_6 octahedra sharing trans-edges, whilst Sc_7Cl_{10} has two parallel chains of octahedra sharing a common edge. Sc_7Cl_{12} is regarded as made of Sc_6Cl_{12} octahedra with isolated scandium atoms in octahedral interstices. Sc_2Br_3 is also known.

7.3 Coordination Compounds of Scandium

As already mentioned, the ionic radius of Sc^{3+} is, at 0.745 Å (for 6-coordination), the largest of the metals in its period, so that high coordination numbers are frequently met with in its chemistry; examples between 3 and 9 are known. ^{45}Sc has $I = 7/2$ and a few NMR studies are being reported.

7.3.1 The Aqua Ion and Hydrated Salts

For many years it was assumed that, like the succeeding d-block metals, scandium formed a $[Sc(H_2O)_6]^{3+}$ ion. Various pieces of evidence indicated this was not the case – salts containing dimeric $[(H_2O)_5Sc(OH)_2Sc(H_2O)_5]^{4+}$ ions with approximately pentagonal

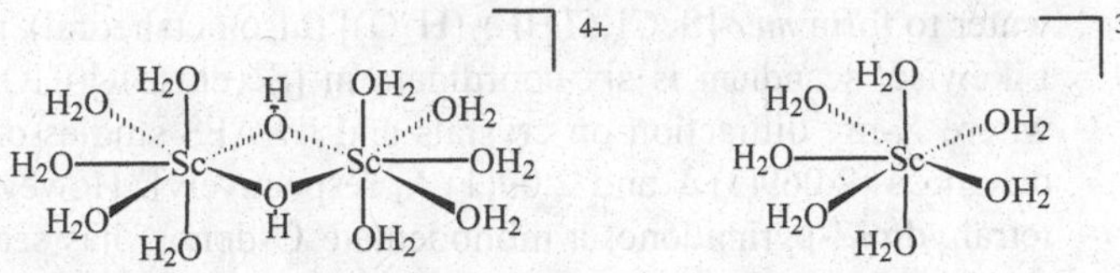

Figure 7.1

bipyramidal seven coordination of scandium were crystallized (Figure 7.1); EXAFS spectra of solutions indicated an Sc–O distance (2.180 Å) longer than that expected for octahedral coordination (2.10–2.12 Å) but in keeping with values obtained for the seven-coordinate scandium ion described above; Raman spectra of aqueous solutions were inconsistent with octahedral $[Sc(H_2O)_6]^{3+}$ ions; size considerations suggested seven coordination. Finally, several salts $[Sc(H_2O)_7] X_3$ were obtained $[X = Cl, Br, I, and C(O_2SCF_3)_3]$ which contain (roughly) pentagonal bipyramidal $[Sc(H_2O)_7]^{3+}$ ions (Figure 7.1) with Sc–O of 2.18–2.19 Å.

Table 7.2 gives the coordination numbers in aqueous solution of a number of $[M(H_2O)_n]^{3+}$ ions, together with their ionic radii (for six coordination, for a fair comparison of size), showing why Sc^{3+} might have been predicted to form a seven-coordinate aqua ion, purely on grounds of size.

A range of solid hydrated salts has been isolated and some structures are known. In addition to the seven-coordinate $[Sc(H_2O)_7] X_3$ $[X = Cl, Br, I, and C(O_2SCF_3)_3]$, $Sc(ClO_4)_3.6H_2O$ contains $[Sc(H_2O)_6]^{3+}$ ions and the triflate $Sc(O_3SCF_3)_3.9H_2O$ has $[Sc(H_2O)_9]^{3+}$ ions – as with the lanthanides (ions with these unusual coordination numbers are stabilized in the solid state by hydrogen-bonding). There are also $Sc(NO_3)_3.4H_2O$ and $Sc_2(SO_4)_3.5H_2O$. The structure of the nitrate is not known, but it should be noted that eight- and nine-coordinate $Sc(NO_3)_3(H_2O)_2$ and $Sc(NO_3)_3(H_2O)_3$ molecules are present in crown ether complexes.

7.3.2 Other Complexes

Scandium forms conventional halide complexes, like the hexafluoroscandates M_3ScF_6 which clearly contain $[ScF_6]^{3-}$ anions, such as Na_3ScF_6; other octahedral anions are found in Na_3ScCl_6 (cryolite structure), $Cs_2LiScCl_6$ (elpasolite structure), and Na_3ScBr_6. Six coordination is the norm; however, a Raman study of $CsCl–ScCl_3$ melts in the range 600–900 °C indicates that though the main species present is $[ScCl_6]^{3-}$ (ν Sc–Cl 275 cm^{-1}) there is evidence for the species $[ScCl_7]^{4-}$ (ν Sc–Cl 260 cm^{-1}).

A range of O-donor ligands (e.g. THF, Me_2SO, R_3PO) forms complexes with scandium, with two main series, ScL_3X_3 (where X is a coordinating anion like chloride) and ScL_6X_3 (with anions like perchlorate); however, at the moment, very few structures are confirmed. Six coordination is confirmed for *mer*-$Sc(THF)_3Cl_3$, which reacts with a small amount of

Table 7.2

M^{3+} ion	Sc	Y	La	Ce	Pr	Nd	Sm	Eu	Gd	Tb	Dy
Ionic radius (Å)	0.745	0.900	1.032	1.01	0.99	0.983	0.958	0.947	0.938	0.923	0.912
n	7	8	9	9	9	9	9	9	8	8	8
M^{3+} ion	Ho	Er	Tm	Yb	Lu	Pu	Ti	V	Cr	Fe	Co
Ionic radius (Å)	0.901	0.89	0.88	0.868	0.861	1.00	0.670	0.640	0.615	0.645	0.545
n	8	8	8	8	8	9	6	6	6	6	6

water to form *mer*-[ScCl$_3$(THF)$_2$(H$_2$O)] (also octahedral), also for [Sc(Me$_3$PO)$_6$] (NO$_3$)$_3$. Likewise, scandium is six coordinate in [Sc(dmso)$_6$](ClO$_4$)$_3$ and also in [Sc(dmso)$_6$]I$_3$, where X-ray diffraction on crystals and EXAFS studies on solutions yield similar Sc–O distances [2.069(3) Å and 2.09(1) Å, respectively]. However, [Sc(pu)$_6$](O$_3$SCF$_3$)$_3$ (pu = tetrahydro-2-pyrimidone, a monodentate O-donor) has scandium in trigonal prismatic 6 coordination.

There are a number of β-diketonate complexes, like Sc(acac)$_3$; this has octahedral 6 co-ordination [like M(acac)$_3$ for M = Ti–Co] and does not show any tendency to form adducts with Lewis bases, as do the lanthanide analogues. When nitrate is involved as a ligand, higher coordination numbers become possible, because a bidentate nitrate group takes up little more space in the coordination sphere than many monodentate ligands (has a small 'bite angle'). Thus in the pentakis(nitrato)scandate ion in (NO$^+$)$_2$[Sc(NO$_3$)$_5$], nine coordination exists with one nitrate being monodentate. Eight coordination is found in Sc(Ph$_3$PO)$_2$(η^2-NO$_3$)$_3$ and [Sc(urea)$_4$(η^2-NO$_3$)$_2$]$^+$ (NO$_3$)$^-$. Similarly, Na$_5$[Sc(CO$_3$)$_4$].2H$_2$O features four bidentate carbonates surrounding scandium. Another way of obtaining high coordination numbers is to use multidentate ligands like EDTA. Crystals of NH$_4$Sc(edta).5H$_2$O contain [Sc(edta)(H$_2$O)$_2$]$^-$ ions, in which edta is (as usual) hexadentate and scandium is thus 8 coordinate (a case where X-ray diffraction studies are essential in determining how many water molecules are bound). Similarly, dtpa is octadentate in Mn[Sc(dtpa)].4H$_2$O, so that scandium is 8 coordinate here too.

Lanthanide nitrate complexes of crown ethers have been studied in detail (Section 4.3.7). Scandium nitrate does not bind directly to crown ethers; complexes like Sc(NO$_3$)$_3$(H$_2$O)$_2$.crown (crown = 15-crown-5, benzo-15-crown-5) do not involve direct bonds between the oxygens in the ether ring and scandium, instead they have hydrated scandium nitrate molecules linked to the crown ether by hydrogen bonds to the ether oxygens. However, reaction of ScCl$_3$ with halogen-extractor SbCl$_5$ and 15-crown-5 in acetonitrile gives [ScCl$_2$(15-crown-5)](SbCl$_6$) in which the scandium is bound to 5 ether oxygens and two chlorines in roughly pentagonal bipyramidal coordination. What has happened is that removal of a chlorine from ScCl$_3$ has effectively generated a linear ScCl$_2$ moiety that is 'threaded' through the crown ether ring. Similar reactions occur with some other crown ethers (benzo-15-crown-5, 18-crown-6; in the latter case the crown ether is only pentadentate – see Figure 7.2). Carrying out the reaction in THF in the absence of crown ether gives [ScCl$_2$(THF)$_4$]$^+$ [SbCl$_5$(THF)]$^-$; in contrast, Y and La form [LnCl$_2$(THF)$_5$]$^+$ [SbCl$_5$(THF)]$^-$.

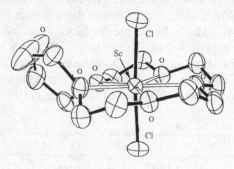

Figure 7.2
Seven-coordinate scandium in [ScCl$_2$(18-crown-6)](SbCl$_6$) (reproduced by permission of the Royal Society of Chemistry from G.R. Willey, M.T. Lakin, and N.W. Alcock, *J. Chem. Soc., Chem. Commun.*, 1992, 1619).

Sc(THF)$_3$Cl$_3$ is a very useful synthon (starting material) for other scandium compounds. It can conveniently be prepared using thionyl dichloride as a dehydrating agent:

$$ScCl_3.x\,H_2O + x\,SOCl_2 \rightarrow ScCl_3(THF)_3 + x\,SO_2 + 2x\,HCl \text{ (reflux with THF)}.$$

Ammine complexes cannot be made by solution methods, as owing to the basicity of ammonia other species (possibly hydroxy complexes) are obtained, but a few ammines have recently been described. (NH$_4$)$_2$[Sc(NH$_3$)I$_5$] is obtained as pink crystals from the reaction of NH$_4$I and metallic scandium in a sealed tube at 500 °C. Sc reacts with NH$_4$Br on heating in a sealed tantalum container, forming [Sc(NH$_3$)$_2$Br$_3$], isotypic with Sc(NH$_3$)$_2$Cl$_3$, which has the structure [Sc$_2$Br$_6$(NH$_3$)$_4$], with isolated dimers of bromide edge-connected [Sc-*mer*-(NH$_3$)$_3$Br$_3$] and [Sc(NH$_3$)Br$_5$] octahedra. [Sc(NH$_3$)Br$_3$], isotypic with Sc(NH$_3$)Cl$_3$, is [Sc$_2$Br$_6$(NH$_3$)$_2$], with zigzag chains of edge-connected [Sc(NH$_3$)Br$_5$] octahedra as [Sc(NH$_3$)$_{1/1}$Br$_{1/1}$Br$_{4/2}$]. Scandium similarly reacts with NH$_4$Cl in the presence of CuCl$_2$ to form [ScCl$_3$(NH$_3$)] and [ScCl$_3$(NH$_3$)$_2$]. Scandium has octahedral coordination in all of these.

7.3.3 Alkoxides and Alkylamides

Scandium resembles the 3d transition metals from titanium through cobalt in forming an unusual three-coordinate silylamide Sc[N(SiMe$_3$)$_2$]$_3$ but unlike them its solid-state structure is pyramidal (Figure 7.3), not planar, in which respect it resembles the lanthanides and uranium. It does not form adducts with Lewis bases, presumably on steric grounds, the resemblance here being to the 3d metals. However, the compound of a less congested amide ligand forms [Sc{N(SiHMe$_2$)$_2$}$_3$(THF)], which has distorted tetrahedral four coordination of scandium, with short Sc...........Si contacts in the solid state; this is in contrast to the 5-coordinate [Ln{N(SiHMe$_2$)$_2$}$_3$(THF)$_2$].

Few scandium alkoxides have been studied in detail, but monomeric 3-coordination exists in [(2,6-Bu$^t_2$-4-MeC$_6$H$_2$O)$_3$Sc] where the bulky ligand enforces steric crowding. It will, however, expand its coordination sphere to form 4-coordinate adducts with Ph$_3$PO and THF – unlike the corresponding three-coordinate silylamide. Sc(OSiBu$^t_3$)$_3$.L (L = THF,

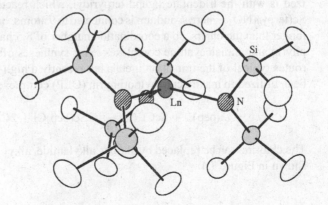

Figure 7.3
Three-coordinate Sc[N(SiMe$_3$)$_2$]$_3$ (where Ln = Sc) (reproduced by permission of the Royal Society of Chemistry from J.S. Ghotra, M.B. Hursthouse, and A.J. Welch, *J. Chem. Soc., Chem. Commun.*, 1973, 669).

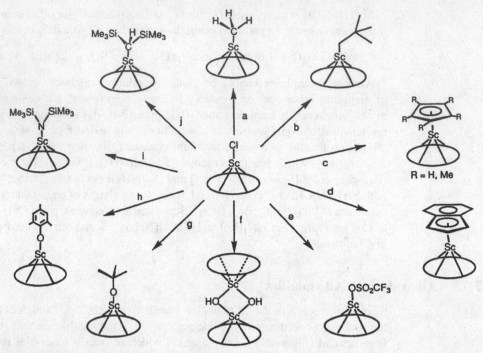

Figure 7.4
Conditions: (a) Me_2Mg in toluene; (b) Np_2Mg in toluene; (c) NaCp in THF, LiCp* in THF, Li(MeCp) in THF; (d) Na(Ind) in THF; (e) TMSOTf in toluene; (f) H_2O in CH_2Cl_2; (g) $LiOCMe_3$ in toluene; (h) $LiO(Me_3C_6H_2)$ in toluene; (i) $NaN(TMS)_2$ in toluene; (j) $LiCH(SiMe_3)_2$ in toluene (reproduced with permission from J. Arnold *et al.*, *Organometallics*, 1993, **12**, 3646 © American Chemical Society 2005).

py, NH_3) have also been prepared (the Lewis base cannot, however, be removed to afford the homoleptic siloxide).

Relatively few complexes of N-donor ligands are known. One of the best characterized is with the tridentate ligand terpyridyl, which reacts with scandium nitrate to form $Sc(terpy)(NO_3)_3$; here scandium is connected to 9 atoms, but one Sc–O bond is considerably longer than the others, so a coordination number of 8.5 has been assigned. A range of porphyrin and phthalocyanine complexes exist; syntheses often involve the high-temperature routes typical of the transition metals but recently a high-yield low-temperature route has been utilized to make octaethylporphyrin (OEP) complexes:

$$Li(oep)^- + ScCl_3(thf)_3 \rightarrow Sc(oep)Cl + 2Cl^- + 3\,THF + Li^+$$

The chloride can be replaced by alkoxy, alkylamide, alkyl, and cyclopentadienyl groups, as shown in Figure 7.4.

7.3.4 Patterns in Coordination Number

Table 7.3 lists in parallel corresponding binary compounds and complexes of scandium, lanthanum, and lutetium. As expected on steric grounds, the smaller scandium generally

Table 7.3

Compound	Sc compound/complex			La compound/complex			Lu compound/complex		
	Formula	C.N.		Formula	C.N.		Formula	C.N.	
Oxide	Sc_2O_3	6		La_2O_3	7		Lu_2O_3	6	
Fluoride	ScF_3	6		LaF_3	9+2		LuF_3	9	
Chloride	$ScCl_3$	6		$LaCl_3$	9		$LuCl_3$	6	
Bromide	$ScBr_3$	6		$LaBr_3$	9		$LuBr_3$	6	
Iodide	ScI_3	6		LaI_3	8		LuI_3	6	
Acetylacetonate	$Sc(acac)_3$	6		$La(acac)_3(H_2O)_2$	8		$Lu(acac)_3(H_2O)$	7	
EDTA complex	$Sc(EDTA)(H_2O)_2^-$	8		$La(EDTA)(H_2O)_3^-$	9		$Lu(EDTA)(H_2O)_2^-$	8	
THF adduct of trichloride	$ScCl_3(THF)_3$	6		$[LaCl(\mu\text{-}Cl)_2(THF)_2]_n$	8		$LuCl_3(THF)_3$	6	
Terpy complex of nitrate	$Sc(NO_3)_3(terpy)$	8.5		$La(NO_3)_3(terpy)(H_2O)_2$	11		$Lu(NO_3)_3(terpy)$	9	
Aqua ion	$[Sc(H_2O)_7]^{3+}$	7		$[La(H_2O)_9]^{3+}$	9		$[Lu(H_2O)_8]^{3+}$	8	
Bis(trimethylsilyl)amide	$Sc\{N(SiMe_3)_2\}_3$	3		$La\{N(SiMe_3)_2\}_3$	3		$Lu\{N(SiMe_3)_2\}_3$	3	
Ph_3PO complex of nitrate	$Sc(\eta^2\text{-}NO_3)_3(Ph_3PO)_2$	8		$La(\eta^1\text{-}NO_3)(\eta^2\text{-}NO_3)_2(Ph_3PO)_4$	9		$[Lu(\eta^2\text{-}NO_3)_2(Ph_3PO)_4]NO_3$	8	

exhibits lower coordination numbers than the lanthanides, although sometimes the value is the same as for lutetium, the smallest lanthanide.

7.4 Organometallic Compounds of Scandium

The organometallic chemistry of scandium is generally similar to that of the later lanthanides. It thus forms a cyclopentadienyl $ScCp_3$ that has mixed mono- and pentahapto-coordination like $LuCp_3$. An anionic methyl $[Li(tmed)]_3[M(CH_3)_6]$ is formed by scandium, as by the lanthanides. However, there are often subtle differences that should be borne in mind. The pentamethylcyclopentadienyl compound $[ScCp*_2Me]$ is a monomer but the lutetium compound is an asymmetric dimer $[Cp*_2Lu(\mu\text{-Me})LuCp*_2Me]$. Similarly, whilst triphenylscandium is obtained as a bis(thf) adduct, $[ScPh_3(thf)_2]$, which has a TBPY structure with axial thf molecules, the later lanthanides form octahedral $[LnPh_3(thf)_3]$. Triphenylscandium and the phenyls of the later lanthanides are made by different routes.

$$ScCl_3(thf)_3 + 3\,C_6H_5Li \rightarrow Sc(C_6H_5)_3(thf)_2 + 3\,LiCl + THF$$

$$2\,Ln + 3\,(C_6H_5)_2Hg + 6\,THF \rightarrow 2\,Ln(C_6H_5)_3(thf)_3 + 3\,Hg\,(Ln = Ho, Er, Tm, Lu)$$

Thus, the synthesis of triphenylscandium is a salt-elimination reaction (or metathesis) whilst the route for the lanthanide phenyls involves a redox reaction. The former has the problem of producing LiCl, which is often significantly soluble in organic solvents and contaminates the desired product, whilst the latter involves disposal of mercury waste, as well as handling toxic organomercury compounds.

7.5 Yttrium

As discussed earlier, yttrium resembles the later lanthanides strongly in its chemistry, so that throughout this book, discussion of lanthanide chemistry includes yttrium, and its compounds will not be examined in this chapter. To give a few, almost random, examples of yttrium compounds strongly resembling those of the later lanthanides:

- YX_3 (X = F, Cl, Br, I) have the same structures as LnX_3 (Ln = Dy–Lu);
- the yttrium aqua ion is $[Y(H_2O)_8]^{3+}$ though solid yttrium triflate $Y(O_3SCF_3)_3.9H_2O$ contains $[Y(H_2O)_9]^{3+}$ ions;
- the acetylacetonate is $[Y(acac)_3(H_2O)]$;
- the bis(trimethylsilyl)amide is $Y[N(SiMe_3)_2]_3$;
- terpyridyl reacts with yttrium nitrate, forming 10 coordinate $[Y(terpy)(NO_3)_3(H_2O)]$.

Yttrium compounds are frequently useful host materials for later Ln^{3+} ions, as mentioned in Section 5.4.4; $Eu:Y_2O_2S$ is the standard material for the red phosphor in virtually all colour and television cathode ray tubes, whilst $Eu:Y_2O_3$ is used for energy-efficient fluorescent tubes. Yttrium oxide is used to stabilize zirconia (YSZ), yttrium iron garnets (YIG) are used in microwave devices, and of course $YBa_2Cu_3O_7$ is the classic 'warm' superconductor. Yttrium, like scandium, is naturally monoisotopic. ^{89}Y has $I = 1/2$; though signals can be difficult to observe, valuable information can be obtained from NMR studies.

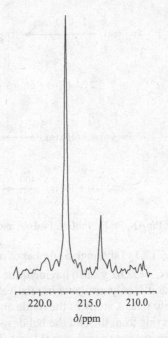

Figure 7.5
^{89}Y NMR spectrum of $[Y_5O(O^iPr)_{13}]$ in C_6D_6 at 25 °C (reproduced with permission of the American Chemical Society from P.S Coan, L.G. Hubert-Pjalzgraf, and H.G. Caulton, *Inorg. Chem.*, 1992, **31**, 1262).

Reaction of YCl_3 with lithium isopropoxide, $LiOCHMe_2$ ($LiOPr^i$), affords an alkoxide $Y_5O(OPr^i)_{13}$, which does not have a simple structure but has a cluster of 5 yttriums arranged round a central oxygen (Fig. 4.12). The ^{89}Y NMR NMR spectrum is shown in Figure 7.5.

Two possible geometries, square pyramidal and trigonal bipyramidal, are possible for the Y_5O core of the molecule; the two NMR signals have relative intensities of 4:1, supporting the square pyramidal structure. Use of bulky R groups, such as $OCBu_3^t$ or $2,6\text{-}Bu_2^tC_6H_3$, would mean that a monomeric $Y(OR)_3$ system is more likely.

7.6 Promethium

Mendeleev's Periodic Table made no provision for a lanthanide series. No one could predict how many of these elements would exist and it was not until Moseley's work on X-ray spectra that resulted in the concept of atomic number (1913) that it was known that an element with atomic number 61, situated between neodymium and samarium, remained to be discovered. Although several claims were made for its discovery in lanthanide ores, it was realized that no stable isotopes of element 61 existed and, from the late 1930s, nuclear chemistry was applied to its synthesis.

Promethium occurs in tiny amounts in uranium ores, thus a sample of Congolese pitchblende was found to contain $(4 \pm 1) \times 10^{-15}$g of ^{147}Pm per kg of ore; it was formed mainly by spontaneous fission of ^{238}U. It is also one of the fission products of uranium-235 and can be obtained from a mixture of lanthanides by ion exchange. The longest-lived isotope is

$$Pm^{3+}(aq) \xrightarrow{H_2C_2O_4(aq)} Pm_2(C_2O_4)_3.xH_2O(s) \xrightarrow[5\,h]{800\,°C} Pm_2O_3(s)$$

$$2\,Pm_2O_3(s) + 6F_2(g) \xrightarrow{300\,°C} 4\,PmF_3(s) + 3\,O_2(g)$$

$$Pm_2O_3(s) + 6HX(g) \xrightarrow{500\,°C} 2\,PmX_3(s) + 3H_2O(g)\ (X = Cl,\ Br)$$

$$PmX_3(s) + 3HI(g) \xrightarrow{400\,°C} PmI_3(s) + 3\,HX(g)\ (X = Cl,\ Br)$$

Figure 7.6
Syntheses of Pm_2O_3, PmF_3, $PmCl_3$, $PmBr_3$, and PmI_3.

^{145}Pm ($t_{1/2} = 17.7$ y) although a number of other isotopes exist, with both ^{146}Pm ($t_{1/2} = 5.5$ y) and ^{147}Pm ($t_{1/2} = 2.62$ y) also having quite long half-lives. Because of its radioactivity and also because it is only available in relatively small amounts, chemical studies are difficult and relatively rare. A study of the oxide and halides began with about 100 μg of the oxide Pm_2O_3. Starting from this, all the halides were synthesized microchemically, carrying out the reactions in quartz capillaries (Figure 7.6).

$$Pm^{3+}(aq) \xrightarrow{H_2C_2O_4(aq)} Pm_2(C_2O_4)_3.xH_2O(s) \xrightarrow[5\,h]{800\,°C} Pm_2O_3(s)$$

$$2\,Pm_2O_3(S) + 6\,F_2(g) \xrightarrow{300\,°C} 4\,PmF_3(s) + 3\,O_2(g)$$

$$Pm_2O_3(s) + 6\,HX(g) \xrightarrow{500\,°C} 2\,PmX_3(s) + 3\,H_2O(g) \quad (X = Cl,\ Br)$$

$$PmX_3(s) + 3\,HI(g) \xrightarrow{400\,°C} PmI_3(s) + 3\,HX(g) \quad (X = Cl,\ Br)$$

On account of the small amounts available, the compounds were characterized by non-destructive techniques – X-ray powder diffraction, Raman spectroscopy, and UV–visible spectra. Purple-pink PmF_3 (mp 1338 °C) has the 11-coordinate LaF_3 structure, lavender $PmCl_3$ (mp 655 °C) has the 9-coordinate UCl_3 structure, and coral-red $PmBr_3$ (mp 624 °C) has the 8-coordinate $PuBr_3$ structure. Red PmI_3 (mp 695 °C) has the eight-coordinate $PuBr_3$ structure at room temperature but the six-coordinate BiI_3 structure at high temperatures. Pm_2O_3 itself can crystallize in one of three modifications; in the case in question, it was obtained in the B-type structure also adopted by the Sm, Eu, and Gd oxides.

Figure 7.7 shows the UV–visible absorption spectra of three promethium halides measured at room temperature. These very sharp absorption bands are characteristic of a lanthanide; if they came from a transition metal, they would be much broader. They also show very little dependence upon halide, showing that ligand-field effects are almost negligible; the position of the maxima would vary much more if promethium were a transition metal. (Although this is not shown in the diagram, the extinction coefficients are also much lower than for a transition metal.)

Other promethium compounds that have been isolated include the hydroxide $Pm(OH)_3$ [which adopts the 9-coordinate structure of $Nd(OH)_3$] and hydrated salts including

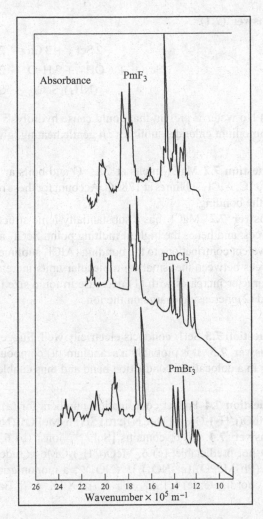

Figure 7.7
Solid-state absorption spectra of PmF$_3$, PmCl$_3$, and PmBr$_3$ (reproduced with permission from W.K. Wilmarth *et al.*, *J. Less Common Metals*, 1988, **141**, 275).

PmCl$_3$.xH$_2$O, Pm(NO$_3$)$_3$.xH$_2$O, and Pm(C$_2$O$_4$)$_3$.10H$_2$O. It would be expected that promethium would form some stable compounds in the +2 oxidation state, though they are unlikely to be made in aqueous solution. No definite evidence has yet been obtained, since studies have been hindered both by the small quantities of the element available and by its radioactivity. The properties of promethium fit neatly into position between neodymium and samarium; it is a microcosm of lanthanide chemistry in general.

Question 7.1 (a) Write equations for the (i) reduction of ScF$_3$ by calcium, (ii) scandium oxide dissolving in potassium hydroxide solution, (iii) the thermal decomposition of (NH$_4$)$_3$ScCl$_6$. (b) Why is the thermal decomposition of (NH$_4$)$_3$ScCl$_6$ a good way of making pure ScCl$_3$?

Answer (7.1)
(a)

$$2\,ScF_3 + 3\,Ca \rightarrow 2\,Sc + 3\,CaF_2$$
$$Sc_2O_3 + 6\,OH^- + 3\,H_2O \rightarrow 2\,[Sc(OH)_6]^{3-}$$
$$(NH_4)_3ScCl_6 \rightarrow ScCl_3 + 3\,NH_4Cl$$

(b) No water is present that could cause hydrolysis and the other product of the reaction, ammonium chloride, sublimes on gentle heating, giving a pure product.

Question 7.2 $MgCl_2$ melts at 712 °C and boils at 1712 °C; $ScCl_3$ normally sublimes at 800 °C; $AlCl_3$ sublimes at 178 °C. Account for these results in terms of the covalent character in the bonding.

Answer 7.2 $MgCl_2$ has a substantially ionic structure, giving it the highest electrostatic forces, and hence the highest melting point. $ScCl_3$ and, even more, $AlCl_3$ have increasing covalent contributions to the bonding ($AlCl_3$ sublimes as Al_2Cl_6 units) so that the attractive forces between the (smaller) molecular units increase. Fajans' rules predict that covalent character increases with (1) decrease in ionic size (aluminium is smaller than scandium) and (2) increasing charge on the ion.

Question 7.3 ScH_2 conducts electricity well. Suggest a reason.

Answer 7.3 It is probably a scandium(III) compound $Sc^{3+}(H^-)_2(e^-)$. The free electrons are in a delocalized conduction band and thus enable it to conduct electricity.

Question 7.4 Predict coordination numbers for (a) Rb_2KScF_6; (b) Ba_2ScCl_7; (c) $[ScCl_3\{MeO(CH_2)_2OMe\}(MeCN)]$; (d) $Sc(Ph_2MePO)_4(NO_3)_3$; (e) $ScX_3(Ph_3PO)_4$ (X = Cl, Br).

Answer 7.4 (a) 6, contains $[ScF_6]^{3-}$ ions; (b) 6, has structure $Ba_2[ScCl_6]Cl$ – possibly not predictable; (c) 6, $MeO(CH_2)_2OMe$ is bidentate and MeCN is N-bonded; (d) 8, $[Sc(Ph_2MePO)_4(\eta^2\text{-}NO_3)_2]^+$ $(NO_3)^-$; a monomeric $Sc(Ph_2MePO)_4(\eta^2\text{-}NO_3)_3$ would be 10 coordinate; (e) 6, $[ScX_2(Ph_3PO)_4]$ X (X = Cl, Br).

Question 7.5 Why are yttrium compounds good host materials for heavier Ln^{3+} ions?

Answer 7.5 Because Y^{3+} has no f (or d electrons), its compounds have no absorptions in the visible region of the spectrum and do not affect the magnetic properties of any added lanthanide ions. Because its ionic radius is similar to that of Ho^{3+}, it can accommodate several of the later lanthanide ions with minimal distortion.

Question 7.6 Suggest why the synthetic routes and spectroscopic methods shown for the binary promethium compounds in Section 7.6 were employed.

Answer 7.6 Only one product is a solid; by-products are gases and thus easily separated without loss of the desired product (important with small quantities). The products in the capillaries can then conveniently be used for X-ray diffraction studies (and UV–visible and Raman spectroscopy).

Question 7.7 Why are the structures of promethium compounds determined by powder methods and not by single-crystal methods?

Answer 7.7 The radiation emitted rapidly destroys the regular lattice in a single crystal.

Question 7.8 The visible spectrum of a promethium(III) compound (Figure 7.7) contains a considerable number of bands. How could you show that none of these were caused by impurity?

Answer 7.8 Wait for a few years (!), then remeasure the spectrum, any bands due to decay products will have become stronger and genuine bands due to Pm^{3+} will have diminished. Next, purify the promethium by ion exchange, then reprepare the compound; the impurity bands should have vanished.

Question 7.9 Explain why it is unlikely that Pm^{2+} compounds could be prepared by reduction in aqueous solution. (Hint: consult the reduction potentials in Table 2.7). Assuming that this were possible, why would $PmSO_4$ be a good choice of compound to isolate? Suggest a solvent that might be a good choice.

Answer 7.9 The reduction potential of $Pm^{2+}(Pm^{3+}/Pm^{2+} = -2.6\,V)$ is such that it is likely to be a very good reducing agent and would react with water. $PmSO_4$ is expected by analogy with $EuSO_4$ to be very insoluble and therefore could be followed by tracer study. A coordinating but non-aqueous solvent like THF would be a good choice.

8 The Lanthanides and Scandium in Organic Chemistry

By the end of this chapter you should be able to:

- recognize key lanthanide compounds used in organic reactions;
- appreciate some of the reactions in which they can be used;
- suggest reagents for certain transformations.

8.1 Introduction

It is a commonplace to say that there has been explosive growth in the use of lanthanides in organic chemistry. For many years, the use of cerium(IV) compounds as oxidants was widespread, but more recently a whole range of other compounds have made their appearance. Thus samarium(II) compounds are now routinely used as one-electron reducing agents and the use of trifluoromethanesulfonate ('triflate') salts of scandium and the lanthanides as water-soluble Lewis acid catalysts is widespread. Beta-diketonate complexes and alkoxides have also come into use; there are even applications of mischmetal in organic synthesis.

8.2 Cerium(IV) Compounds

Cerium is the most stable lanthanide in the (+4) oxidation state, so Ce^{IV} compounds have long been used as oxidants, most often as the nitrate complex, $(NH_4)_2[Ce(NO_3)_6]$ [cerium(IV) ammonium nitrate; CAN], but also $(NH_4)_4[Ce(SO_4)_4]$ [cerium(IV) ammonium sulphate; CAS] in particular, along with cerium trifluoracetate and others. The redox potential depends on the medium; in aqueous solution it is pH sensitive, getting more positive (i.e., more strongly oxidizing) the more acidic the solution. Mixtures of cerium salts with sodium bromate have been used, since the bromate is an oxidizing agent that can regenerate the Ce^{IV} state, permitting the use of cerium in catalytic, rather than stoichiometric, quantities.

8.2.1 Oxidation of Aromatics

When working in acidic solution, arenes are generally oxidized to quinones; reasonable yields are obtained with symmetric hydrocarbons as starting molecules, but

Lanthanide and Actinide Chemistry   S. Cotton
© 2006 John Wiley & Sons, Ltd.

non-symmetrical arenes lead to several products. CAS is the reagent of choice for these oxidations, as use of CAN may lead to introduction of nitro groups into the aromatic rings.

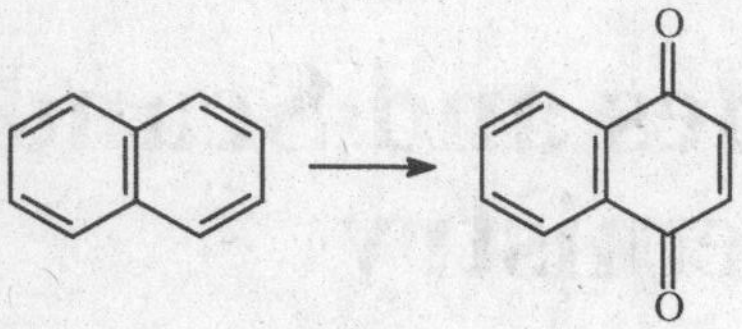

(CAS, MeCN–H₂O–H₂SO₄; RT; 90–95%)

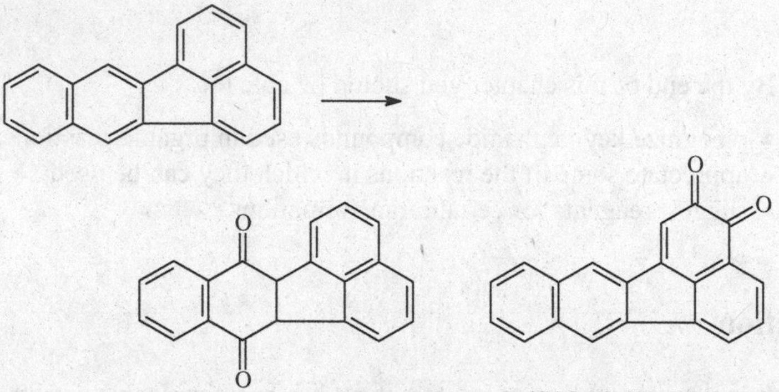

74:26   (CAS, MeCN–H₂O–H₂SO₄; RT; 38%)

Working in other media, the rings can be functionalized instead of being converted into quinones.

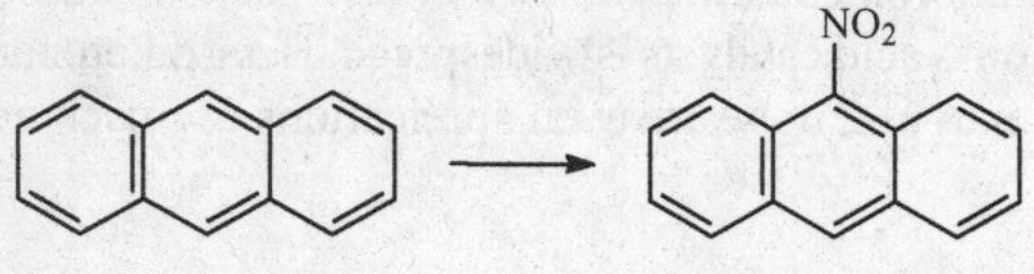

(CAN/silica gel; RT; 55%)

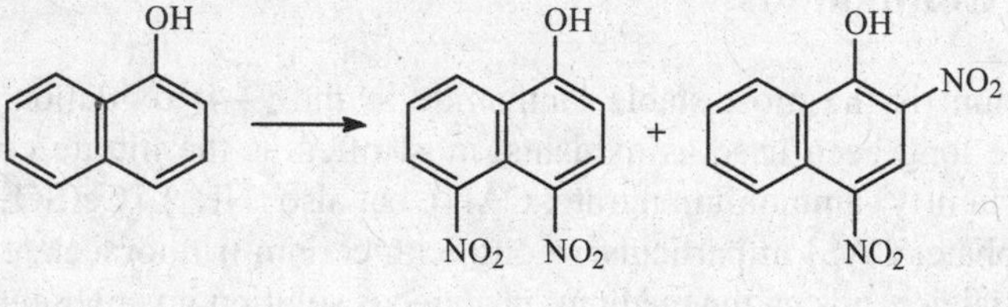

(CAN; CH₃COOH)

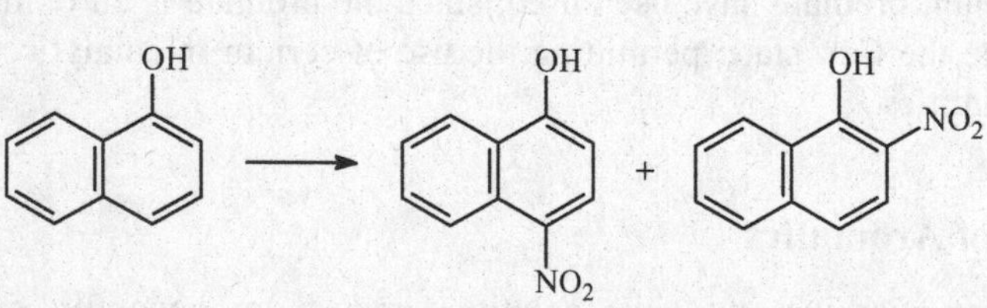

(CAN/silica gel)

Ce^IV Compounds oxidize side chains of aromatic compounds effectively and selectively, methylene carbons at the benzylic positions being oxidized to carbonyl groups. Polymethylated aromatics generally are oxidized to a single aldehyde group.

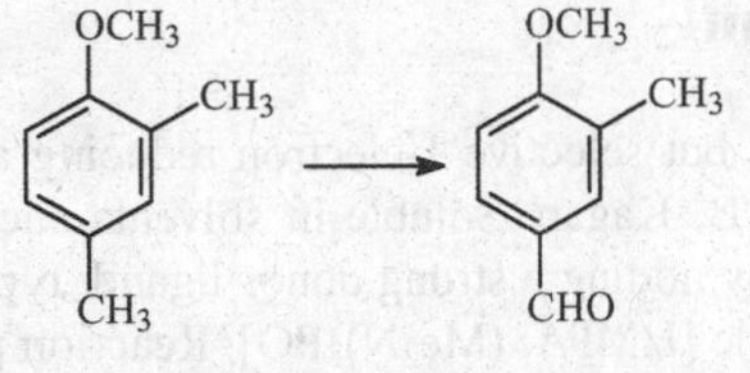

(CAN; MeOH 83%)

If alkylbenzenes are oxidized in non-aqueous media, other products are obtained; working in ethanoic acid, acetates are formed; in alcohol, ethers result.

8.2.2 Oxidation of Alkenes

Alkenes are readily oxidized, products depending strongly upon the solvent and also upon nucleophiles present.

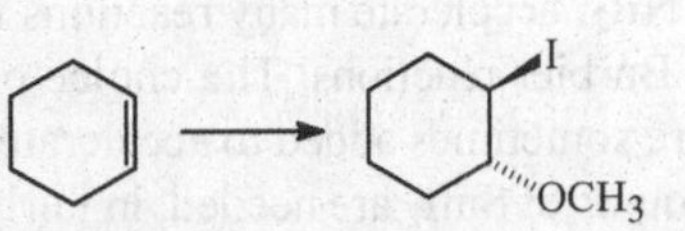

(CAN–I$_2$; MeOH)

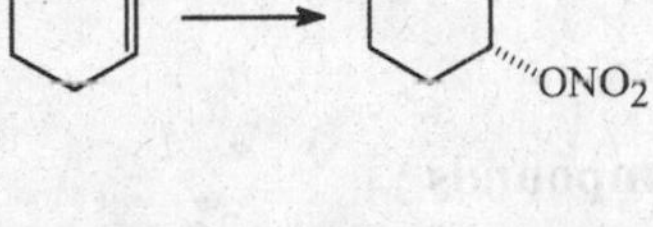

(CAN–I$_2$; Bu^{t}OH)

8.2.3 Oxidation of Alcohols

There are a variety of products, depending upon the alcohol. Allylic and benzylic alcohols are easily oxidized under mild conditions. Secondary alcohols are oxidized under rather stronger conditions. Simple primary alcohols (i.e., not 'activated' benzylic alcohols) are not oxidized. The oxidation of tertiary alcohols is accomplished with C–C bond fission.

(CAN/NaBrO$_3$; H$_2$O–MeCN; 80 °C; 88%)

PhCH$_2$OH ⟶ PhCHO   (CAN; AcOH–H$_2$O; 90 °C; 94%)

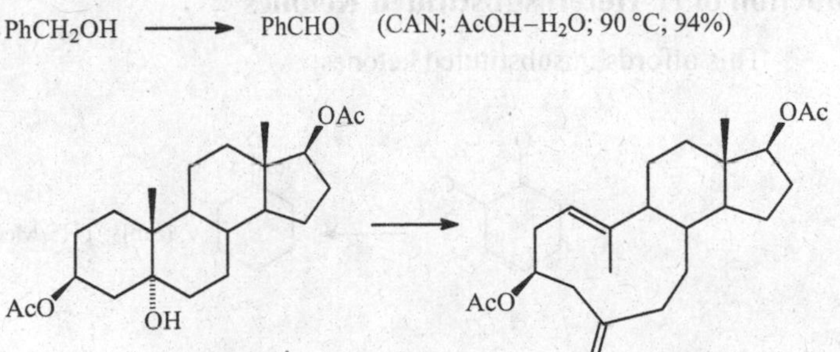

(CAN; MeCN–H$_2$O; 80 °C; 80%)

The last named is an example of a Grob oxidative fragmentation.

8.3 Samarium(*II*) Iodide, SmI$_2$

This is a powerful but selective 1-electron reducing agent, whose use was pioneered especially by Henri B. Kagan. Soluble in solvents such as alcohols and THF, its activity can be improved by adding a strong donor ligand, typically a Lewis base such as hexamethylphosphoramide [HMPA; (Me$_2$N)$_3$PO]. Reaction proceeds at an optimum HMPA: Sm ratio of 4:1, the acceleration probably involving a complex like SmI$_2$(hmpa)$_4$ (which has been isolated). Many reactions may involve radical intermediates.

A convenient synthesis is:

$$Sm + ICH_2CH_2I \rightarrow SmI_2 + H_2C = CH_2$$

Carried out in THF at room temperature, this affords a deep blue solution. The effectiveness of SmI$_2$ for some reactions can also be enhanced with certain catalysts. Thus, nickel halides, especially NiI$_2$, accelerate many reactions in THF, whilst iron(III) salts are used in samarium-assisted Barbier reactions. The choice of solvent is sometimes important; thus water or alcohols are sometimes added to accelerate reactions. A disadvantage is that large, stoichiometric, amounts of SmI$_2$ are needed, in fairly dilute mixtures, making its use costly, not least as large volumes of solvent are consumed; interest has developed in catalytic reactions, for example in the use of Mischmetal as a co-reducing agent for *in-situ* regeneration of SmI$_2$.

8.3.1 Reduction of Halogen Compounds

Halogen compounds are readily reduced, converting RX into alkanes in good yields. The reactivity order is I > Br > Cl. The process is very solvent-dependent; in THF only primary halides are reduced, but in HMPA primary, secondary, tertiary, and aryl halides are all reduced.

$$n\text{-}C_{10}H_{21}Br \longrightarrow n\text{-}C_{10}H_{22} \quad (2.5 \text{ mol SmI}_2 \text{ THF/HMPA, Pr}^i\text{OH, 10 min, RT, 95\%})$$

Sometimes coupling occurs:

$$PhCH_2Br \longrightarrow PhCH_2CH_2Ph \quad (SmI_2; THF; RT; 82\%)$$

8.3.2 Reduction of α-Heterosubstituted Ketones

This affords unsubstituted ketones

(SmI$_2$; THF/MeOH; –78 °C; 100%)

8.3.3 Reductions of Carbonyl Groups

Two outcomes are possible. In the presence of a proton source (e.g., H$_2$O) they are reduced to alcohols, as in E.J. Corey's approach to the synthesis of atractyligenin

(SmI$_2$; THF–H$_2$O; RT; 97%)

In this synthesis, the reactant was chosen to control stereochemistry and give an equatorial alcohol, due to an intermediate axial radical; use of hydride-transfer reagents would tend to afford predominantly the axial alcohol.

Asymmetric reduction of the carbonyl group occurs in the presence of a chiral base.

PhCOCOPh ⟶ (SmI$_2$/quinidine; THF–HMPA; RT; 78%; 56% ee)

In the absence of a proton source, coupling to give pinacols occurs:

C$_6$H$_{13}$C(O)CH$_3$ ⟶ (SmI$_2$; THF; RT; 80%)

This can be used to create cyclic systems containing *cis*-vicinal diol units.

R = Bu^{t}Ph$_2$Si   96 : 4   (SmI$_2$; THF; −78 to 0 °C; 81%)

Carboxylic acids and esters (and other acid derivatives) are generally inert to SmI$_2$. However, in strongly basic or acidic conditions, reduction occurs: Carboxylic acids and amides are reduced to primary alcohols (amides afford some amine as secondary product)

$$CH_3CH_2CH_2CH_2CH(C_2H_5)CO_2H \longrightarrow CH_3CH_2CH_2CH_2CH(C_2H_5)CH_2OH$$

(SmI$_2$; THF$-$H$_2$O$-$NaOH; RT; 94%)

$$PhCONH_2 \longrightarrow PhCH_2OH + PhCH_2NH_2$$

91 : 9

(SmI$_2$; THF$-$H$_2$O$-$KOH; RT; 90%)

Nitriles are also reduced to amines

$$PhCN \longrightarrow PhCH_2NH_2 \quad (SmI_2; THF-H_3PO_4; RT; 99\%)$$

8.3.4   Barbier Reactions

These involve addition of a halogenoalkane to a carbonyl compound.

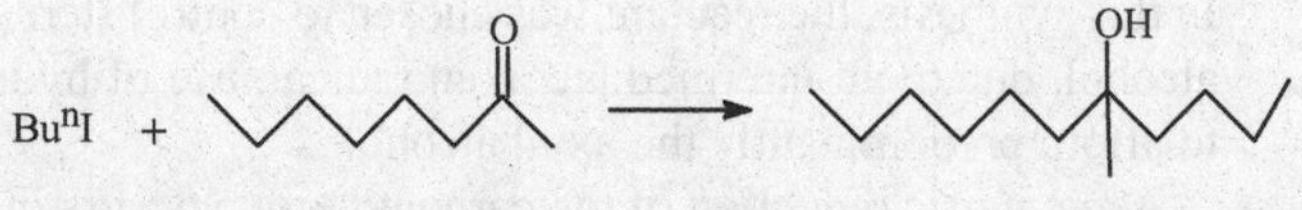

(SmI$_2$; THF; reflux; 97%)

This reaction can be carried out rapidly at room temperature by using small amounts of Fe^{3+} catalyst. Considerable acceleration can be achieved by using hexamethylphosphoramide as a reaction medium (compare the following two reactions).

$$C_4H_9Br + C_6H_{13}COCH_3 \longrightarrow C_6H_{13}C(OH)(C_4H_9)CH_3$$

(2 moles SmI$_2$; THF; reflux; 96%; 12 h)

$$C_4H_9Br + C_6H_{13}COCH_3 \longrightarrow C_6H_{13}C(OH)(C_4H_9)CH_3$$

(2 moles SmI$_2$; THF$-$HMPA; RT; 92%; 1 min)

Intramolecular Barbier-type reactions occur between carbonyl and halogen groups in the same molecule:

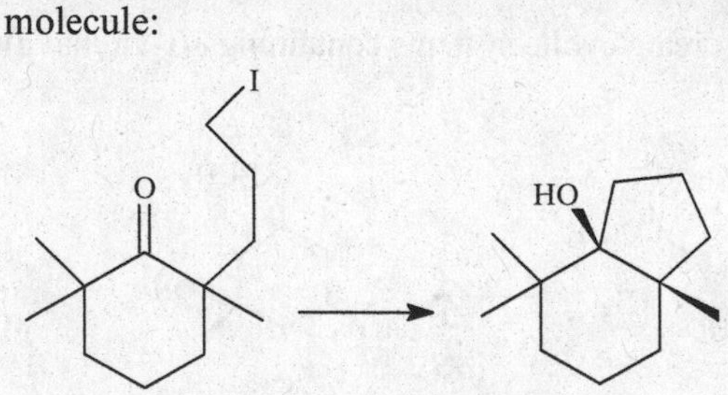

(SmI$_2$; THF; RT; Fe^{3+} catalyst; 90%)

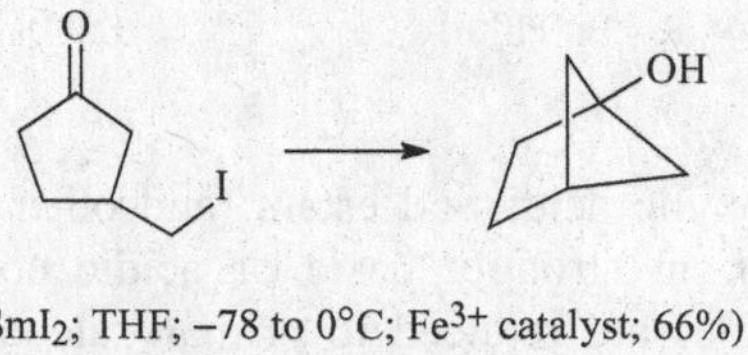

(SmI$_2$; THF; -78 to 0°C; Fe^{3+} catalyst; 66%)

8.3.5 Reformatsky Reactions

These occur between α-bromo esters and carbonyl compounds and are often used in ring-formation reactions.

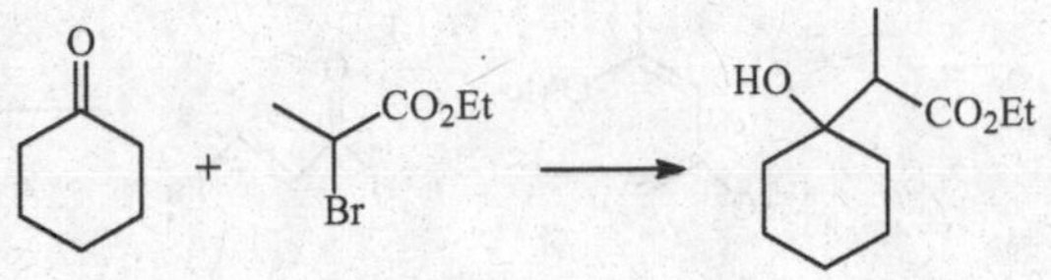

(SmI$_2$; THF; 90%)

8.4 Lanthanide β-Diketonates as Diels–Alder Catalysts

In the Diels–Alder reaction a diene (which must be capable of being s-*cis*) reacts with a dienophile, which must have an electron-withdrawing group conjugated to an alkene.

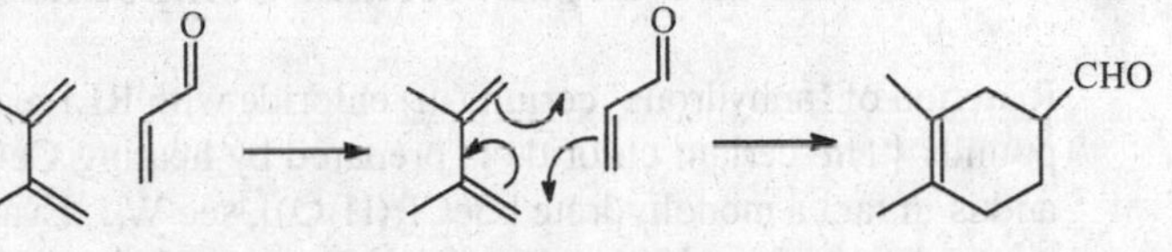

Lanthanide β-diketonate complexes, Ln(fod)$_3$, widely used formerly as NMR shift reagents (Section 5.5.1) are employed (fod = 6,6,7,7,8,8,8-heptafluoro-2,2-dimethyloctane-3,5-dionato). They have the advantages of working well under mild conditions (so that acid-sensitive groups are unaffected), as well as displaying good regio- and stereo-selectivity, with *endo*-adducts usually obtained.

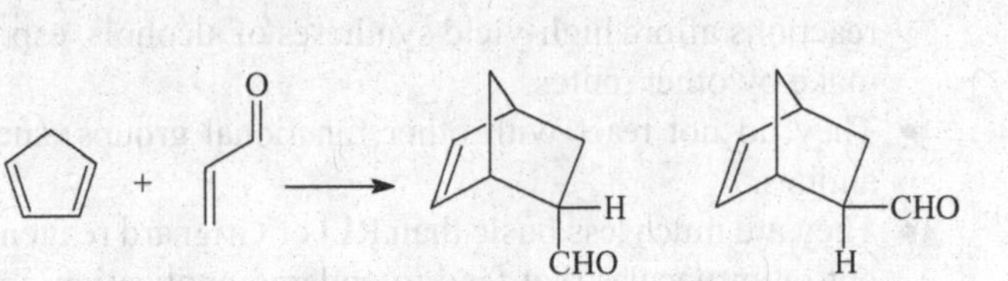

81%   5%

[Yb(fod)$_3$; RT; 24–48 h]

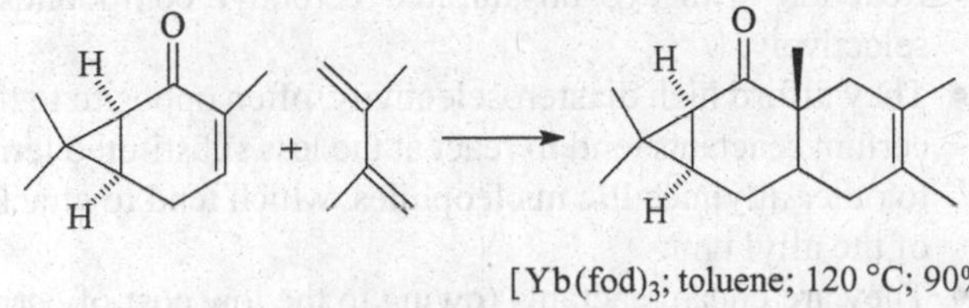

[Yb(fod)$_3$; toluene; 120 °C; 90%]

8.4.1   Hetero-Diels–Alder Reactions

Electron-rich alkenes, like the diene referred to in the first example, are very acid labile, so need a gentle catalyst. Aldehydes are the most usual dienophiles used in these syntheses.

[Eu(fod)$_3$; CH$_2$Cl$_2$; RT, 48h; 84%]

[Eu(fod)$_3$; CHCl$_3$; RT, 66%]

8.5   Cerium(*III*) Chloride and Organocerium Compounds

Reaction of 'anhydrous' cerium(III) chloride with RLi reagents affords organocerium compounds {The cerium chloride is prepared by heating CeCl$_3$(H$_2$O)$_7$ *in vacuo* up to 140 °C and is in fact a monohydrate [CeCl$_3$(H$_2$O)], see W.J. Evans, J.D. Feldman, and J.W. Ziller, *J. Am. Chem. Soc.*, 1996, **118**, 4581.} Reaction is carried out at −78 °C, as decomposition is rapid at 0 °C, especially if a β-hydrogen is present in the R group. The exact nature of the cerium species is uncertain.

Some characteristics and advantages of these species are:

- They readily react with carbonyl groups by nucleophilic attack, in the same way as RLi or Grignard species, but generally react more smoothly (as they are less basic reagents than the alternatives, so reactions are not complicated by deprotonation side-reactions). These reactions afford high-yield syntheses of alcohols, especially those that can be difficult to make by other routes.
- They do not react with other functional groups, such as halides, esters, epoxides, or amines.
- They are much less basic than RLi or Grignard reagents, and the reactions can be carried out on molecules that tend to undergo enolization, metal-halogen exchange, reduction, or pinacol coupling when RLi is used.
- They often work in highly hindered situations, where other reagents do not avail.
- Reaction with α,β-unsaturated carbonyl compounds gives 1,2-addition compounds selectively.
- They afford high diastereoselectivity, often opposite to that obtained by use of RLi. Allyl-cerium reactants tend to react at the less substituted terminus of the allyl unit, in contrast to other allylmetallic nucleophiles, which tend to attack at the most substituted terminus of the allyl unit.
- They are cheap reactants (owing to the low cost of cerium compounds).

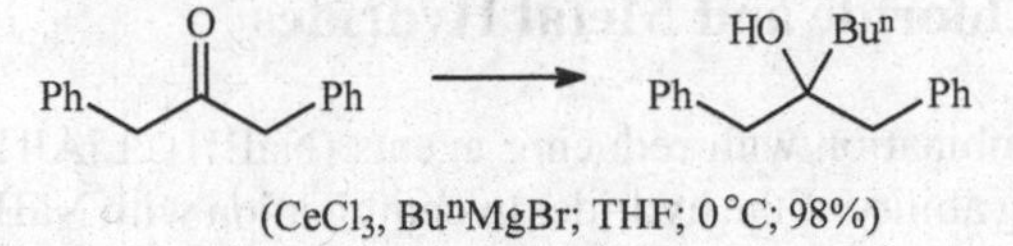

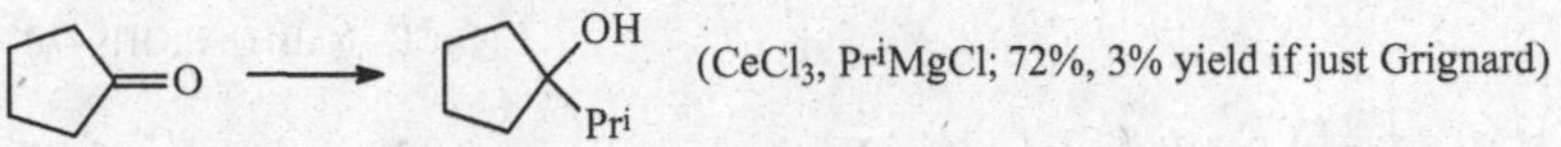

$Ph-CH_2-CO-CH_2-Ph \longrightarrow Ph-CH_2-C(OH)(Bu^n)-CH_2-Ph$

(CeCl₃, BunMgBr; THF; 0°C; 98%)

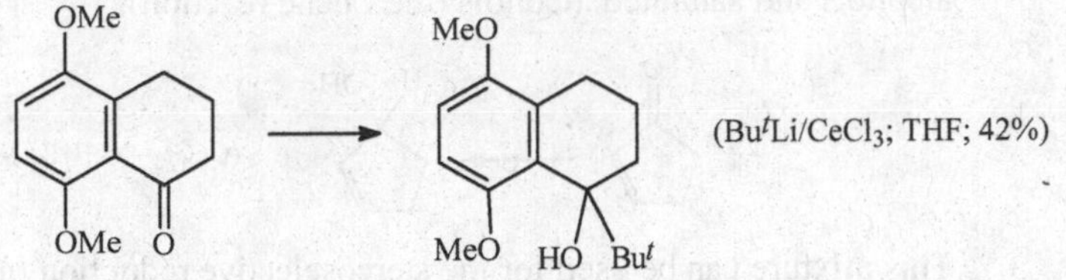

(CeCl₃, Bu^{n}MgBr; THF; 0°C; 98%)

(Bu^{t}Li/CeCl₃; THF; 42%)

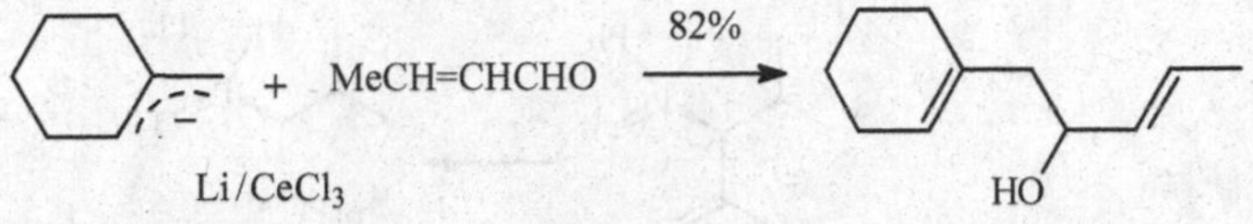

82%

Li/CeCl₃

(CeCl₃, Pr^{i}MgCl; 72%, 3% yield if just Grignard)

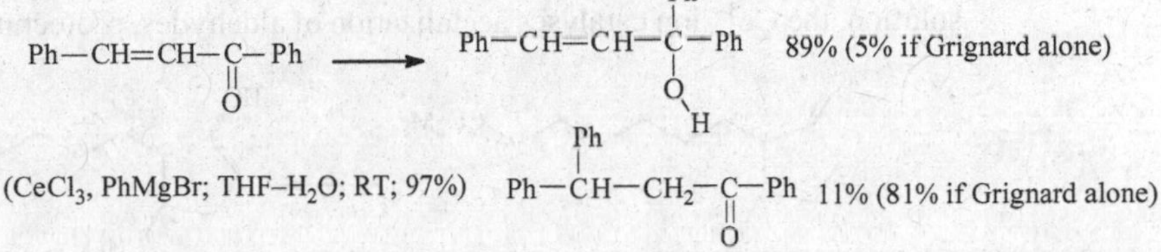

$Ph-CH=CH-C(O)-Ph \longrightarrow$

$Ph-CH=CH-C(Ph)(OH)-Ph$ 89% (5% if Grignard alone)

$Ph-CH(Ph)-CH_2-C(O)-Ph$ 11% (81% if Grignard alone)

(CeCl₃, PhMgBr; THF–H₂O; RT; 97%)

Reactions of chiral cycloalkenyl reagents with rigid ketones affords products with a high degree of stereodifferentiation:

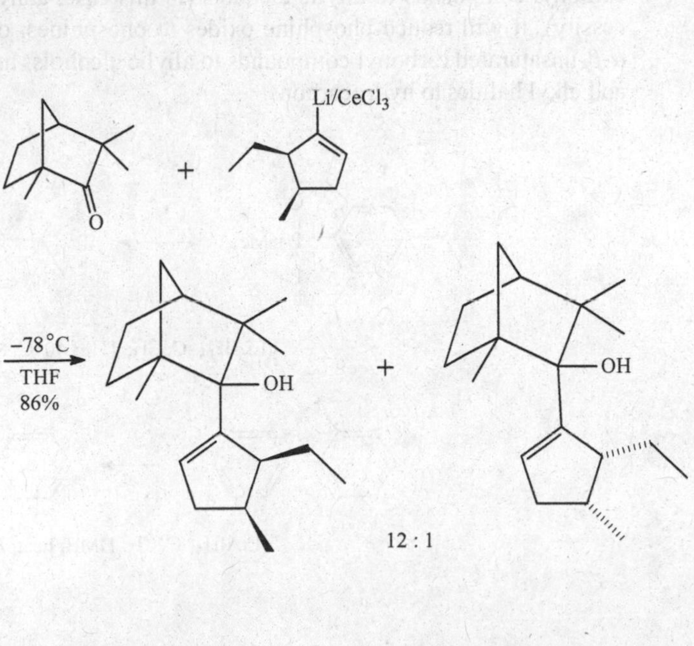

Li/CeCl₃

−78°C
THF
86%

12 : 1

8.6   Cerium(*III*) Chloride and Metal Hydrides

In combination with reducing agents (NaBH$_4$, LiAlH$_4$), cerium chloride modifies the reducing ability of the hydride. In combination with NaBH$_4$, CeCl$_3$.7H$_2$O selectively reduces the C=O group in enones, when use of NaBH$_4$ by itself would give a mixture of allylic alcohols and saturated alcohols (the Luche reaction).

(CeCl$_3$–NaBH$_4$; MeOH; 0°C; 88%)

This mixture can be used for the stereoselective reduction of saturated ketones

>95%   <5%

(CeCl$_3$–NaBH$_4$; EtOH; −78°C; ~100%)

and for the selective reduction of ketones in the presence of aldehydes, since, in alcoholic solution, the Ce^{3+} ion catalyses acetalization of aldehydes, protecting them from reduction.

(CeCl$_3$–NaBH$_4$; EtOH/H$_2$O; 75%)

A mixture of LiAlH$_4$ and CeCl$_3$ is a powerful reducing agent, reducing unsaturated carbonyl compounds to allylic alcohols (in this case, anhydrous cerium chloride is a necessity). It will reduce phosphine oxides to phosphines, oximes to primary amines, and α,β-unsaturated carbonyl compounds to allylic alcohols. In particular, it reduces both aryl and alkyl halides to hydrocarbons.

(LiAlH$_4$–CeCl$_3$; THF; 40°C; 89%)

(LiAlH$_4$–CeCl$_3$; DME; heat, *hv*; 94%)

8.7   Scandium Triflate and Lanthanide Triflates

These reagents are associated especially with S. Kobayashi. $Ln(O_3SCF_3)_3$ [$Ln(OTf)_3$] are synthesized in aqueous solution, crystallized as nonahydrates (Section 4.3.2) and the anhydrous triflates can be prepared by heating the hydrates. They are very effective Lewis acid catalysts, which perform well under mild conditions and can be used in both aqueous and organic solutions. Whilst most Lewis acids are decomposed in water (strictly dry organic solvents are employed in these reactions, cf. $AlCl_3$ in Friedel–Crafts reactions), triflates are not affected. The ability to avoid the use of organic solvents gives triflates powerful 'green' credentials. They are catalysts for many reactions (Michael, Diels–Alder, allylation, Friedel–Crafts, imino-Diels–Alder, etc.). Scandium triflate has attracted especial attention, though the use of the cheaper ytterbium compound has also been well investigated.

8.7.1   Friedel–Crafts Reactions

Small amounts of $Sc(OTf)_3$ [as well as many $Ln(OTf)_3$] catalyse some Friedel–Crafts reactions. Unlike conventional Friedel–Crafts catalysts, there is no need to use stoichiometric amounts. The yield does depend upon the lanthanide; for example, in the following reaction, 89% (Sc), 55% (Yb), 28% (Y).

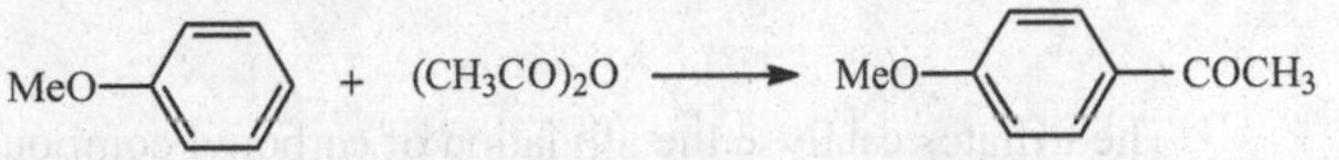

$$[M(OTf)_3 \text{ cat}; CH_3NO_2; 50\,°C]$$

$Yb(OTf)_3$ is an excellent catalyst for the aldol reactions of silylenol ethers with aldehydes in aqueous solution, working better than in organic solvents like THF and MeCN, though the reactions can also be performed in organic solvents, and, after the reaction has been quenched by the addition of water, the triflate catalyst may be recovered from the aqueous layer.

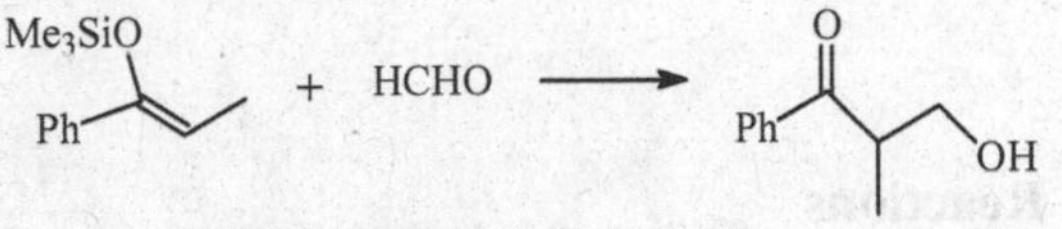

$$[Yb(OTf)_3 \text{ cat}; H_2O\text{–THF}; 94\%]$$

Even greater rates have been achieved in the $Yb(OTf)_3$-catalysed aldol reactions of silylenol ethers with aldehydes in micelles, by adding a small quantity of surfactant, such as sodium dodecyl sulfate.

A study of the reaction of benzaldehyde with 1-(trimethylsilyloxy)cyclohexane shows a remarkable dependence of the yield upon the lanthanide used, being as high as 91% for $Yb(OTf)_3$ and as low as 8% with $La(OTf)_3$.

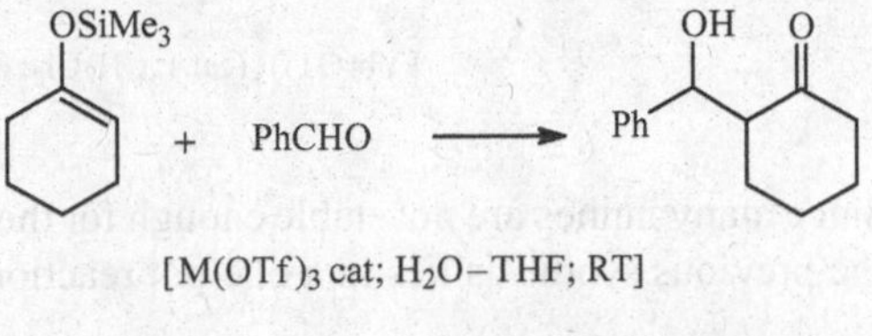

$$[M(OTf)_3 \text{ cat}; H_2O\text{–THF}; RT]$$

8.7.2 Diels–Alder Reactions

Lanthanide triflates catalyse Diels–Alder reactions, with the scandium complex as the most effective catalyst, and, again, the catalyst can be recovered and reused, being just as effective in subsequent runs.

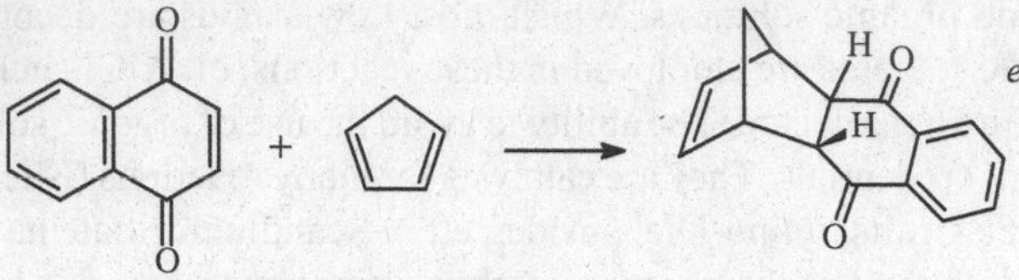

endo/exo = 100/0

[Sc(OTf)$_3$ cat.; THF–H$_2$O; RT; 93%]

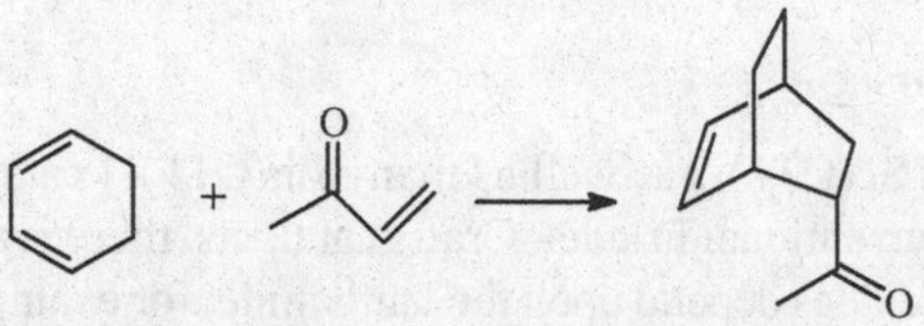

endo/exo = 95/5

[Sc(OTf)$_3$ cat.; THF–H$_2$O; RT; 83%]

The triflates catalyse the allylation of carbonyl compounds with tetraallyltin, producing intermediates in the synthesis of higher sugars.

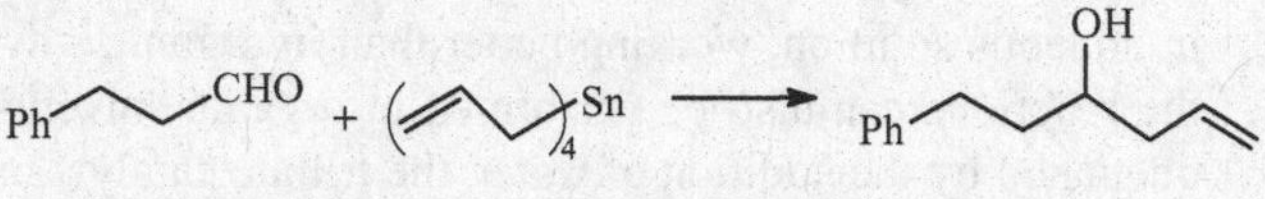

[Yb(OTf)$_3$ cat.; H$_2$O–EtOH–toluene; RT; 90%]

8.7.3 Mannich Reactions

Another synthesis to which they have been applied is Mannich-type reactions between imines and enolates (especially silyl enolates) to afford β-amino ketones or esters.

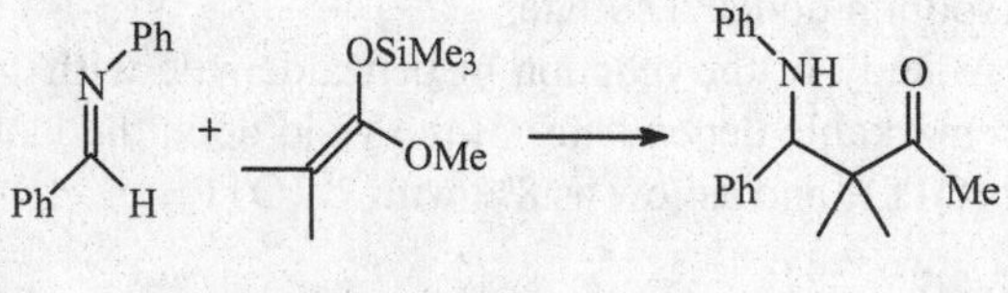

[Yb(OTf)$_3$ (cat.); CH$_2$Cl$_2$; 0°C; 97%]

Since many imines are not stable enough for their isolation and purification, an extension of the previous synthesis lies in a one-pot reaction that reacts with *in situ* prepared imines with silyl enolates to produce β-amino esters.

PhCHO + BuNH₂ + [structure: ketene silyl acetal with OSiMe₃ and OMe] → [product: β-amino ester with Bu-NH, Ph, two methyls, C=O, OMe]

[Yb(OTf)₃ (cat.) + dehydrating agent; CH₂Cl₂; RT; 85%]

β-Amino ketones can be made from an aldehyde, an amine, and a vinyl ether in a one-pot synthesis, an aqueous Mannich-type reaction.

PhCHO + *p*-ClC₆H₄NH₂ + [structure: isopropenyl, Me, Me] → [product: β-amino ketone with *p*-ClC₆H₄-NH, Ph, two methyls, C=O, OMe]

[Yb(OTf)₃ (cat.); THF–H₂O; RT; 90%]

8.7.4 Imino-Diels–Alder Reactions

Lanthanide triflates are catalysts for a powerful imino-Diels–Alder reaction for building N-containing six-membered heterocycles.

[reaction scheme: PhCH=N-phenyl + cyclopentadiene → fused tricyclic product II with H, H, N, Ph]

[Yb(OTf)₃ (cat.); CH₃CN; RT; 69%]

[reaction scheme: *o*-Cl-phenyl imine of PhCHO + ethyl vinyl ether (OEt) → tetrahydroquinoline product with H, OEt, N, H, Cl, Ph]

[Yb(OTf)₃ (cat.); CH₃CN; RT; 95%]

The yields again depend upon the choice of the lanthanide triflate. In a study of the reaction:

[reaction scheme: PhCH=N-phenyl + cyclopentadiene → fused tricyclic product with H, H, N, H, Ph]

[M(OTf)₃ (cat.); CH₃CN; RT]

best yields were obtained with heavy lanthanides such as Er (97%) and Yb (85%), in contrast to La (45%) and Pr (60%).

Another one-pot synthesis uses reaction of an aldehyde, an amine, and an alkene to give pyridine and quinoline derivatives via the imino-Diels–Alder route.

[Yb(OTf)$_3$(cat.); CH$_3$CN; RT; 65%]

Extensions of the triflate catalyst include the use of versions with fluorinated 'ponytails' such as [Sc{C(SO$_2$C$_8$F$_{17}$)$_3$}$_3$] for use in fluorous phase Diels–Alder reactions (fluorinated solvent.)

8.8   Alkoxides and Aryloxides

Complexes LnM$_3$tris(binapthoxide) (M = alkali metal) have a good deal of utility as catalysts.

The lanthanide complexes are soluble in ethers like Et$_2$O or THF. They are catalysts for the enantioselective reduction of aldehydes, with best results for the lanthanum complex. Thus the yield is 72% when M = La, but 50% when M = Yb; more strikingly, the enantiomeric excess is 69% for the La complex but only 3% when the Yb complex was used.

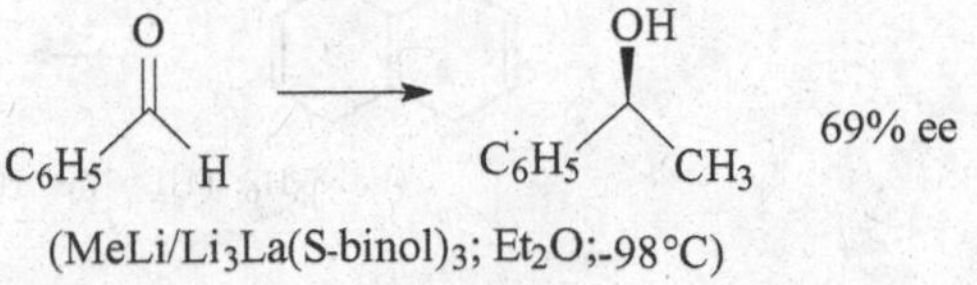

$$C_6H_5-CHO \longrightarrow C_6H_5-CH(OH)-CH_3 \quad 69\% \; ee$$

(MeLi/Li$_3$La(S-binol)$_3$; Et$_2$O;-98°C)

The research group of Shibasaki has studied similar compounds, LnM$_3$(binapthoxide)$_3$. x THF.yH$_2$O (M = Li, Na, K) as catalysts for a wide range of reactions, including epoxidation of enones, hydrophosphonylation of imines and aldehydes, and a range of asymmetric C—C bond-forming reactions (Diels–Alder, Michael addition, aldol and nitroaldol formation). Yields and enantioselectivity can depend markedly upon the alkali metal. Thus, in the nitroaldol reaction shown below, the yields are very similar (90, 92%) when LnM$_3$(binapthoxide)$_3$ are used (M = Li, Na), however the ee is 94% when using the Li compound but only 2% if the sodium compound is employed.

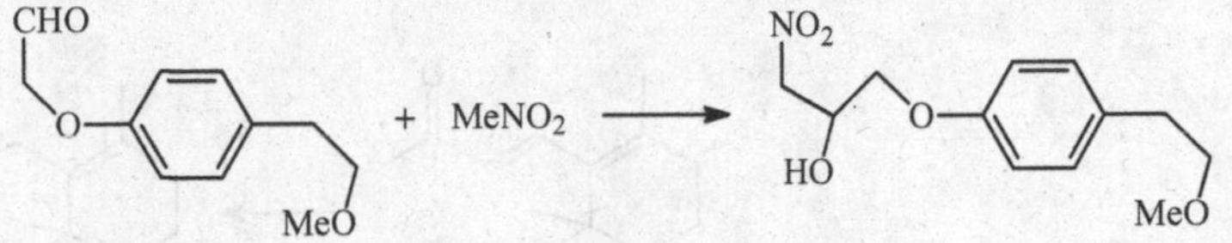

Lanthanide isopropoxides, usually written Ln(OPr^i)$_3$, but more likely to be oxo-centred clusters Ln$_5$O(OPr^i)$_{13}$, are used, not just as starting materials for the synthesis of catalysts such as the naphthoxides but also as catalysts in their own right. They have been used in the Meerwein–Ponndorf–Verley reaction, where carbonyl compounds are reduced to alcohols, recent studies having shown that the reaction takes place exclusively by a carbon-to-carbon hydrogen transfer.

8.9 Lanthanide Metals

Apart from the role of mischmetal as a co-reductant in SmI$_2$-catalysed reactions such as pinacolization (Section 8.3.3) mischmetal has been found to be a useful reactant in its own right. Thus a combination of mischmetal and additives (1,2-diiodoethane or iodine) has been active in Reformatsky-type reactions. Though giving slightly lower yields than SmI$_2$, they are less air-sensitive; there is the further handicap of requiring slow (dropwise) addition of reactants.

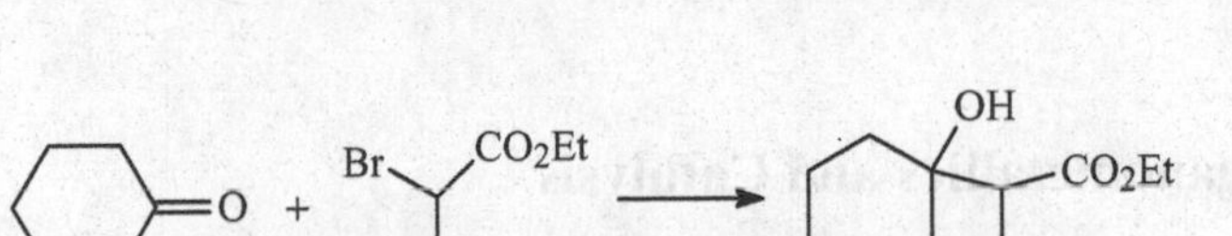

(mischmetal + additive; THF; RT; then acidify)

Lanthanum–nickel alloys are useful hydrogenation catalysts. The alloy LaNi$_5$ readily absorbs large amounts of hydrogen. The catalysts are robust, hard to poison, are capable of repeated reuse, and operate under mild conditions.

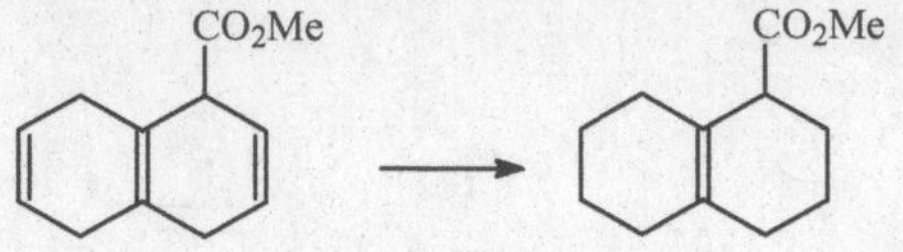

(LaNi$_5$H$_6$; THF−MeOH; RT; 78%)

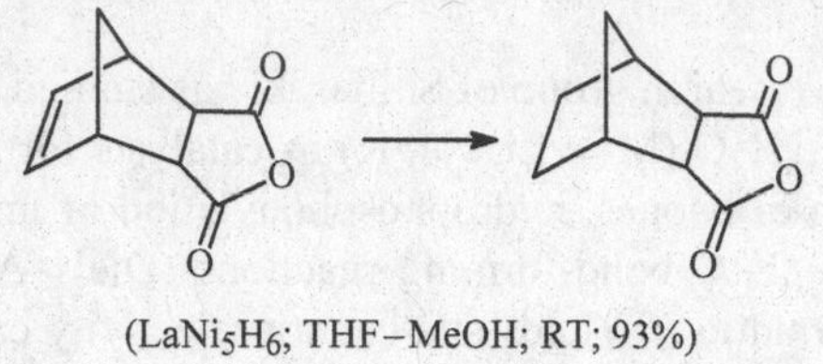

(LaNi$_5$H$_6$; THF−MeOH; RT; 93%)

It selectively catalyses hydrogenation of C=C units (especially conjugated double bonds) but under more forcing conditions it also reduces other functional groups such as carbonyl, nitro, and nitrile.

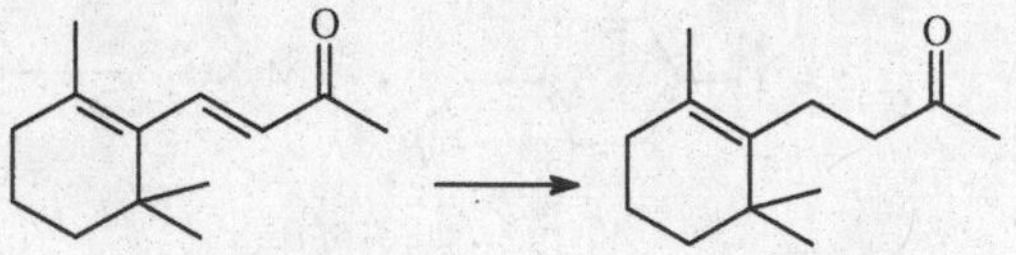

(LaNi$_5$H$_6$; THF−MeOH; 0 °C; 6 h; 89%)

Compare

(LaNi$_5$H$_6$; THF−MeOH; 0 °C; 3 h; 89%)

and

(LaNi$_5$H$_6$; THF−MeOH; RT; 12 h; 95%)

8.10   Organometallics and Catalysis

Considerable study is being made of the ability of organometallic compounds of the lanthanides (and actinides) to catalyse reactions of organic substrates, much of this work associated with the name of Tobin J. Marks. The hydride [Cp*$_2$Lu(μ-H)$_2$LuCp*$_2$] (Cp* = C$_5$Me$_5$) is an exceptionally active catalyst for the homogeneous hydrogenation of alkenes and alkynes. A suggested mechanism, involving dissociation to the coordinatively unsaturated monomer [Cp*$_2$LuH], is shown below (Figure 8.1) – readers will note a resemblance to the mechanism of hydrogenation using [RhCl(PPh$_3$)$_3$].

A whole range of reactions are catalysed by such complexes. Figure 8.2 shows the mechanism proposed in a recent study of hydroamination, catalysed by $[Cp*_2La\{CH(SiMe_3)_2\}_2]$. It indicates a two-stage mechanism, cyclization to form La–C and C–N bonds, followed by La–C protonolysis. The alkene inserts into the La–C bond via a four-centre transition state, followed by protonolysis by a second substrate molecule and dissociation of the cyclized amine, which regenerates the catalyst. The rate-determining step involves a highly organised seven-membered chair-like cyclic transition state.

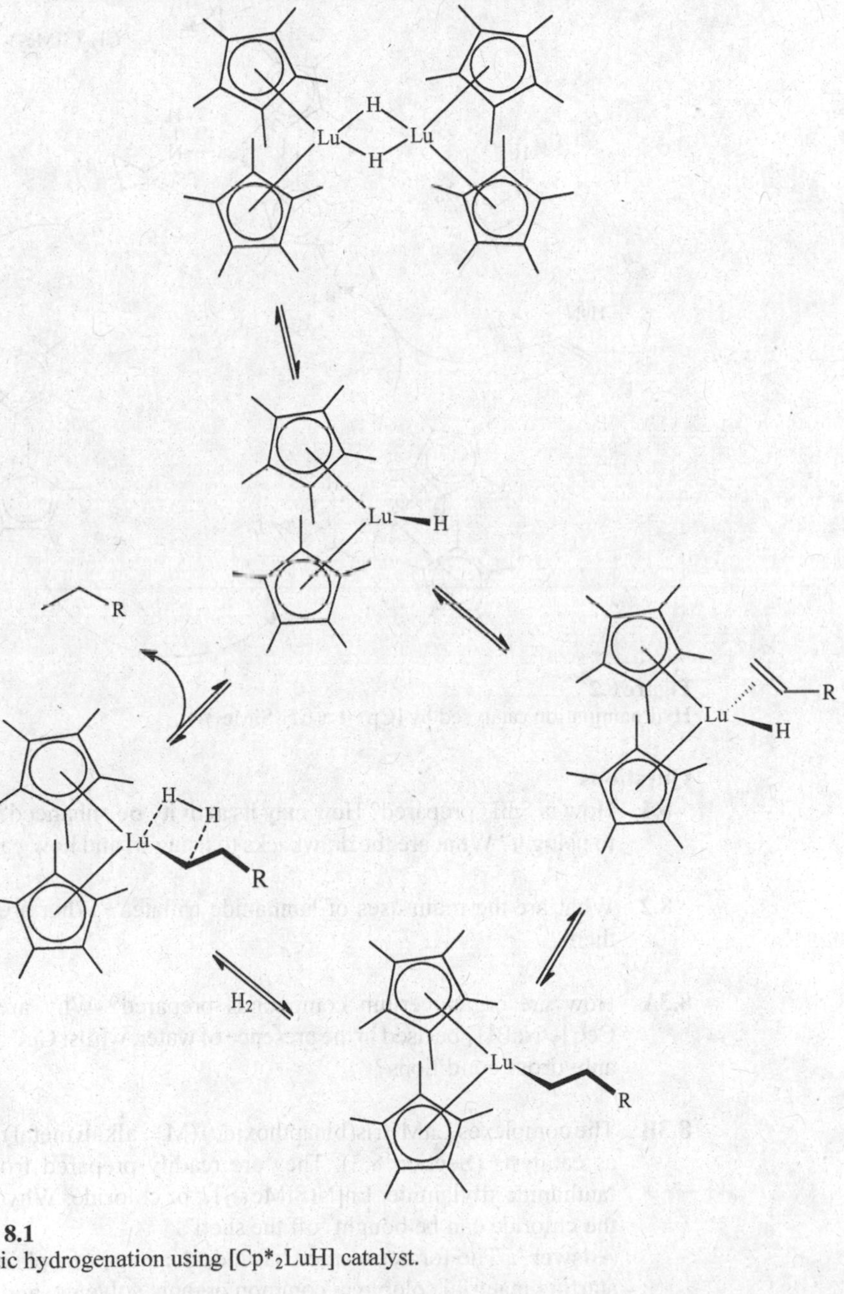

Figure 8.1
Catalytic hydrogenation using $[Cp*_2LuH]$ catalyst.

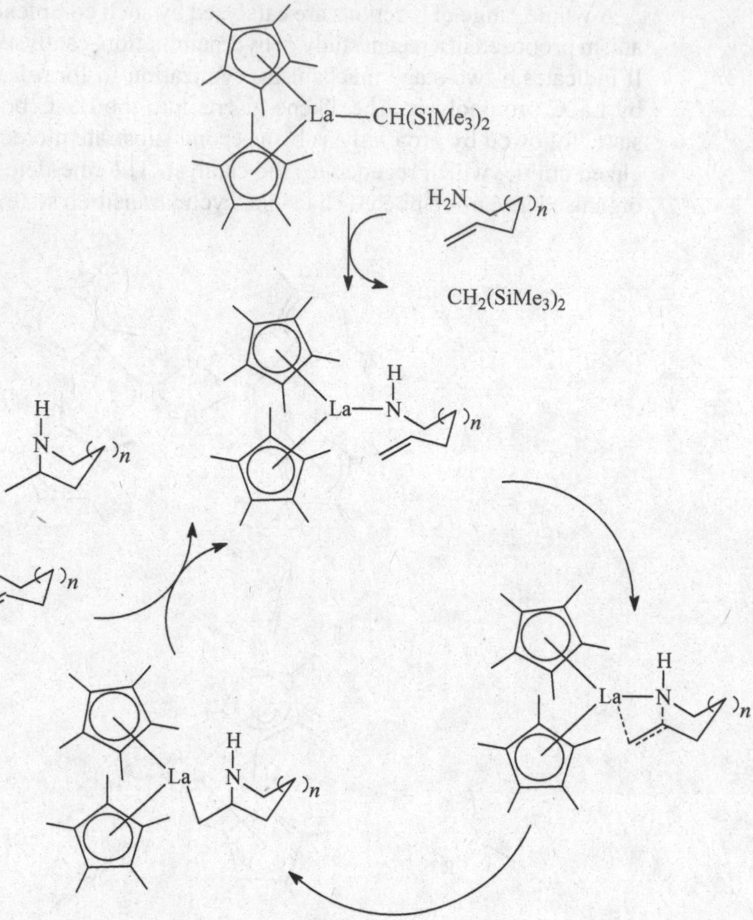

Figure 8.2
Hydroamination catalysed by [Cp*$_2$La(CH(SiMe$_3$)$_2$)].

Questions

8.1 How is SmI$_2$ prepared? How may its activity be enhanced? What are the advantages to using it? What are the drawbacks to using it, and how can these be circumvented?

8.2 What are the main uses of lanthanide triflates? What are the advantages to using them?

8.3A How are organocerium compounds prepared? Why are they useful? Why can CeCl$_3$/NaBH$_4$ be used in the presence of water, whilst CeCl$_3$/LiAlH$_4$ requires strictly anhydrous conditions?

8.3B The complexes LnM$_3$tris(binapthoxide) (M = alkali metal) have a good deal of utility as catalysts (Section 8.3). They are readily prepared from either the appropriate lanthanide silylamide, Ln[N(SiMe$_3$)$_2$]$_3$, or chloride. Why choose the former, when the chloride can be bought 'off the shelf'.
Answer : The former method has the advantage of a cleaner reaction, a lanthanide starting material soluble in common organic solvents, and chloride-free products.

Problems

8.4 Suggest a product for:

CH3
OAc
O
O
O
AcO
OAc
CH3
(CAN; MeCN–H2O)

Answer:

CH3
OAc
O
O
O
AcO
OAc
CHO

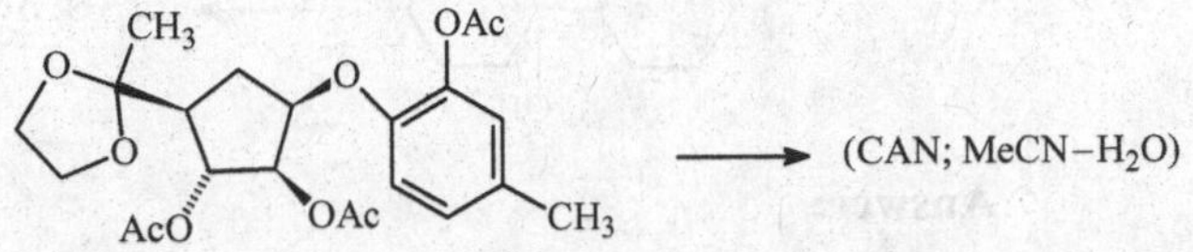

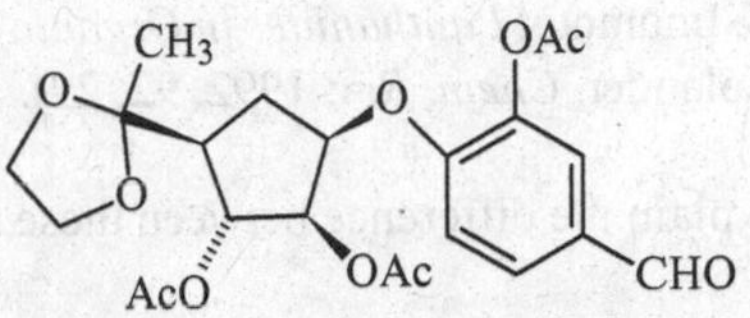

8.5 Suggest a product for:

OH
OH
[CAN (10%)–NaBrO3; MeCN–H2O; 80°C]

Answer:

O
OH

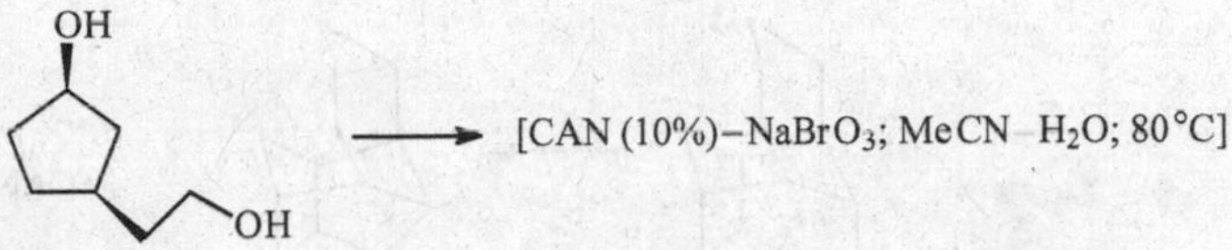

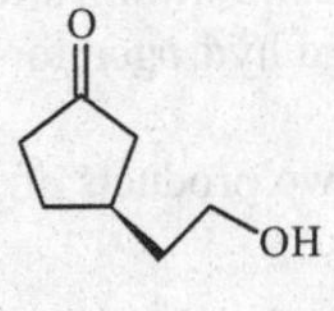

8.6 Suggest a product for this reaction:

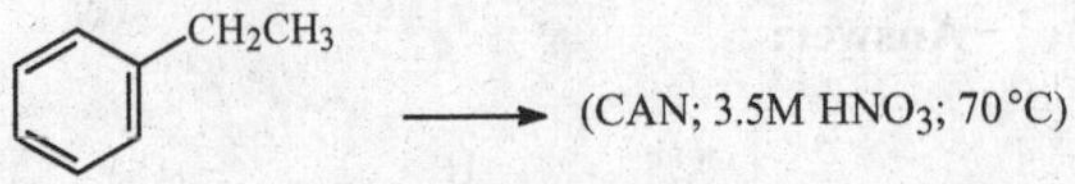

CH2CH3
(CAN; 3.5M HNO3; 70°C)

Answer:

O
CH3

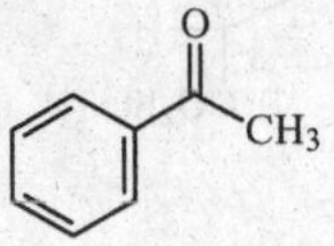

[G.A. Molander, *Chem. Rev.*, 1992, **92**, 29 (in particular, p. 31)].

8.7 Suggest a product of this reaction:

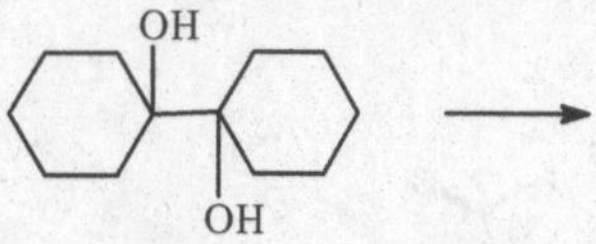

→ (CAN; AcOH–H$_2$O; 25 °C)

Answer:

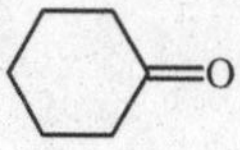

(2 molecules)

(T. Imamoto, *Lanthanides in Organic Synthesis*, Academic Press, 1994, p. 130; G.A. Molander, *Chem. Rev.* 1992, **92**, 29).

8.8 Explain the difference between these reactions:

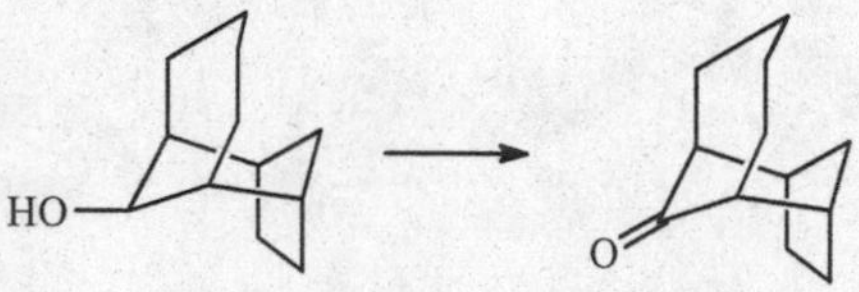

(CAN; MeCN–H$_2$O; 60 °C; 90%)

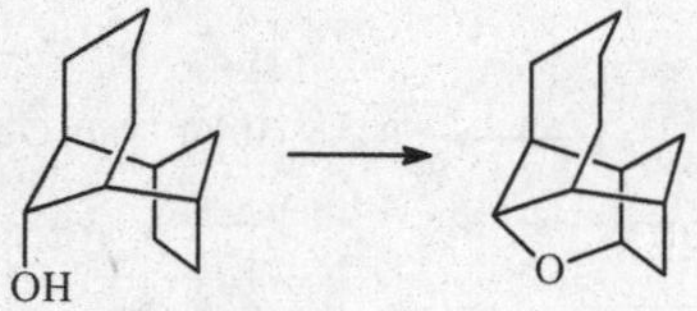

(CAN; MeCN–H$_2$O; 60 °C; 85%)

Answer: In one isomer, there is the possibility of delta-hydrogen abstraction by an alkoxyl radical, forming a furan ring; in the other, the alcohol group is too far away from the delta hydrogen for this to happen.

8.9 Predict the two products of:

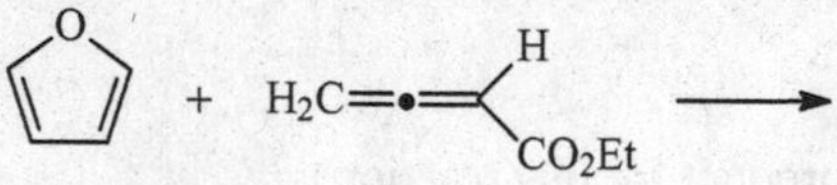

+ H$_2$C=•=C (H) (CO$_2$Et) → [Eu(fod)$_3$; RT; 20 h]

Answer:

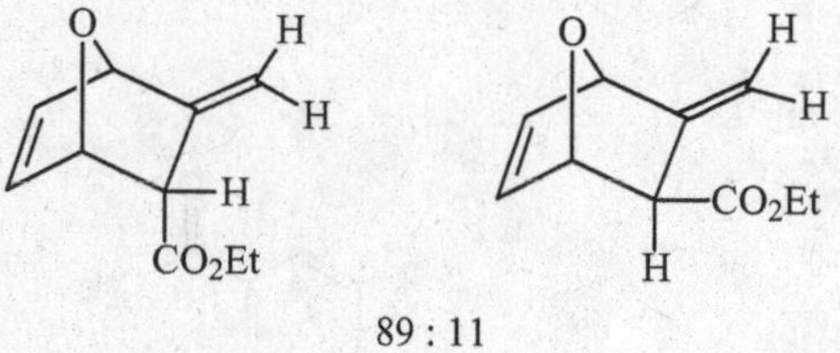

75% yield

89 : 11

[G.A. Molander, *Chem. Rev.*, 1992, **92**, 29 (in particular, p. 60)].

8.10 Predict the product of this reaction:

[Yb (fod)$_3$; neat; RT, 60–80%]

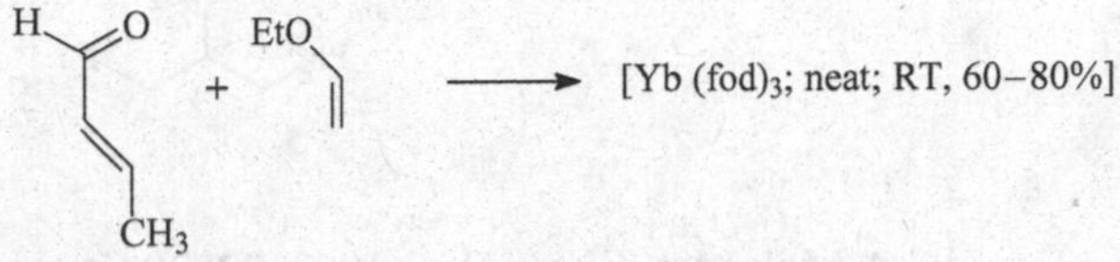

Answer:

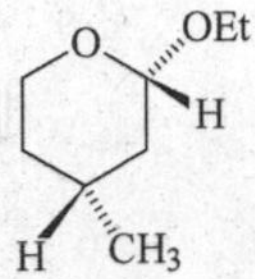

(T. Imamoto, *Lanthanides in Organic Synthesis*, Academic Press, 1994, 107).

8.11 Predict the product of this reaction:

(CeCl$_3$, Pr^{i}MgBr; THF; 0 °C)

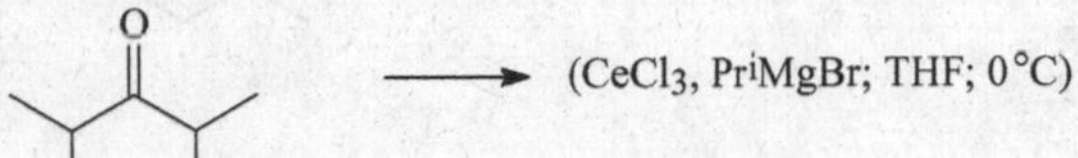

Answer: (Pr^i)$_3$COH
G.A. Molander, *Chem. Rev.*, 1992, **92**, 29 [(in particular, p. 44)].

8.12 Predict the product of this reaction:

CH$_3$(CH$_2$)$_{11}$F $\longrightarrow$ (LiAlH$_4$–CeCl$_3$; THF; heat)

Answer: C$_{12}$H$_{26}$
[G.A Molander, *Chem. Rev.*, 1992, **92**, 29 (in particular, p. 37)].

8.13 Predict the product of this reaction:

(LiAlH$_4$–CeCl$_3$; THF; RT)

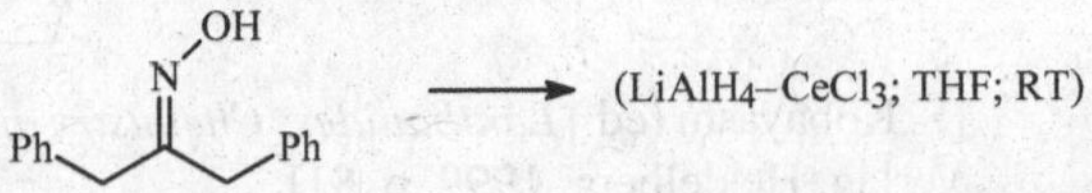

Answer:

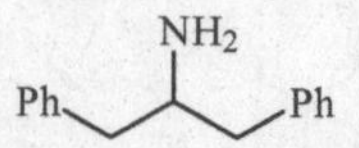
NH2
Ph
Ph

[G.A Molander, *Chem. Rev.*, 1992, **92**, 29 (in particular, p. 39)].

8.14 Predict the product of this reaction:

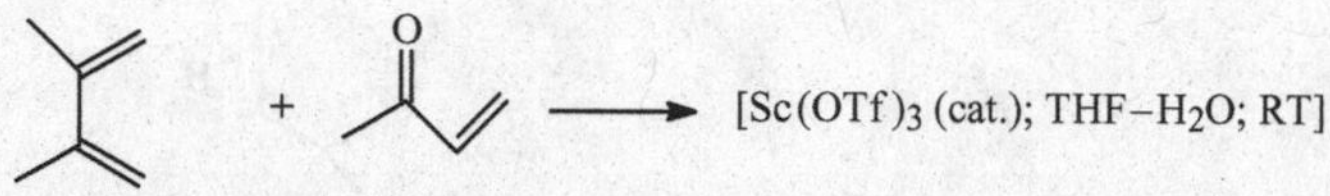
O
+
[Sc(OTf)3 (cat.); THF–H2O; RT]

Answer:

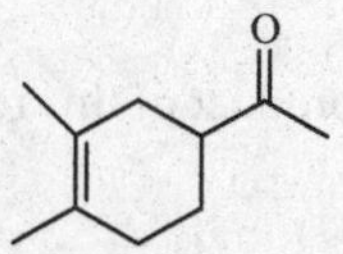
O

[S. Kobayashi (ed.) *Lanthanides: Chemistry and use in Organic Synthesis,* Springer-Verlag, Heidelberg, 1999, p. 78].

8.15 Predict the product of this reaction:

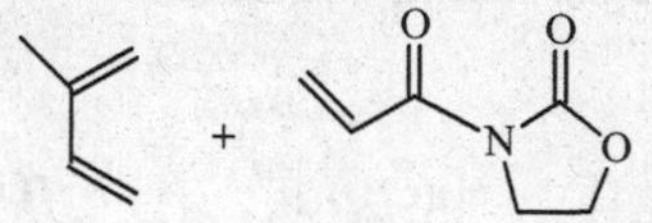
O
O
+
N
O
[Sc(OTf)3 (cat.); THF–H2O;RT]

Answer:

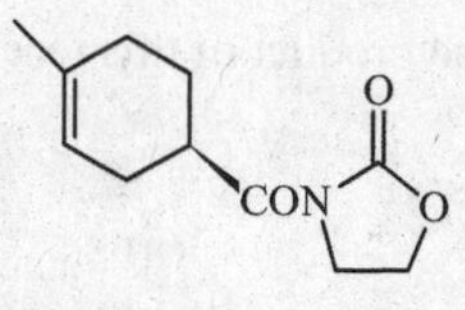
O
CON
O

[S. Kobayashi (ed.) *Lanthanides: Chemistry and Use in Organic Synthesis,* Springer-Verlag, Heidelberg, 1999, p. 81].

8.16   Predict the product of this reaction:

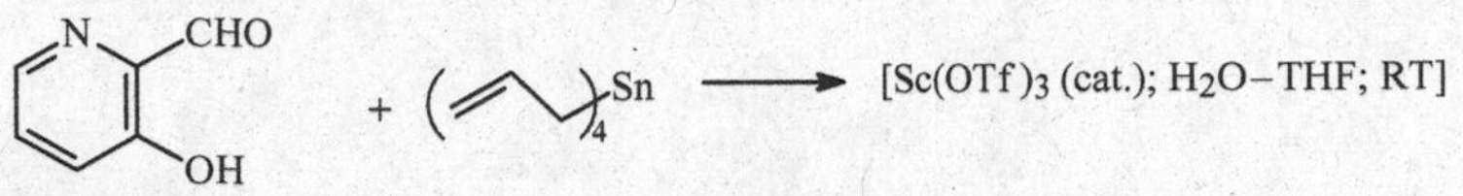

$$+ \quad \left(\underset{4}{\diagup\!\!\!\diagdown\!\!\!\diagup} \right)\!Sn \quad \longrightarrow \quad [Sc(OTf)_3 \text{ (cat.); } H_2O-THF; \text{ RT}]$$

Answer:

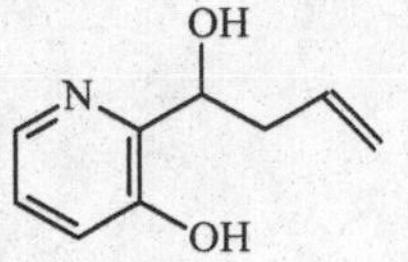

[S. Kobayashi (ed.) *Lanthanides: Chemistry and Use in Organic Synthesis,* Springer-Verlag, Heidelberg, 1999, p. 81].

8.17   Predict the product of this reaction:

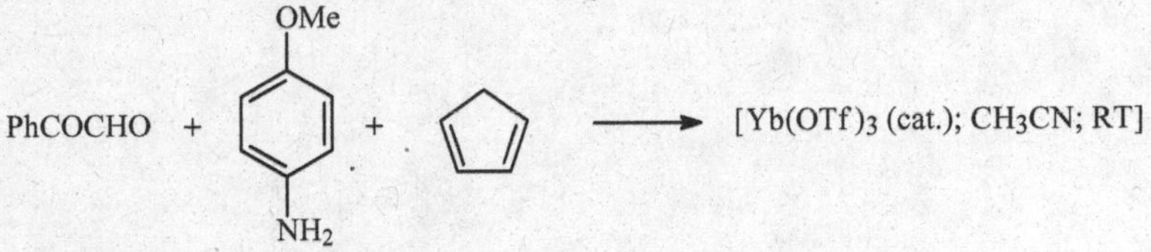

$$PhCOCHO \quad + \quad \text{OMe-C}_6H_4\text{-}NH_2 \quad + \quad \bigcirc \quad \longrightarrow \quad [Yb(OTf)_3 \text{ (cat.); } CH_3CN; \text{ RT}]$$

Answer:

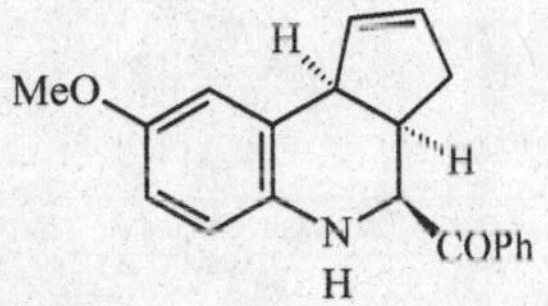

[S. Kobayashi (ed.) *Lanthanides: Chemistry and Use in Organic Synthesis,* Springer-Verlag, Heidelberg, 1999, p. 95].

9 Introduction to the Actinides

By the end of this chapter you should be able to:

- recall that most of the actinides do not occur in nature, that all are radioactive, and that they have increasingly short half-lives as the series is traversed;
- understand the methods used to extract and synthesize them;
- understand isotopic separation processes and the need for them;
- recall that the early actinides exhibit transition-metal-like behaviour and that the later actinides resemble the lanthanides in their chemistry;
- explain the stability of the different oxidation states for the actinides;
- recognize +4 and +6 as the main oxidation states for uranium, and +4 as the only important state for thorium;
- recall that relativistic effects are significant in the chemistry of these elements.

9.1 Introduction and Occurrence of the Actinides

Although the first compounds of uranium and thorium were discovered in 1789 (Klaproth) /1828 (Berzelius), most of these elements are man-made products of the 20th century. Thorium and uranium are both long-lived and present in the earth in significant amounts, but for practical purposes the others should be regarded as man-made, though actinium and protactinium occur in nature in extremely small amounts, as decay products of ^{235}U and ^{238}U, whilst microscopic amounts of plutonium, generated through neutron capture by uranium, have been reported to occur naturally. The principal thorium ore is monazite, a phosphate ore also containing large amounts of the lanthanides, whilst the main uranium ore is U_3O_8, usually known as pitchblende. Elements beyond uranium are man-made and their synthesis is described in section 9.2. All the actinides are radioactive, sometimes intensely; this requires special care in their handling, and the radioactivity often plays a part in their chemistry, as in causing radiation damage in solutions and in dislocating the regular arrangements of particles in crystals.

9.2 Synthesis

Apart from thorium, protactinium, and uranium, the actinides are obtained by a bombardment process of some kind. Thus for actinium (and also protactinium):

$$^{226}Ra + {}^1n \rightarrow {}^{227}Ra + \gamma; \qquad {}^{227}Ra \rightarrow {}^{227}Ac + \beta^-$$

$$^{230}Th + {}^1n \rightarrow {}^{231}Th + \gamma; \qquad {}^{231}Th \rightarrow {}^{231}Pa + \beta^-$$

Lanthanide and Actinide Chemistry S. Cotton
© 2006 John Wiley & Sons, Ltd.

Similarly, for the elements directly beyond uranium:

$$^{238}U + {}^1n \rightarrow {}^{239}U + \gamma; \qquad {}^{239}U \rightarrow {}^{239}Np + \beta^-; \qquad {}^{239}Np \rightarrow {}^{239}Pu + \beta^-$$

$$\text{and} \quad {}^{238}U + {}^1n \rightarrow {}^{237}U + 2{}^1n; \qquad {}^{237}U \rightarrow {}^{237}Np + \beta^-$$

For elements beyond plutonium, successive neutron capture is required. This is, obviously, a slower process; thus starting with 1 kg of ^{239}Pu, using a neutron flux of 3×10^{14} neutrons $cm^{-2}s^{-1}$ (a high but experimentally feasible value), around 1 mg of ^{252}Cf is obtained after 5–10 y.

$$^{239}Pu \xrightarrow{n} {}^{240}Pu \xrightarrow{n} {}^{241}Pu \xrightarrow{n} {}^{242}Pu \xrightarrow{n} {}^{243}Pu \xrightarrow{n,\beta^-} {}^{243}Am \xrightarrow{n} {}^{244}Am$$

$$\downarrow \beta^- \qquad\qquad\qquad\qquad\qquad\qquad\qquad\qquad\qquad\qquad\qquad \downarrow \beta^-$$

$$^{241}Am \xrightarrow{n,\gamma} {}^{242}Am \xrightarrow{\beta^-} {}^{242}Cm \qquad\qquad\qquad {}^{244}Cm$$

$$^{242}Pu \text{ or } {}^{243}Am \text{ or } {}^{244}Cm \xrightarrow{\text{multiple n, } \beta^-} {}^{249}Bk, {}^{252}Cf, {}^{253}Es, {}^{257}Fm$$

The limit of this process in a reactor is ^{257}Fm; the product of the next neutron absorption is ^{258}Fm, which undergoes spontaneous fission ($t_{1/2} = 0.38$ ms). This point can be passed in two ways. One is to utilize a more intense neutron flux than can be obtained in a reactor, in the form of a thermonuclear explosion, so that a product such as ^{258}Fm undergoes further neutron absorption before fission can occur. Here, in the synthesis of ^{255}Fm in 'Ivy Mike', the world's first thermonuclear test, at Eniwetok atoll on 1st November 1952, the initial product of multiple neutron capture, ^{255}U, underwent a whole series of rapid decays, yielding ^{255}Fm.

$$^{238}U \xrightarrow{17n} {}^{255}U \xrightarrow{\beta^-} {}^{255}Np \xrightarrow{\beta^-} {}^{255}Pu \xrightarrow{\beta^-} {}^{255}Am \xrightarrow{\beta^-} {}^{255}Cm \xrightarrow{\beta^-} {}^{255}Bk$$

$$\downarrow \beta^-$$

$$^{255}Fm \xleftarrow{\beta^-} {}^{255}Es \xleftarrow{\beta^-} {}^{255}Cf$$

For obvious reasons, this route is not likely to be followed in the future; instead, heavy-ion-bombardment, using particles such as ^{11}B, ^{12}C, and ^{16}O, will be used, though more recently heavier ones like ^{48}Ca and ^{56}Fe have been utilized. This route is reliable but has the twin drawbacks (from the point of view of yield) of requiring a suitable actinide target and also being an atom-at-a-time route. Examples include:

$$^{244}Cm + {}^4He \quad \rightarrow \quad {}^{247}Bk + {}^1H$$

$$^{253}Es + {}^4He \quad \rightarrow \quad {}^{256}Md + {}^1n$$

$$^{248}Cm + {}^{18}O \quad \rightarrow \quad {}^{259}No + 3{}^1n + {}^4He$$

$$^{249}Cf + {}^{12}C \quad \rightarrow \quad {}^{255}No + 2{}^1n + {}^4He$$

$$^{248}Cm + {}^{15}N \quad \rightarrow \quad {}^{260}Lr + 3{}^1n$$

$$^{249}Cf + {}^{11}B \quad \rightarrow \quad {}^{256}Lr + 4{}^1n$$

The most important and longest-lived isotopes are listed in Table 9.1.

Table 9.1 Longest-lived isotopes of the actinide elements

	Mass	Half-life	Means of decay
Ac	227	21.77 y	$\beta-$
Th	232	1.41×10^{10} y	α
Pa	231	3.28×10^4 y	α
U	234	2.45×10^5 y	α
	235	7.04×10^8 y	α
	238	4.47×10^9 y	α
Np	236	1.55×10^5 y	$\beta-$, EC
	237	2.14×10^6 y	α
Pu	239	2.14×10^4 y	α
	240	6.57×10^3 y	α
	242	3.76×10^5 y	α
	244	8.26×10^7 y	α
Am	241	432.7 y	α
	243	7.38×10^3 y	α
Cm	244	18.11 y	α
	245	8.5×10^3 y	α
	246	4.73×10^3 y	α
	247	1.56×10^7 y	α
	248	3.4×10^5 y	α
	250	1.13×10^4 y	α, SF
Bk	247	1.38×10^3 y	α
	249	320 d	$\beta-$
Cf	249	351 y	α
	250	13.1 y	α
	251	898 y	α
	252	2.64 y	α
Es	252	472 d	α
	253	20.47 d	α
	254	276 d	α
	255	39.8 d	$\beta-$
Fm	257	100.4 d	α
Md	258	55 d	α
No	259	1 h	α, EC
Lr	260	3.0 min	α

9.3 Extraction of Th, Pa, and U

9.3.1 Extraction of Thorium

Thorium makes up around 10% of monazite; one process for its extraction relies on treating the ground ore with concentrated aq. NaOH at 240 °C for some hours, then adding hot water. This dissolves out Na_3PO_4, leaving a mixture of the hydrated oxides and hydroxides of the lanthanide and Th. On adjusting the pH to 3.5 with boiling HCl, the lanthanide oxides dissolve, leaving behind only hydrated ThO_2. This can be converted into $Th(NO_3)_4$, and purified by solvent extraction into kerosene with tributyl phosphate.

In the acid route, the phosphate ore is digested with sulfuric acid at 230 °C for some hours and thorium phosphate is precipitated on adjusting the pH to 1. This is purified by first converting it into $Th(OH)_4$, thence into thorium nitrate, purified as above.

9.3.2 Extraction of Protactinium

First identified in 1913 (the first compound, Pa_2O_5, was isolated in 1927 by von Grosse, who isolated the element in 1931), protactinium is not generally extracted. Most of what is known about the chemistry of protactinium ultimately results from the extraction in 1960 by the UK Atomic Energy Authority of some 125 grams of Pa, from 60 tons of waste material left over from the extraction of uranium, at a cost of about $500,000 (very roughly, £1,250,000 at today's exchange rate).

9.3.3 Extraction and Purification of Uranium

Uranium is deposited widely in the Earth's crust, hence it has few ores, notably the oxides uraninite and pitchblende. The ores are leached with H_2SO_4 in the presence of an oxidizing agent such as $NaClO_3$ or MnO_2, to oxidize all the uranium to the (+6) state as a sulfate or chloride complex. On neutralization with ammonia a precipitate of 'yellow cake', a yellow solid with the approximate composition $(NH_4)_2U_2O_7$ is formed. This is converted into UO_3 on ignition at 300 °C. This can be purified further by conversion into uranyl nitrate, followed by solvent extraction using tributyl phosphate in kerosene as the extractant.

9.4 Uranium Isotope Separation

Having purified the uranium, it is then treated to separate the ^{235}U and ^{238}U isotopes for nuclear fuel purposes (any uranium compounds purchased commercially are already depleted in ^{235}U). In practice, nuclear fuel requires enrichment from the natural abundance of 0.71 % ^{235}U to around 5%, so what follows details a degree of enrichment not usually required.

The uranium compound usually used is UF_6. It is chosen on account of its volatility (sublimes at 56.5 °C) and low molecular mass (M_r), despite its extreme sensitivity to moisture (and toxicity of the HF produced) requiring the use of scrupulously sealed and water-free conditions, as well as fluorine-resistant materials.

9.4.1 Gaseous Diffusion

UF_6 vapour diffuses through barriers (F_2-resistant Al or Ni) with pores of diameter *ca* 10–25 nm at 70–80 °C.

Applying Graham's Law dictates a separation factor, α^*, where

$$\alpha^* = \sqrt{M_r(^{238}UF_6)/M_r(^{235}UF_6)} = \sqrt{352/349} = 1.00429$$

Consecutive stages are linked in a cascade to provide the desired degree of enrichment, with 3000 stages giving up to 90% enrichment in ^{235}U. A great deal of energy is needed to pump the UF_6 round the system.

9.4.2 Gas Centrifuge

On centrifuging UF_6 vapour in a gas centrifuge, $^{238}UF_6$ tends to concentrate towards the outside of the centrifuge whilst the lighter $^{235}UF_6$ predominates near the axis of the centrifuge. Better separations are achieved at high speeds and temperatures, and an advantage of the process is that the separation is in direct proportion to M_r.

9.4.3 Electromagnetic Separation

Ionized UCl_4 was separated using a cyclotron-like system to obtain the first enriched samples used in the Manhattan project. In practice, this was a difficult method to utilize and was rejected in favour of gaseous diffusion.

9.4.4 Laser Separation

One process has involved selectively ionizing ^{235}U in uranium vapour using a tuneable laser. In another process (Figure 10.6), $^{235}UF_6$ molecules in a gaseous mixture are selectively ionized using an IR laser (the isotope shift of ν U-F is around 0.5 cm^{-1}); the excited $^{235}UF_6$ molecules are decomposed by a UV laser to give solid UF_5; the $^{238}UF_6$ molecules are unaffected. Low temperatures are used to give sharp absorption bands so that the vibration of the $^{235}UF_6$ molecules but not that of the $^{238}UF_6$ molecules is activated. Decomposition of $U(OMe)_6$ has also been studied.

9.5 Characteristics of the Actinides

Table 9.2 depicts the oxidation states of the actinides. For the early actinides, Ac–Np, the highest (though not necessarily the most stable) oxidation state reflects the total number of electrons (6d and 5f) that can be removed from the outer shell. This resemblance to the transition metals was noted for Ac–U nearly a century ago and initially made people think that the actinides were another block of transition metals. Later, starting round about Bk, most elements tend to exhibit one stable oxidation state, +3 in nearly all cases, thus resembling the lanthanides.

Table 9.2 Oxidation states of the actinides

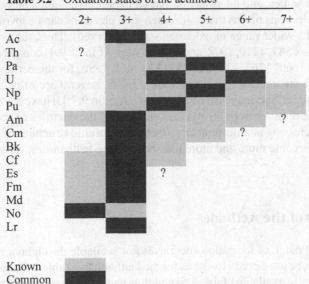

Table 9.3 Electron configurations of the actinides and their ions[a]

	M	M^{3+}	M^{4+}
Ac	$6d^1\,7s^2$		
Th	$6d^2\,7s^2$	$5f^1$	
Pa	$5f^2\,6d^1\,7s^2$	$5f^2$	$5f^1$
U	$5f^3\,6d^1\,7s^2$	$5f^3$	$5f^2$
Np	$5f^4\,6d^1\,7s^2$	$5f^4$	$5f^3$
Pu	$5f^6\,7s^2$	$5f^5$	$5f^4$
Am	$5f^7\,7s^2$	$5f^6$	$5f^5$
Cm	$5f^7\,6d^1\,7s^2$	$5f^7$	$5f^6$
Bk	$5f^9\,7s^2$	$5f^8$	$5f^7$
Cf	$5f^{10}\,7s^2$	$5f^9$	$5f^8$
Es	$5f^{11}\,7s^2$	$5f^{10}$	$5f^9$
Fm	$5f^{12}\,7s^2$	$5f^{11}$	$5f^{10}$
Md[b]	$5f^{13}\,7s^2$	$5f^{12}$	$5f^{11}$
No[b]	$5f^{14}\,7s^2$	$5f^{13}$	$5f^{12}$
Lr[b]	$5f^{14}\,6d^1\,7s^2$	$5f^{14}$	$5f^{13}$

[a]Configurations are in addition to a radon 'core' $1s^2\,2s^2\,2p^6\,3s^2\,3p^6\,3d^{10}\,4s^2\,4p^6$ $4d^{10}\,4f^{14}\,5s^2\,5p^6\,5d^{10}\,6s^2\,6p^6$
[b]Predicted

Early in the actinide series, electrons in the 6d orbitals are lower in energy than there is 5f orbitals, This is clear from the ground-state electronic configurations (Table 9.3) of the atoms, which show that the 6d orbitals are filled before 5f. The 5f orbitals are starting to be filled at protoactinium, and with the exception of curium, the 6d orbitals are not occupied again.

The 5f orbitals are not shielded by the filled 6s and 6p subshells as the 4f orbitals of the lanthanides are (by the corresponding 5s and 5p subshells). Moreover, the energy gap between $5f^n\,7s^2$ and $5f^{n-1}\,6d\,7s^2$ configurations is less than for the corresponding lanthanides. Not only are the 5f orbitals less 'inner orbitals' in the sense that the 4f orbitals are for the lanthanides, and thus more perturbed in bonding, but the near-degeneracy of the 5f, 6d, and 7s electrons means that more outer-shell electrons can be involved in compound formation (and a wider range of oxidation states observed). Thus, the first four ionization energies of Th are 587, 1110, 1978, and 2780 kJ mol^{-1} (Table 9.4) compared with respective values for Zr of 660, 1267, 2218, and 3313 kJ mol^{-1}. So, for the earlier actinides, higher oxidation states are available and, as for the d block, several are often available for each metal [relativistic effects may also be important (Section 9.7)]. However, as the 5f electrons do not shield each other from the nucleus effectively, the energies of the 5f orbitals drop rapidly with increasing atomic number, so that the electronic structures of the later actinides and their ions become more and more like those of the lanthanides, whose chemistry they thus resemble.

9.6 Reduction Potentials of the Actinides

Although a full range of ionization energies is not available throughout the actinide series and thus cannot be used predictively, as for the lanthanides (Table 2.2), electrode potentials are more generally available (Table 9.5) and thus can be so used, as follows:

Table 9.4 Ionization energies of the actinides (kJ/mol)

	I_1	I_2	I_3	I_4
Ac	499	1170	1900	4700
Th	587	1110	1978	2780
Pa	568	1128		2991
U	584	1420	1900	3145
Np	597	1128	1997	3242
Pu	585	1128	2084	3338
Am	578	1158	2132	3493
Cm	581	1196	2026	3550
Bk	601	1186	2152	3434
Cf	608	1206	2267	3599
Es	630	1216	2334	3734
Fm	627	1225	2363	3792
Md	635	1235	2470	3840
No	642	1254	2643	3956
Lr	444	1428	2228	4910

Table 9.5 Reduction potentials of the actinides (V)

	$M^{3+} + 3e \rightarrow M$	$M^{4+} + 4e \rightarrow M$	$M^{3+} + e \rightarrow M^{2+}$	$M^{4+} + e \rightarrow M^{3+}$
Ac	−2.13		−4.9	
Th		−1.83	−4.9	−3.7
Pa		−1.47	−4.7	−2
U	−1.8	−1.38	−4.7	−0.63
Np	−1.79	−1.30	−4.7	0.15
Pu	−2.03	−1.25	−3.5	0.98
Am	−2.07	−0.90	−2.3	2.3
Cm	−2.06		−3.7	3.1
Bk	−1.96		−2.8	1.64
Cf	−1.91		−1.6	3.2
Es	−1.98		−1.6	4.5
Fm	−2.07		−1.1	4.9
Md	−1.74		−0.15	5.4
No	−1.26		1.45	6.5
Lr	−2.1			7.9

1. The negative reduction potentials for the M^{3+}/M potentials indicate that the process

$$M(s) + 3H^+(aq) \rightarrow M^{3+}(aq) + 3/2\, H_2(g)$$

is energetically favourable, suggesting that the metals should react well with dilute acid (and possibly water too), as is in fact observed.

2. The large negative reduction potentials for the M^{3+}/M^{2+} process early in the series indicate that for these metals no aqueous chemistry of M^{2+} ions is to be expected. The potentials become less negative with increasing atomic number, indicating increasing stability of M^{2+}, and the positive value for No is in keeping with No^{2+} being the most stable ion for this metal in aqueous solution (Table 9.2). A number of compounds have been isolated in the solid state for Am^{2+} ($5f^7$), Cf^{2+}, and Es^{2+} (as might be expected from the reduction potentials) and the (+2) ion is known in solution for Fm and Md too (Table 9.2).

3. The large negative values of E^o for M^{4+}/M^{3+} for Th and Pa indicate that reduction to form (+3) species for these metals will be difficult, whilst for U, Np, and Pu the smaller E^o values indicate that both the +3 and +4 states will have reasonable stability. However, from Am onwards, $E^o > 2V$ for all these elements (except Bk), suggesting that for all these metals the (+3) state will be more favoured (as observed). The tendency for the E^o values for both M^{4+}/M^{3+} and M^{3+}/M^{2+} to become more positive, and for the reduction to be more favoured, on crossing the series from left to right, shows that overall the lower oxidation states become more stable the higher the atomic number.

9.7 Relativistic Effects

For the lighter chemical elements, the velocity of the electrons is negligible compared with the velocity of light. However, for the actinides and to a lesser extent the lanthanides this is not the case; as the velocity of the electrons increases towards c, then their mass increases too.

For a 1s electron in a uranium atom, the average radial velocity, $V_{rad} = 92c/137$, $\sim 0.67c$. The mass increase is given by $m = m_e/\sqrt{(1 - 0.67^2} = 1.35\ m_e$. This produces a contraction of the 1s orbital and stabilization of 1s electrons. Electrons in other s shells also tend to be stabilized owing to their orthogonality with 1s. Similar but smaller effects occur for p electrons. In contrast, d and f electrons tend to be expanded and destabilized (compared with imaginary, 'non-relativistic' atoms), due to the increased shielding of the nucleus by the (increasingly stabilized) outer core s and p electrons. Since f electrons are poor at screening other electrons from the nuclear charge, then as the 4f shell is filled, there is a contraction of the 5p and 6s orbitals.

A important 'relativistic effect' is that 5f orbitals of actinides are larger and their electrons more weakly bound than predicted by non-relativistic calculations, hence the 5f electrons are more chemically 'available'. This leads to:

(a) a bigger range of oxidation states than with the lanthanides;
(b) a greater tendency to covalent bond formation (but maybe involving 6d rather than 5f orbitals) in ions like MO_2^+ and MO_2^{2+} (most notably the uranyl ion, UO_2^{2+}).

Question 9.1 4.5×10^9 tonnes of uranium is present in the Earth's oceans, at a concentration of 3.3 mg m^{-3} (about 1/1000th that in the crust). How might it be extracted?
Answer 9.1 Ion-exchange technology is a possibility. The main cost would be that of pumping all the water needed; wave energy technology (as part of a power station?) could be an answer. Recovering other valuable metals (e.g. gold) at the same time might also improve the economics.

Question 9.2 What are the advantages of using UF_6 for isotope separation?
Answer 9.2 It vapourizes at low temperatures, so that little energy is used for that; it has a low molecular mass, so that, since separation factors are proportional to the difference in mass between the ^{235}U- and ^{238}U-containing molecules, easier separations are achieved; since fluorine is monoisotopic (^{19}F), only molecules of two different masses are involved, minimizing overlap between ^{235}U- and ^{238}U-containing species (this would not be the case if, say, Cl or Br were involved).

Question 9.3 What are the patterns in oxidation states seen in (a) the early actinides Ac–Pu and (b) the later ones, Am–Lr. With which parts of the Periodic Table are resemblances most pronounced? How does this relate to the electronic structure?

Answer 9.3 (a) The maximum oxidation state observed for Ac–Np corresponds to the number of 'outer-shell' electrons. Similar behaviour is seen in transition-metal chemistry, for example for the metals Sc–Mn in the 3d series.

(b) Especially from Bk onwards, the elements exhibit one common oxidation state, +3 in almost all cases, resembling the lanthanides in that respect.

10 Binary Compounds of the Actinides

By the end of this chapter you should be able to:

- suggest syntheses for typical compounds;
- recognize patterns in formulae and explain them;
- recognize patterns in the stability of halides;
- recognize and explain trends in the volatility and coordination number in uranium halides;
- state why the properties of UF_6 make it suitable for isotope separation;
- suggest properties for compounds;
- appreciate problems in the synthesis of halides of the heavier actinides;
- understand the structures of the actinide halides.

10.1 Introduction

This chapter examines some of the most important binary compounds of the actinides, especially the halides. Despite the problems caused by their radioactivity, some binary compounds of most of these elements have been studied in considerable detail, and form a good vehicle for understanding trends in the actinide series.

The actinides are reactive metals, typified by the reactions of thorium and uranium shown in Figures 10.1 and 10.2. These binary compounds frequently have useful properties. Thus, choosing examples from thorium chemistry, Th_2S_3 is a high-temperature crucible material, ThN is a superconductor, and Th_3P_4 and Th_3As_4 are semiconductors.

10.2 Halides

The compounds known are summarized in Table 10.1. The only compound of an early actinide in the +2 state is ThI_2, a metallic conductor which is probably Th^{4+} $(e^-)_2$ $(I^-)_2$. Certain heavier actinides form MX_2 (Am, Cf, Es), which usually have the structure of the corresponding EuX_2 and are thus genuine M^{2+} compounds. All four trihalides exist for all the actinides as far as Es, except for thorium and protactinium. Tetrafluorides exist for Th–Cm and the other tetrahalides as far as NpX_4 (and in the gas phase in the case of $PuCl_4$). Pentahalides are only known for Pa, U, and Np; whilst there are a few MF_6 (M = U–Pu), uranium is the only actinide to form a hexachloride. The known actinide halides are generally stable compounds; most are soluble in (and hydrolysed by) water.

Lanthanide and Actinide Chemistry S. Cotton
© 2006 John Wiley & Sons, Ltd.

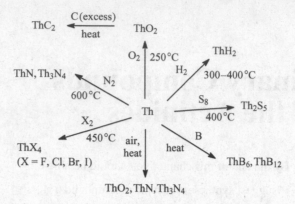

Figure 10.1
Reactions of thorium.

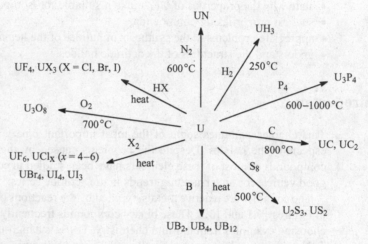

Figure 10.2
Reactions of Uranium.

The maximum oxidation state found in compounds of the early actinides corresponds to the total number of electrons available in the 7s, 6d, and 5f orbitals. Beyond uranium this is no longer the case, and the stability of the high oxidation states decreases sharply; this probably reflects the decreasing availability of the 5f electrons for bonding. By the second half of the series, one oxidation state (+3) dominates, in the way observed for the lanthanides. Fluorine, the strongest oxidizing agent among the halogens, supports the highest oxidation states, as is generally observed elsewhere.

10.2.1 Syntheses of the halides

A considerable number of synthetic routes are available, many involving reaction of the oxides with HX or X_2 (according to the oxidation state desired) though sometimes more unusual halogenating agents like AlX_3 or hexachloropropene have been used. A selection of methods follow.

Table 10.1 Oxidation states and halides of the actinides

Oxidation state	2	3	4	5	6
Ac		F Cl Br I			
Th	I	I	F Cl Br I		
Pa		I	F Cl Br I	F Cl Br I	
U		F Cl Br I	F Cl Br I	F Cl Br	F Cl
Np		F Cl Br I	F Cl Br I	F	F
Pu		F Cl Br I	F		F
Am	Cl Br I	F Cl Br I	F		F?
Cm		F Cl Br I	F		F?
Bk		F Cl Br I	F		
Cf	Cl Br I	F Cl Br I	F		
Es	Cl Br I	F Cl Br I	F?		
Fm					
Md					
No					
Lr					

Fluorides

$$Ac(OH)_3 \rightarrow AcF_3 \quad (HF, 700\,°C)$$
$$AmO_2 \rightarrow AmF_3 \quad (HF, heat)$$
$$PaO_2 \rightarrow PaF_4 \quad (HF/H_2, 600\,°C)$$
$$PaF_4 \rightarrow PaF_5 \quad (F_2, 800\,°C)$$

In some cases, where the +3 state is easily oxidized, a reductive route is used:

$$MO_2 \rightarrow MF_3 \quad (M = Pu, Np; HF/H_2, 500\,°C)$$
$$UF_4 + Al \rightarrow UF_3 + AlF \text{ (Unstable) } (900\,°C)$$

Chlorides

In the case of AnCl$_3$, the route used for lanthanide trihalides involving heating the oxide or hydrated chloride with ammonium chloride works well:

$$Ac_2O_3 \rightarrow AcCl_3 \quad (NH_4Cl, 250\,°C)$$
$$AmCl_3.6H_2O \rightarrow AmCl_3 \quad (NH_4Cl, 250\,°C)$$
$$NpO_2 \rightarrow NpCl_4 \quad (CCl_4, 500\,°C)$$

Again, reductive methods are sometimes necessary:

$$UH_3 \rightarrow UCl_3 \quad (HCl, 350\,°C)$$
$$NpCl_4 \rightarrow NpCl_3 \quad (H_2, 600\,°C)$$

Bromides

$$Ac_2O_3 \rightarrow AcBr_3 \quad (AlBr_3, 750\,°C)$$
$$NpO_2 \rightarrow NpBr_4 \quad (AlBr_3, heat)$$
$$M_2O_3 \rightarrow MBr_3 \quad (M = Cm, Cf; HBr, 600–800\,°C)$$

Table 10.2 Coordination number and structure types of the actinide trihalides

	C.N.	Ac	Th	Pa	U	Np	Pu	Am	Cm	Bk	Cf	Es	Fm	Md	No	Lr
Fluorides	11	LaF$_3$			LaF$_3$	LaF$_3$	LaF$_3$	LaF$_3$	LaF$_3$	LaF$_3$	LaF$_3$					
	9									YF$_3$	YF$_3$					
Chlorides	9	UCl$_3$			UCl$_3$	UCl$_3$	UCl$_3$	UCl$_3$	UCl$_3$	UCl$_3$	UCl$_3$					
	8									PuBr$_3$	PuBr$_3$	PuBr$_3$				
	6															
Bromides	9	UCl$_3$			UCl$_3$	UCl$_3$										
	8					PuBr$_3$	PuBr$_3$	PuBr$_3$	PuBr$_3$	PuBr$_3$						
	6									AlCl$_3$	AlCl$_3$	AlCl$_3$				
	6									BiI$_3$	BiI$_3$					
Iodides	8			PuBr$_3$	PuBr$_3$	PuBr$_3$	PuBr$_3$									
	6							BiI$_3$	BiI$_3$	BiI$_3$	BiI$_3$	BiI$_3$				

Iodides

$$Th \rightarrow ThI_4 \quad (I_2, 400\ °C)$$
$$CfCl_3 \rightarrow CfI_3 \quad (HI, 540\ °C)$$
$$CfI_3 \rightarrow CfI_2 \quad (H_2, 520\text{--}570\ °C)$$

10.2.2 Structure Types

Trihalides are known for most of the actinides and form the basis for comparison with the lanthanides. Table 10.2 lists the known structures adopted by the trihalides.

Of these structure types, the LaF$_3$ structure has a fully capped trigonal prismatic structure with 11 coordination (9 + 2); the YF$_3$ structure has rather distorted tricapped trigonal prismatic 9 coordination; the UCl$_3$ structure is a tricapped trigonal prism, from which the 8-coordinate PuBr$_3$ structure is derived by removing one of the face-capping halogens; the AlCl$_3$ and BiI$_3$ structures both have octahedral six-coordination.

The actinide trihalides display a similar pattern of structure to those of the lanthanide trihalides. However, comparing the coordination numbers for Ln^{3+} and An^{3+} ions with the same number of f electrons ('above one another in the Periodic Table'), it can be seen that the coordination number of the lanthanide halides decreases sooner than in the actinide series, a reflection of the fact that the larger actinide ions allow more halide ions to pack around them. Table 10.3 gives comparative coordination numbers for the trihalides of the lanthanides and actinides.

The tetrahalides have coordination numbers of 8 in the solid state whilst the pentahalides have 6 and 7 coordination. Thus PaF$_5$ and PaCl$_5$ are both seven coordinate (Figure 10.3),

Table 10.3 Comparison of coordination for trihalide structures

	La	Ce	Pr	Nd	Pm	Sm	Eu	Gd	Tb	Dy	Ho	Er	Tm	Yb	Lu
LnF$_3$	11	11	11	11	11	11, 9	11, 9	9	9	9	9	9	9	9	9
LnCl$_3$	9	9	9	9	9	9	9	9	8	6	6	6	6	6	6
LnBr$_3$	9	9	9	8	8	8	8	6	6	6	6	6	6	6	6
LnI$_3$	8	8	8	8	8	6		6	6	6	6	6	6	6	6

	Ac	Th	Pa	U	Np	Pu	Am	Cm	Bk	Cf	Es	Fm	Md	No	Lr
AnF$_3$	11			11	11	11	11	11	11, 9	11, 9					
AnCl$_3$	9			9	9	9	9	9	9, 8	9, 8	8				
AnBr$_3$	9			9	9, 8	8	8	8	8, 6	6	6				
AnI$_3$			8	8	8	8	6	6	6	6	6				

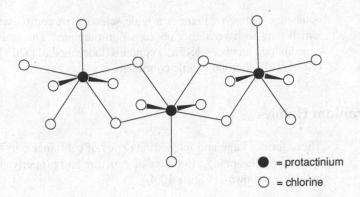

● = protactinium
○ = chlorine

Figure 10.3
The chain structure adopted by PaCl$_5$.

whilst PaBr$_5$ is six coordinate; in contrast UF$_5$ is seven coordinate (as is NpF$_5$), whilst UBr$_5$ and UCl$_5$ are both six coordinate. Hexafluorides are formed by U–Pu (and just possibly Am), UCl$_6$ is the only hexachloride; all MX$_6$ are octahedral.

The properties and structures of the halides are mainly exemplified in this chapter by a consideration of the halides of thorium and uranium.

10.3 Thorium Halides

Apart from two rather ill-defined iodides in the +2 and +3 oxidation states, thorium forms exclusively halides in the +4 state. Heating thorium metal with ThI$_4$ affords ThI$_2$ and ThI$_3$, the latter not well characterized. ThI$_2$ exists in black and gold forms; the high conductivity suggests that, like some of the apparent lanthanide(II) iodides (Section 3.3.3), it contains free electrons and is in fact Th^{4+} (e$^-$)$_2$ (I$^-$)$_2$.

The Th^{IV} halides can be made by various routes including:

$$ThO_2 + 4\ HF \rightarrow ThF_4 + 2\ H_2O$$

$$ThH_2 + 3\ Cl_2 \rightarrow ThCl_4 + 2\ HCl$$

$$Th + 2\ Br_2 \rightarrow ThBr_4\ (700\ ^\circ C)$$

$$Th + 2\ I_2 \rightarrow ThI_4\ (400\ ^\circ C)$$

All are white solids. Table 10.4 summarizes the structures of the thorium(IV) halides (the chloride and the bromide exist in two different forms in the solid state).

Thorium halides form many complexes with neutral donors, discussed in chapter 11. Here it may be noted that there are a number of anionic complexes, including Cs$_2$ThCl$_6$, (pyH)$_2$ThBr$_6$, and (Bu$_4$N)$_2$ThI$_6$, which generally seem to contain isolated [ThX$_6$]$^{2-}$

Table 10.4 Thorium(IV) halides

	Coord. No.	Solid state Bond length (Å)	Closest polyhedron	Gas phase Coord. No.	Bond length (Å)
ThF$_4$	8	2.30–2.37	Dodecahedron	4	2.14
ThCl$_4$	8	2.72–2.90 (β); 2.85–2.89 (α)	Sq. antiprism	4	2.58
ThBr$_4$	8	2.85–3.12 (β); 2.91–3.02 (α)	Sq. antiprism	4	2.72
ThI$_4$	8	3.13–3.29	Dodecahedron		

octahedra. However, there is a wide selection of complexes of the smaller fluoride ion, which tend to have higher coordination numbers. The stoichiometry is no guide to the coordination number – K_5ThF_9 contains dodecahedral $[ThF_8]^{4-}$ ions whilst in Li_3ThF_7 and $(NH_4)_3ThF_7$ thorium is nine coordinate

10.4   Uranium Halides

These form a large and interesting series of substances, in oxidation states ranging form +3 to +6, illustrating principles of structure and property. The structures of the uranium halides are shown in Figure 10.4.

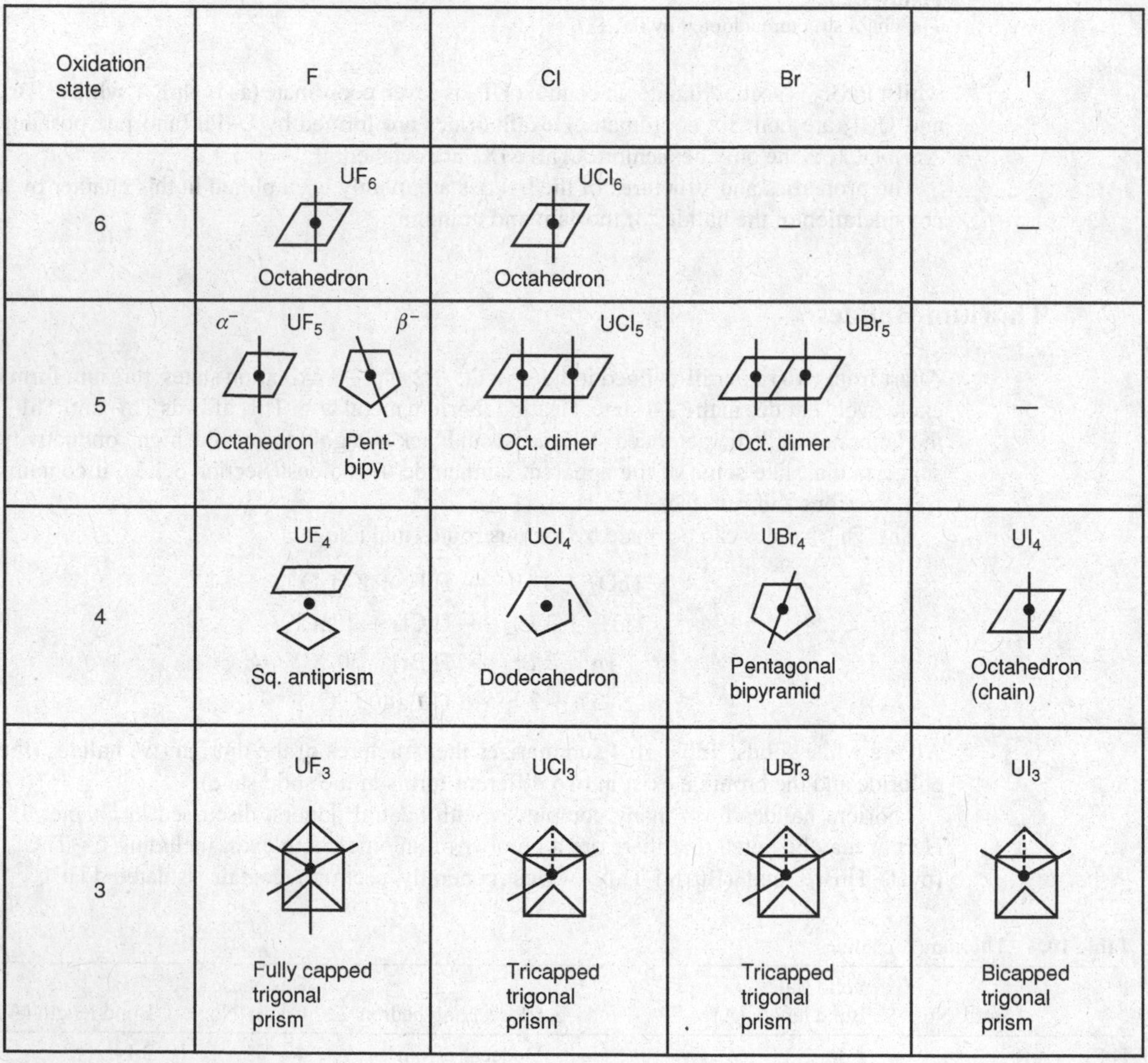

Oxidation state	F	Cl	Br	I
6	UF_6 Octahedron	UCl_6 Octahedron	—	—
5	α^- UF_5 β^- Octahedron / Pent-bipy	UCl_5 Oct. dimer	UBr_5 Oct. dimer	—
4	UF_4 Sq. antiprism	UCl_4 Dodecahedron	UBr_4 Pentagonal bipyramid	UI_4 Octahedron (chain)
3	UF_3 Fully capped trigonal prism	UCl_3 Tricapped trigonal prism	UBr_3 Tricapped trigonal prism	UI_3 Bicapped trigonal prism

Figure 10.4
Coordination polyhedra in the uranium halides [adapted from J.C. Taylor, *Coord. Chem. Rev.* 1976, **20**, 203; reprinted with permission of Elsevier Science publishers].

10.4.1 Uranium(*VI*) Compounds

UF_6 and UCl_6 are two remarkable substances, the former on account of its application in isotopic enrichment of uranium (Sections 9.4 and 10.4.5); the latter as the only actinide hexachloride. Colourless UF_6 can be made by a wide variety of routes, including:

$$UF_4 + F_2 \rightarrow UF_6 \quad (250\text{–}400\ °C)$$
$$3\,UF_4 + 2ClF_3 \rightarrow 3UF_6 + Cl_2$$
$$2\,UF_4 + O_2 \rightarrow UF_6 + UO_2F_2 \quad (600\text{–}900\ °C)$$

The latter is the Fluorox process, which has the considerable advantage of not requiring the use of elemental fluorine. Dark green, hygroscopic, UCl_6 may be synthesized by halogen exchange or disproportionation:

$$UF_6 + 2\,BCl_3 \rightarrow UCl_6 + 2\,BF_3 \quad (-107\ °C)$$
$$2\,UCl_5 \rightarrow UCl_4 + UCl_6 \quad (120\ °C)$$

Both have octahedral molecular structures ($U\text{–}F = 1.994$ Å in UF_6) and are volatile under reduced pressure at temperatures below 100 °C.

10.4.2 Uranium(*V*) Compounds

These are among the best characterized compounds in this somewhat rare oxidation state, although they do tend to be unstable (see the preparation of UCl_6) and UI_5 does not exist.

$$2\,UF_6 + 2\,HBr \rightarrow 2\,UF_5 + Br_2 + 2\,HF$$
$$2\,UO_3 + 6\,CCl_4 \rightarrow 2\,UCl_5 + 6\,COCl_2 + Cl_2 \ (20\,atm., 160\ °C)$$
$$2\,UBr_4 + Br_2 \rightarrow 2\,UBr_5 \quad (\text{Soxhlet extraction; brown solid})$$

Grey UF_5 has a polymeric structure with 6- and 7-coordinate uranium, whilst red-brown UCl_5 and brown UBr_5 have a dimeric structure in which two octahedra share an edge.

10.4.3 Uranium (*IV*) Compounds

In normal laboratory work, these are the most important uranium halides, especially UCl_4.

$$UO_2 + 4\,HF \rightarrow UF_4 + 2\,H_2O \ (550\ °C)$$
$$UO_2 + 2\,CCl_4 \rightarrow UCl_4 + 2\,COCl_2 \quad (250\ °C)$$
$$U_3O_8 + C_3Cl_6 \rightarrow 3\,UCl_4 + Cl_2C{=}CClCOCl \text{ etc. (reflux)}$$
$$U + 2\,Br_2 \rightarrow UBr_4 \quad (He/Br_2; 650\ °C)$$
$$U + 2\,I_2 \rightarrow UI_4 \quad (I_2;\ 500\ °C, 20\ kPa)$$

HF is used in the synthesis of the tetrafluoride since obviously the use of fluorine in this reaction would tend to produce UF_6. Although Peligot first prepared UCl_4 in 1842 by the reaction of uranium oxide with chlorine and charcoal, nowadays it is conveniently made by refluxing the oxide with organochlorine compounds such as hexachloropropene and CCl_4. U^{4+} does not have reducing tendencies, and UI_4 is stable, though not to hydrolysis.

Table 10.5 lists some properties of these compounds. As usual in the lower oxidation states, the fluorides have significantly lower volatilities. This can alternatively be explained

Table 10.5 Uranium(IV) halides

	UF$_4$	UCl$_4$	UBr$_4$	UI$_4$
Description	Air-stable green solid	Deliquescent green solid	Deliquescent brown solid	Deliquescent black solid
Mp (°C)	1036	590	519	506
Bp (°C)		789	761	
Solubility		H$_2$O and most org.solvents	H$_2$O and most org.solvents	H$_2$O and most org.solvents
Coordination geometry	Square antiprism	Dodecahedron	Pentagonal bipyramid	Octahedron
C.N. of uranium	8	8	7	6
U–X distance (Å)	2.25–2.32	2.64–2.87	2.61 (term.) 2.78–2.95 (bridge)	2.92(term.) 3.08–3.11 (bridge)

in terms of high lattice energies, on account of the small size of the fluoride ion, or by the greater ionic character in the bonding. In fact, like the other MF_4 and MCl_4, UF_4 vapourizes as MF_4 molecules. As usual, the coordination number C.N. of the metal decreases as the halogen gets larger whilst the bond lengths increase.

10.4.4 Uranium(*III*) Compounds

Because of the ease of oxidation of the U^{3+} ion, these are all made under reducing conditions.

$$2\,UF_4 + H_2 \rightarrow 2\,UF_3 + 2\,HF \quad (950\,°C)$$
$$2\,UH_3 + 6\,HCl \rightarrow 2\,UCl_3 + 3\,H_2 \quad (350\,°C)$$
$$2\,UH_3 + 6\,HBr \rightarrow 2\,UBr_3 + 3\,H_2 \quad (300\,°C)$$
$$2\,UH_3 + 3\,HI \rightarrow 2\,UI_3 + 3\,H_2 \quad (300\,°C)$$

These compounds have structures typical of the actinide trihalides (Section 10.2.2). Green UF_3 has the 11-coordinate LaF_3 structure whilst UCl_3 and UBr_3, both red, have the tricapped trigonal prismatic 'UCl_3' structure. Nine iodide ions cannot pack round uranium, so black UI_3 adopts the 8-coordinate $PuBr_3$ structure (Figure 10.5).

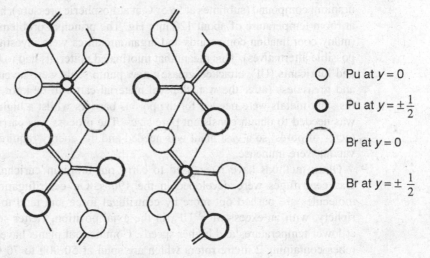

Pu at $y = 0$

Pu at $y = \pm\frac{1}{2}$

Br at $y = 0$

Br at $y = \pm\frac{1}{2}$

Figure 10.5
The layer structure of $PuBr_3$ [after A.F. Wells, *Structural Inorganic Chemistry*, Clarendon Press, Oxford (3rd edn, 1962) and reproduced by permission of the Oxford University Press]. Note that 9-coordination is prevented by non-bonding Br...Br interaction. The notation $y = 0$ and $y = \pm\frac{1}{2}$ indicates the relative height of atoms with respect to the plane of the paper. The planes of the layers are normal to the plane of the paper.

10.4.5 Uranium Hexafluoride and Isotope Separation

^{235}U is the only naturally occurring fissionable nucleus, but it makes up only 0.72% of natural uranium. Scientists working on the Manhattan project to build the first atomic bombs in the early 1940s were faced with the problem of producing uranium enriched in ^{235}U. In principle various routes were available, such as deflection in electric and magnetic fields, diffusion across thermal or osmotic pressure barriers, and ultracentrifugation. E.O. Lawrence developed a modified cyclotron ('calutron') that relied on the fact that $^{235}UCl_4^+$

and $^{238}UCl_4{}^+$ ions would follow slightly different paths in a magnetic field and thus be separated. Although eventually gaseous diffusion became the preferred route, the calutron did produced significant amounts of enriched uranium by 1945 that were used to make the first atomic bomb. The preferred technique was eventually gaseous diffusion of UF_6 through porous membranes, usually of nickel or aluminium, with a pore size ca. 10–25 nm. Graham's law of diffusion indicates that the rate of diffusion of a compound $\propto \sqrt{1/M_r}$. A separation factor, α^*, can be defined as:

$$\alpha^* = \sqrt{M_r(^{238}UF_6)/M_r(^{235}UF_6)} = \sqrt{352/349} = 1.00429.$$

Using a 'cascade' process with up to 3000 stages, an enrichment of up to 90% is obtained.

Desirable qualities for the compound used included high volatility, so that it was easy to convert into a vapour and minimize energy requirements; as low an M_r as possible, in order to enhance the separation factor; ease of synthesis. UF_6 met these criteria quite well; in addition, fluorine has only one isotope, so that the UF_6 molecules of any given uranium isotope all have the same mass and this does not complicate the separation. Although UF_6 is actually a solid at room temperature and atmospheric pressure, it is by far the most volatile uranium compound (sublimes at 56.5 °C at atmospheric pressure), having a vapour pressure at room temperature of about 120 mm Hg. The principal problem was its high reactivity (many coordination compounds and organometallics were investigated unsuccessfully as possible alternatives). This meant that fluorinated materials had to be developed for valves and lubricants (UF_6 attacked grease); new pump seals were invented that were gas-tight and greaseless (after the war, the seal material came to be known as Teflon). Fluorine-resistant metals were needed for the porous barriers whilst a high quality of engineering was needed to obtain consistent pore sizes. The process was carried out on a huge scale (acres of pores) so a vast plant was needed and the energy requirements for pumping the vapour were immense.

Other methods have been used to carry out uranium enrichment, again using UF_6. Gas centrifuges were developed in the 1960s. On centrifugation, the heavier $^{238}UF_6$ molecules are pushed out more by centrifugal force and tend to accumulate in the periphery, with an excess of $^{235}UF_6$ in the axial position. Better separations are obtained at lower temperatures and higher speeds. Commercial plants have used series of vacuum tubes containing 2 metre rotors which are spun at 50 000 to 70 000 rpm. Like gaseous diffusion, it runs as a cascade process, but requiring many fewer stages than gaseous diffusion.

Another technique that shows great promise and is still being developed is laser enrichment, which can be applied to either atoms or molecules (Figure 10.6). The molecular laser process exploits the fact that $^{238}UF_6$ and $^{235}UF_6$ molecules have slightly different vibrational frequencies (of the order of 0.5 cm^{-1}). The $^{235}UF_6$ molecules in UF_6 vapour (supercooled in order to produce sharper absorption bands) are selectively excited with a tuneable IR laser, then irradiated with a high-intensity UV laser, whereupon the excited $^{235}UF_6$ molecules are photodecomposed into $^{235}UF_5$ (the $^{238}UF_6$ molecules are unaffected). Under these conditions, the $^{235}UF_5$ is a solid, and is separated from the $^{238}UF_6$ (using a sonic impactor).

The leading process of this type is the Australian SILEX system (which is also being used to carry out isotopic separation for other elements such as silicon and zirconium).

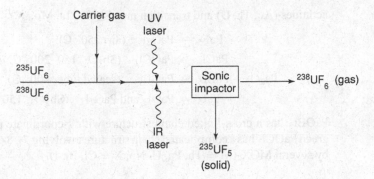

Figure 10.6
Laser separation method for $^{235}UF_6$ and $^{238}UF_6$ (reproduced with permission from S.A. Cotton, *Lanthanides and Actinides*, Macmillan, 1991, p. 95).

10.5   The Actinide Halides (Ac–Am) excluding U and Th

10.5.1   Actinium

Actinium forms halides only in the (+3) oxidation state, AcX_3 and $AcOX$ (X = F, Cl, Br). The trihalides have the LaF_3 (X = F) and UCl_3 (X = Cl, Br) structures, respectively.

10.5.2   Protactinium

Protactinium has a wealth of halides, with all PaX_4 and PaX_5 known, also PaI_3. There are also several oxyhalides.

Pa_2O_5 is the usual starting material for the synthesis of the halides:

$$Pa_2O_5 \rightarrow PaCl_5 \quad (C/Cl_2/CCl_4; \ 400\text{–}500 \ ^\circ C, \ in \ vacuo)$$

$$PaCl_5 \rightarrow PaCl_4 \quad (H_2; \ 800 \ ^\circ C \ or \ Al; \ 400 \ ^\circ C)$$

$$Pa_2O_5 \rightarrow PaF_4 \quad (HF/H_2; \ 500 \ ^\circ C)$$

$$PaF_4 \rightarrow PaF_5 \quad (F_2; \ 700 \ ^\circ C)$$

$$Pa_2O_5 \rightarrow PaBr_5 \quad (C/Br_2; \ 600\text{–}700 \ ^\circ C)$$

$$PaBr_5 \rightarrow PaBr_4 \quad (Al \ or \ H_2; \ 400 \ ^\circ C \ in \ vacuo)$$

$$Pa_2O_5 \rightarrow PaI_5 \quad (SiI_4; 600 \ ^\circ C, \ in \ vacuo)$$

$$PaI_5 \rightarrow PaI_4 \quad (Al \ or \ H_2; \ 400 \ ^\circ C, \ sealed \ tube)$$

$$PaI_5 \rightarrow PaI_3 \quad (360 \ ^\circ C, \ in \ vacuo)$$

Seven coordination is found in both colourless $PaF_5(\beta\text{-}UF_5$ structure) and yellow $PaCl_5$ (Figure 10.3; chain structure), whilst two forms of dark red $PaBr_5$ adopt the α- and β-UCl_5 structures, dimeric with six-coordinate protactinium. PaX_4(X = F, Cl, Br) have the eight-coordinate UF_4 (X = F; square antiprismatic) and UCl_4 (X = Cl, Br; dodecahedral) structures, all eight coordinate; the fluoride is brown, the chloride green-yellow, and the bromide orange-red). Dark brown PaI_3 is also eight coordinate ($PuBr_3$ structure). Protactinium shares the same ability to form oxyhalides as several other actinides (Ac, Th, U) and transition metals (Nb, Ta, Mo, W).

$$PaF_5 \rightarrow PaO_2F \quad (air; 250\ °C)$$
$$PaCl_4 \rightarrow PaOCl_2 \quad (Sb_2O_3; 150–200\ °C,\ in\ vacuo)$$
$$Pa_2O_5 + PaBr_5 \rightarrow PaOBr_3 \quad (sealed\ tube; 400\ °C)$$
$$PaI_5 \rightarrow PaOI_3\ and\ PaO_2I \quad (Sb_2O_3; 150–200\ °C,\ in\ vacuo)$$

$PaOBr_3$ has a cross-linked chain structure with 7-coordinate protactinium, whilst yellow-green $PaOCl_2$ has a complicated chain structure involving 7-, 8- and 9- coordination, adopted by several $MOX_2(M = Th, Pa, U, Np; X = Cl, Br, I)$.

10.5.3   Neptunium

Neptunium maintains the flavour of a range of halides in different oxidation states, but only fluorides are known in oxidation states above +4. All NpX_3 are known, with assorted pleasant shades (purple fluoride, green chloride and bromide, brown iodide), together with $NpX_4(X = F, Cl, Br$; green, red-orange, and dark red, respectively), NpF_5 (blue-white) and NpF_6 (orange). Attempts to make the plausible NpF_7 using fluorinating agents like NpF_5 have not met with success. Np^{IV} has a wide chemistry, but there is no NpI_4, though here there are parallels with the inability of iodide to coexist with a stable 'high' oxidation state (*cf.* Cu^{2+}, Fe^{3+} given the instability of the known FeI_3).

$$NpF_3, NpF_4,\ NpO_2 \rightarrow NpF_6 \quad (F_2; 500\ °C)$$
$$NpF_4 \rightarrow NpF_5 \quad (KrF_2/HF)$$
$$NpO_2 \rightarrow NpF_4 \quad (HF; 500\ °C)$$
$$NpO_2 \rightarrow NpCl_4 \quad (CCl_4; 500\ °C)$$
$$NpO_2 \rightarrow NpBr_4 \quad (AlBr_3; 350\ °C)$$
$$NpO_2 \rightarrow NpF_3 \quad (H_2/HF; 500\ °C)$$
$$NpCl_4 \rightarrow NpCl_3 \quad (H_2; 450\ °C)$$
$$NpO_2 \rightarrow NpBr_3 \quad (HBr; 500\ °C)$$
$$NpO_2 \rightarrow NpI_3 \quad (HI; 500\ °C)$$

These have the same structures as the uranium analogues, except that a second form of $NpBr_3$ has the $PuBr_3$ structure. Like UF_6 and PuF_6, NpF_6 forms a volatile, toxic vapour (bp 55.2 °C). Oxyhalides $NpOF_3$, NpO_2F_2, $NpOF_4$, $NpOCl$, and $NpOI$ have been described.

10.5.4   Plutonium

The tendency towards lower stability of high oxidation state noted with Np is continued here. Though all PuX_3 exist, pale brown PuF_4 is the only tetrahalide, and there are no pentahalides, just red-brown PuF_6. The trihalides exhibit pastel colours (fluoride violet-blue; chloride blue-green; bromide light green; iodide bright green). The (+2) state is not accessible; with the next actinide, americium, the reaction with mercury(II) iodide yields AmI_2.

$$2\ PuO_2 + H_2 + 6HF \rightarrow 2PuF_3 + 4H_2O \quad (600°C)$$
$$Pu_2(C_2O_4)_3.10H_2O + 6HCl \rightarrow 2PuCl_3 + 3CO + 3CO_2 + 13H_2O$$

$$2PuH_3 + 6HBr \rightarrow 2PuBr_3 + 3H_2 \quad (600°C)$$

$$2Pu + 3HgI_2 \rightarrow 2PuI_3 + 3Hg \quad (500°C)$$

$$Pu^{4+}(aq) + 4HF(aq) \rightarrow PuF_4 \cdot 2\tfrac{1}{2}H_2O + HF(g) \rightarrow PuF_4$$

$$PuF_4 + F_2 \rightarrow PuF_6 \quad (400°C)$$

Two further points merit attention. PuF_6 is a very volatile substance, like the other actinide hexafluorides (mp 51.5 °C; bp 62.15 °C); very reactive (it undergoes controlled hydrolysis to PuO_2F_2 and $PuOF_4$), it tends to be decomposed by its own α-radiation, and is best kept in the gaseous state (resembling NO_2 in appearance).

As already noted, $PuCl_4$ does not exist in the solid state; however, it has been detected in the gas phase above 900 °C by its UV–visible absorption spectrum, so the equilibrium below moves to the right on heating:

$$PuCl_3 + \tfrac{1}{2}Cl_2 \rightleftharpoons PuCl_4$$

Halide complexes in the (+4) state are stable for all but iodides. Apart from the two mentioned above, other oxyhalides include $PuO_2Cl_2 \cdot 6H_2O$ and $PuOX$ (X = F, Cl, Br, I), the absence of any in the (+4) and (+5) states is again notable.

10.5.5   Americium

At americium, dihalides become isolable for the first time.

$$Am + HgX_2 \rightarrow AmX_2 + Hg \quad (X = Cl, Br, I; heat)$$

They are black solids with the $PbCl_2$ (X = Cl); $EuBr_2$ (X = Br); and EuI_2 (X = I) structures. As with plutonium, only fluorides occur in the higher (>3) oxidation states. The trihalides are important; they have the usual pastel colours, pink in the case of the fluoride and chloride, white–pale yellow (bromide), and pale yellow (iodide).

$$AmO_2 + 4HF \text{ (aq) (with HNO}_3) \rightarrow AmF_4 \cdot xH_2O$$

$$AmF_4 \cdot xH_2O \rightarrow AmF_3 \quad (NH_2HF_2; 125°C)$$

$$AmO_2 \rightarrow AmCl_3 \quad (CCl_4; 800°C)$$

$$AmO_2 \rightarrow AmBr_3 \quad (AlBr_3; 500°C)$$

$$AmO_2 \rightarrow AmI_3 \quad (AlI_3; 500°C)$$

$$AmF_3 + \tfrac{1}{2}F_2 \rightarrow AmF_4$$

There is an unsubstantiated report of the isolation of AmF_6 (Iu. V. Drobyshevshii *et al.*, *Radiokhimiya*, 1980, **22**, 591):

$$AmF_4 \rightarrow AmF_6 \quad (KrF_2/HF; 40–60 °C)$$

The product was a volatile dark brown solid, with an IR absorption at 604 cm^{-1} (compare ν U–F at 624 cm^{-1} in UF_6).

Halides of the Heavier Transactinides

Earlier mention (Section 10.2.2 and Tables 10.1–10.3) of the trihalides of the transplutonium elements does not do justice to the synthetic expertise of experimentalists who have studied these compounds. Some examples of the techniques involved follow.

10.6.1 Curium(III) Chloride

The first synthesis, reported in the mid 1960s, was made difficult by the use of the short-lived ^{244}Cm ($t_{1/2} = 18.1$ y) isotope, which caused significant damage to the sample through irradiation. However, by 1970 microgram quantities of ^{248}Cm ($t_{1/2} = 3.4 \times 10^5$ y) were available, albeit on a microgram scale. Ion-exchange beads were loaded with 1 μg of ^{248}Cm, then heated to 1200 °C to convert the curium into the oxide; the Cm_2O_3 was placed in capillaries and chlorinated using HCl at 500 °C, forming pale yellow $CmCl_3$. Slow cooling of molten $CmCl_3$ from over 600 °C afforded a single crystal of the curium chloride, which when studied by X-ray diffraction revealed its adoption of the UCl_3 structure and afforded the Cm–Cl bond lengths (J. R. Peterson and J. H. Burns, *J. Inorg. Nucl. Chem.*, 1973, **35**, 1525).

10.6.2 Californium(III) Chloride, Californium(III) Iodide, and Californium(II) Iodide

$CfCl_3$ was first reported by Burns *et al.*, in 1973, synthesized by the ion-exchange bead route described above. As more ^{249}Cf ($t_{1/2} = 351$ y) became available on the μg scale, another route was employed. A precipitate of californium oxalate was filtered in a quartz capillary using a piece of ashless filter paper. Heating at 800–900 °C destroyed the filter paper and turned the oxalate into Cf_2O_3, which reacted with CCl_4 at 520–550 °C to form lime-green $CfCl_3$. X-ray diffraction showed that this had the UCl_3 structure. The $CfCl_3$ was purified by sublimation at 800 °C, then employed in the synthesis of two iodides:

$$CfCl_3 \rightarrow CfI_3 \ (HI; 540 °C)$$

The yellow-orange triiodide forms the lavender-violet diiodide on reduction:

$$CfI_3 \rightarrow CfI_2 \ (H_2; 520–570 °C)$$

These compounds were identified using X-ray diffraction and UV-visible spectroscopy.

10.6.3 Einsteinium(III) Chloride

The first synthesis of this compound was carried out on the nanogram scale in 1968 by Peterson's group in the face of several difficulties. The ^{253}Es isotope is so short lived ($t_{1/2} = 20$ d) that it releases some 1400 kJ mol^{-1} min^{-1}, sufficient to dislocate the crystal lattice within 3 minutes of formation and it is also capable of destroying ion-exchange resins. Around 200 nanograms of einsteinium were absorbed by a charcoal chip, which was heated to turn the Es into Es_2O_3, then chlorinated using CCl_4 to form $EsCl_3$. A crystal of $EsCl_3$ was just heated to below its melting point, so that defects in the lattice produced by the radiation were continuously annealed out. One decade later, in 1978, a different route was employed by Peterson to make Es_2O_3, first precipitating the oxalate, then converting it into the oxide by calcination. Treatment of the oxide first with HF, then fluorine at 300 °C, yielded the fluoride EsF_3. The absorption spectrum of EsF_3 was examined over a period of 2 years, so that additionally the spectra of the $^{249}BkF_3$ daughter and $^{249}CfF_3$ granddaughter were also obtained, enabling their assignment of the LaF_3 structure.

10.7 Oxides

Table 10.6 summarizes the formulae of the oxides. It should be emphasized that this represents a simplification and, especially for U and Pu, a number of phases exist between the compositions shown.

The pattern of oxidation state here is very different to that of the lanthanides, where the common oxides are Ln_2O_3. The early actinides in particular show much more resemblance to the transition metals, where the maximum oxidation state corresponds to the number of 'outer shell' electrons; this reflects the greater availability of d and f electrons in the actinide elements. The +4 state in particular is more stable than in the lanthanide series, where it is only encountered in a few ions.

10.7.1 Thorium Oxide

ThO_2, is a white solid that adopts the fluorite structure (as do the MO_2 phases of the other actinides). When heated it gives off a rather bluish light; if about 1% cerium is added, the light is both whiter and more intense so that the mixture came to be used in making incandescent gas mantles, widely used for lighting until comparatively recently.

10.7.2 Uranium Oxides

The uranium–oxygen phase diagram is very complicated, with some of the 14 reported phases not being genuine and several phases showing variable composition. The important phases are UO_2, U_4O_9, U_3O_8, and UO_3. Brown-black UO_2 has the fluorite structure. It is best made by reduction of higher uranium oxides (e.g. UO_3 with H_2 or CO at 300–600 °C).

Additional oxygen can be incorporated into interstitial sites in the basic fluorite structure until the composition reaches U_4O_9 (black $UO_{2.25}$). Green-black U_3O_8 is the result of heating uranyl salts at around 650–800 °C; above 800 °C, it tends to lose oxygen.

$$3\ UO_2(NO_3)_2 \rightarrow U_3O_8 + 6\ NO_2 + 2\ O_2$$

This is a mixed-valence compound with pentagonal bipyramidal coordination of uranium. Addition of more oxygen eventually results in orange-yellow UO_3. This has several crystalline forms, most of which contain uranyl groups in $2 + 4$ coordination, and can be made by heating, e.g., $(NH_4)_2U_2O_7$ or $UO_2(NO_3)_2$ at 400–600 °C (above which temperature, oxygen loss to U_3O_8 tends to occur). When uranium oxides are heated with M_2CO_3 or $M'CO_3$

Table 10.6 Oxides of the actinide elements

Ac	Th	Pa	U	Np	Pu	Am	Cm	Bk	Cf	Es
Ac_2O_3					Pu_2O_3	Am_2O_3	Cm_2O_3	Bk_2O_3	Cf_2O_3	Es_2O_3
	ThO_2		UO_2	NpO_2	PuO_2	AmO_2	CmO_2	BkO_2	CfO_2	
		Pa_2O_5		Np_2O_5						
			UO_3	NpO_3 (?)						

(M, M' are group I or II metals), uranates with formulae such as Na_2UO_4 or $K_2U_2O_7$ result; unlike d-block metal compounds with similar formulae, these contain UO_6 polyhedra.

10.8 Uranium Hydride UH_3

The +3 oxidation state of uranium is strongly reducing, so it is not altogether surprising to find it forms a stable hydride. Massive uranium reacts with hydrogen gas on heating to around 250 °C (at lower temperatures if powdered metal is used), swelling up to a fine black powder, which is pyrophoric in air. It decomposes on further heating (350–400 °C under a hydrogen atmosphere) to hydrogen and uranium powder, so that making then decomposing the hydride can be a useful way of preparing a reactive form of the metal (a strategy that can be followed with other actinides, e.g. Th). Some reactions of UH_3 are summarized in Figure 10.7. The crystal structure shows that each uranium has 12 hydrogen neighbours.

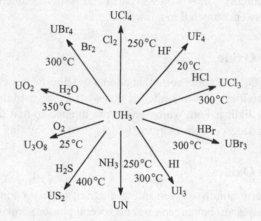

Figure 10.7
Reactions of UH_3.

10.9 Oxyhalides

Although not binary compounds, a discussion of actinide oxyhalides is not out of place here. A number of these are formed by the earlier actinides, another difference between the 4f and 5f metals.

Thorium forms $ThOX_2$ (X = Cl, Br, I) but there is a greater variety with protactinium:

$$PaF_5.2H_2O \rightarrow PaO_2F \text{ (air; 250 °C)}$$

$$PaCl_4 \rightarrow PaOCl_2 \text{ (Sb}_2O_3; 150–200 °C)}$$

$$Pa_2O_5 + PaBr_5 \rightarrow PaOBr_3 \text{ (sealed tube; 400 °C)}$$

$$PaI_5 \rightarrow PaOI_3 \text{ and } PaO_2I \text{ (Sb}_2O_3; 150–200 °C)}$$

Of these, yellow $PaOBr_3$ has a cross-linked chain structure with seven-coordinate Pa (bound to 3 O and 4 Br). Yellow-green $PaOCl_2$ has an important structure adopted by many oxyhalides MOX_2 (X = Cl, Br, I; M = Th, Pa, U, Np), which contains seven-, eight-, and nine-coordinate Pa.

Neptunium forms $NpOF_3$, NpO_2F_2, $NpOF_4$, and NpOX (X = Cl, I), whilst the plutonium compounds PuOX (X = F, Cl, Br, I), $PuO_2Cl_2.6H_2O$, $PuOF_4$, and PuO_2F_2 are known, as well as AmO_2F_2 (and certain americium oxychloride complexes).

The most important oxyhalides, though, are the uranium compounds. Yellow UO_2F_2 and UO_2Cl_2, and red UO_2Br_2, have been known for a long while, but the existence of beige UO_2I_2 was doubted for many years, only being confirmed very recently (2004) by Berthet *et al.*

$$UO_3 \rightarrow UO_2F_2 \quad (HF; 400\,°C)$$

$$U_3O_8 \rightarrow UO_2Cl_2.2\,H_2O \quad (HCl/H_2O_2)$$

$$UO_2Cl_2.2\,H_2O \rightarrow UO_2Cl_2 \quad (Cl_2/HCl; 450\,°C)$$

$$UCl_4 \rightarrow UO_2Cl_2 \quad (O_2; 350\,°C)$$

$$UBr_4 \rightarrow UO_2Br_2 \quad (O_2; 170\,°C)$$

$$UO_2(OTf)_2 \rightarrow UO_2I_2 \quad (Me_3SiI; OTf = CF_3SO_3)$$

In the solid state, UO_2F_2 has a structure in which a uranyl unit is bound to six fluorides (all bridging); whilst in UO_2Cl_2 uranium is bound to the two 'oxo' oxygens, four chlorines, and another oxygen, from another uranyl unit. UO_2I_2 has not been obtained in crystalline form, but it is soluble in pyridine and THF (L), forming $UO_2I_2L_3$, which have pentagonal bipyramidal coordination of uranium, with the usual *trans-* O=U=O linkage.

A number of adducts of the oxyhalides are well characterized, such as *all-trans-* $[UO_2X_2L_2]$ [X = Cl, L = Ph_3PO; X = Br, L = $(Me_2N)_3PO$; X = I, L = $(Me_2N)_3PO$, Ph_3PO, Ph_3AsO]. Others are anionic, containing ions such as $[UO_2F_5]^{3-}$ and $[UO_2X_4]^{2-}$ (X = Cl, Br). Hydrates have recently been reported of the bromide and iodide, *all-trans-*$UO_2I_2(OH_2)_2 \cdot 4Et_2O$, $[UO_2Br_2(OH_2)_3]$ (*cis*-equatorial bromides), and the dimer $[UO_2Br_2(OH_2)_2]_2$, which is $[Br(H_2O)_2O_2U(\mu\text{-}Br)_2UO_2Br(OH_2)_2]$.

Question 10.1 Discuss the halides formed by the actinides, commenting on the trends in oxidation state as the series is crossed.

Answer 10.1 Early on in the series, as far as uranium, the maximum oxidation state corresponds to the total number of 'outer shell' electrons. Uranium forms a hexachloride, in addition to the MF_6 also formed by Np and Pu. After uranium, neptunium forms the full range of tetrahalides, but, from plutonium onwards, the (+3) state dominates the chemistry of the binary halides, which strongly resemble those of the lanthanides. This may reflect decreased availability of 5f (and 6d?) electrons for bonding. As usual, F supports the highest oxidation states.

Question 10.2 Study the information about the thorium tetrahalides in Table 10.4 and comment on the trends and patterns in it.

Answer 10.2 The bond lengths increase as the atomic number of the halogen increases, since the atomic (ionic) radius of the halogen increases. The bond length is less for the ThX_4 molecules in the gas phase than for the corresponding 8-coordinate species in the solid state; again this is expected since there will be more interatomic repulsion in the eight-coordinate species. Since there is no obvious pattern in which compounds adopt particular polyhedra in the solid state, this suggests that the dodocahedron and square antiprism have very similar energies. [It may be parenthetically noted that there is no decrease in coordination number as the size of the halogen increases, as seen for uranium (Figure 10.4).]

Table 10.7   Melting points and boiling points of uranium chlorides

	UCl_3	UCl_4	UCl_5	UCl_6
Mp (°C)	835	590	287	177.5 (dec.)
Bp (°C)	dec.	790		

Question 10.3 Table 10.7 shows the melting and boiling points of the chlorides of uranium. Comment on how these relates to their structures.

Answer 10.3   There is a clear relationship with the degree of molecularity. UCl_6 has a monomeric molecular structure with relatively weak intermolecular forces and is therefore the most volatile (and on account of its high oxidation state, the least stable); UCl_5 is dimeric U_2Cl_{10}, therefore as a larger molecule will have stronger intermolecular forces and be less volatile than UCl_6. Continuing to UCl_4 and UCl_3, the more halogens surrounding the uranium, the more energy will be needed to break the lattice into small mobile units, and the higher will be the melting and boiling points.

Question 10.4 Thorium(IV) oxide has a melting point of 3390 °C. Explain why.

Answer 10.4 It is composed of Th^{4+} and O^{2-} ions. The strong electrostatic attraction between these highly charged ions (a very high lattice energy) means that a great deal of energy has to be supplied to overcome the attraction in order to free the ions and melt it.

Question 10.5 Write equations for the reduction of UO_3 by both hydrogen and carbon monoxide.

Answer 10.5 $UO_3 + H_2 \rightarrow UO_2 + H_2O; UO_3 + CO \rightarrow UO_2 + CO_2$

Question 10.6 UH_3 is high melting and has a high conductivity, similar to uranium metal. Comment on its bonding and structure.

Answer 10.6 The high conductivity suggests the presence of delocalized electrons. One possible structure is $U^{4+}(H^-)_3(e^-)$; the ionic nature would explain the high melting point.

11 Coordination Chemistry of the Actinides

By the end of this chapter you should be able to:

- recognize +4 and +6 as the main oxidation states for uranium and +4 as the only important oxidation state for thorium;
- appreciate that coordination numbers of >6 are the norm;
- be familiar with the main features of the chemistry of the other actinides, and appreciate the transition in chemical behaviour across the series;
- appreciate the importance of nitrate complexes;
- recognize the characteristic structures of uranyl complexes;
- suggest structures for compounds;
- explain the separation of nuclear material and the recovery of U and Pu in terms of the chemistry involved;
- appreciate the problems presented in nuclear waste disposal and understand the solutions;
- suggest safe ways to handle actinide compounds.

11.1 Introduction

Studies of the coordination chemistry of the actinides have been limited by a number of factors – the care needed in handling radioactive materials and the possibility of damage to human tissue from the radiation; toxicity (especially Pu); the very small quantities available and very short half-lives of the later actinides; radiation and heating damage to solutions; and radiation damage (defects and dislocations) to crystals.

The vast majority of the studies reported have concerned the metals thorium and uranium, particularly the latter, due to accessibility of raw materials, ease of handling, and the long lifetimes of the relatively weakly α-emitting elements Th and U. In many cases, compounds of neptunium and plutonium with similar formulae to U and Th analogues have been made and found to be isomorphous and thus presumably isostructural. This chapter will therefore commence with, and concentrate largely on, the chemistry of complexes of these elements, followed by sections on the other actinides.

11.2 General Patterns in the Coordination Chemistry of the Actinides

As noted earlier (Table 9.2), the patterns of oxidation states of the early actinides resemble those of d-block metals, with the maximum oxidation state corresponding to the number of 'outer shell' electrons. Thus the chemistry of thorium is essentially confined to the +4

Lanthanide and Actinide Chemistry   S. Cotton
© 2006 John Wiley & Sons, Ltd.

state (to an even greater degree than Zr or Hf) but uranium exhibits oxidation states of $+3$, $+4$, $+5$, and $+6$, with most compounds being in either the $+4$ or $+6$ state. The reason for this is that uranium(III) compounds tend to be easy to oxidize:

$$U^{4+}(aq) + e^- \rightarrow U^{3+}(aq) \qquad E = -0.63\,V$$

whilst under aqueous conditions the UO_2^+ ion readily disproportionates to a mixture of U^{4+} and UO_2^{2+} [though there is a chemistry of uranium(V) in non-aqueous solvents].

As the actinide series is crossed, it becomes harder to involve the f electrons in compound formation, so that the later actinides increasingly show a chemistry in the $+3$ state, resembling the lanthanides, but with a more prominent $+2$ state.

11.3　Coordination Numbers in Actinide Complexes

As for the lanthanides, actinide complexes display high coordination numbers. A study of the aqua ions of early actinides makes an interesting comparison (Table 11.1 lists numbers of water molecules and bond lengths).

Table 11.1　Actinide aqua ions – numbers of bound water molecules and metal–water distances[a]

Ox state		Ac	Th	Pa	U	Np	Pu	Am	Cm	Bk	Cf
6	MO_2^{2+}				5.0	5.0	6.0				
	$M–OH_2$ (Å)				2.40	2.42	2.40–2.45				
5	MO_2^+					5.0	4.0				
	$M–OH_2$ (Å)					2.50	2.47				
4	M^{4+}		10.0		9.0; 10.0	11.2	8 or 9				
	$M–OH_2$ (Å)		2.45		2.51; 2.42	2.40	2.39				
3	M^{3+}				9 or 10	9 or 10	10.2	10.3	10.2		8.5 ± 1.5
	$M–OH_2$ (Å)				2.61	2.52	2.51	2.48	2.45		2.4

[a] The uncertainty in the hydration number is generally in the region of 1.

The best characterized aqua ion is the hydrated uranyl ion $[UO_2(OH_2)_5]^{2+}$, which has been isolated in several salts studied by diffraction methods as well as X-ray absorption methods (EXAFS) in solution. However, similar ions are firmly believed to exist in the cases of $[MO_2(aq)]^{2+}$ (M = Np, Pu) and $[MO_2(aq)]^+$ (M = U, Np, Pu). In studying Table 11.1, two features stand out. Ions like UO_2^{2+} are unprecedented in lanthanide chemistry, both in respect of the $+6$ oxidation state and also in the presence of strong and non-labile U=O bonds. Secondly, the coordination numbers of the hydrated $+3$ actinide ions appear to be higher than those of the Ln^{3+} ions (9 for early lanthanides, 8 for later ones), explicable on account of the slightly higher ionic radii of the actinide ions. One of the few other characterized aqua ions is the trigonal prismatic $[Pu(H_2O)_9]^{3+}$, strongly resembling the corresponding lanthanide series.

11.4　Types of Complex Formed

Table 11.2 gives some stability constants for complexes of Th^{4+} and UO_2^{2+}. Thorium forms stronger complexes with fluoride, the 'hardest' halide ion, than with chloride and bromide; this is the behaviour expected of a 'hard' Lewis acid. Thorium also forms quite

Table 11.2 Stability constants of Th^{4+} and UO_2^{2+} complexes

	Stability constants for complexes of Th^{4+}					Stability constants for complexes of UO_2^{2++}		
Ligand	I (mol/dm^3)	Log K_1	Log K_2	Log K_3	Log K_4	Ligand	I (mol/dm^3)	Log K_1
F^-	0.5	7.56	5.72	4.42		F^-	1	4.54
	1	7.46				Cl^-	1	−0.10
Cl^-	1	0.18				Br^-	1	−0.3
Br^-	1	−0.13				NO_3^-	1	−0.3
NO_3^-	1	0.67				$EDTA^{4-}$	0.1	7.4
SO_4^{2-}	2	3.3	2.42					
NCS^-	1	1.08						
$EDTA^{4-}$	0.1	25.3						
$acac^-$	0.1	8.0	7.5	6.0	5.3			

strong complexes with oxygen-donor ligands like nitrate. The hexadentate ligand $EDTA^{4-}$ forms very stable complexes, due largely to the favourable entropy change when six water molecules are replaced.

11.5 Uranium and Thorium Chemistry

11.5.1 Uranyl Complexes

The great majority of uranium(VI) compounds contain the UO_2 group and are known as uranyl compounds; exceptions are a few molecular compounds, such as the halides UOF_4, UF_6, and UCl_6, and some alkoxides such as $U(OMe)_6$. Uranyl compounds result eventually from exposure of compounds of uranium in other oxidation states to air. They characteristically have a yellow fluorescence under UV light; from the early 19th century, glass manufacturers added uranium oxide when making yellow and green glass (it is sometimes known as Vaseline glass).

Uranyl complexes can be thought of as derivatives of the UO_2^{2+} ion. There is a very wide range of them; they may be cationic, such as $[UO_2(OH_2)_5]^{2+}$ ions; neutral, e.g. $[UO_2(OPPh_3)_2Cl_2]$; or anionic, such as $[UO_2Cl_4]^{2-}$, yet all feature a *trans*-UO_2 grouping with the characteristic short U–O bonds (1.7–1.9 Å), quite comparable with those found in 'osmyl' compounds, which have the OsO_2 grouping. The presence of the uranyl group can readily be detected in the IR spectrum of a uranium compound through the presence of a strong band in the region 920–980 cm^{-1} caused by the asymmetric O–U–O stretching vibration; a corresponding band around 860 cm^{-1} caused by the symmetric O–U–O stretching vibration is seen in the Raman spectrum. Fine structure due to symmetric uranyl stretching vibrations can be seen on an absorption peak in the spectrum of uranyl complexes around 450 nm (Figure 12.1). There has been considerable speculation on the bonding in the uranyl ion. The essentially linear geometry of the UO_2 unit is an invariable feature of uranyl complexes; no other atoms can approach the uranium nearer than ~2.2 Å. Uranium d–p and f–p π bonding have both been invoked to explain the bonding (Figure 11.1).

An energy level diagram for the uranyl ion is shown in Figure 11.2. An electron count takes 6 electrons from uranium, four from each oxygen, deducting two for the positive charges; alternatively, if the uranyl ion is thought of as a combination of U^{6+} and two O^{2-}, taking six electrons from each oxide and none from U^{6+}, again giving 12. These completely occupy the six σ_u, σ_g, π_u, and π_g molecular orbitals ($\sigma_u^2 \sigma_g^2 \pi_u^4 \pi_g^4$). It has been suggested

X
X
Z
(a) π u
(b) π g
σg
σu
(c)

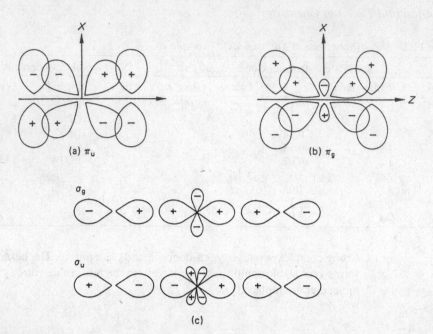

Figure 11.1
π-bonding in the uranyl, $[UO_2]^{2+}$ ion: (a) d_{xz}–p_x overlap; (b) f_{xz^2}–p_x overlap; (c) σ-bonding in the uranyl ion (reproduced with permission from Figure 3.24 of S.A. Cotton, *Lanthanides and Actinides*, Macmillan, 1991).

6d
5f
2p
Occupied in UO2 2+
The relative ordering of the bonding MOs is uncertain

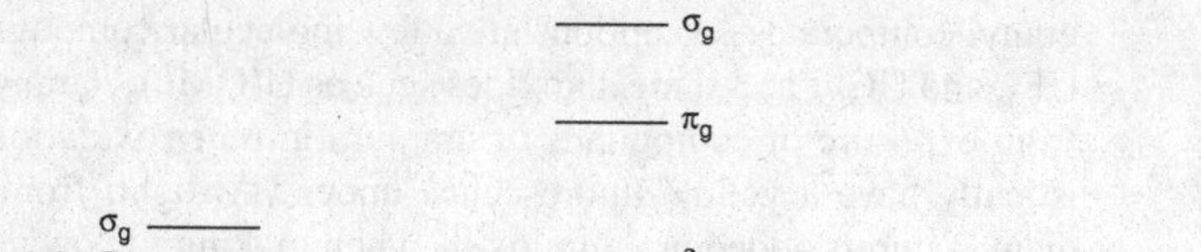

Figure 11.2
M.O. scheme for the uranyl ion, UO_2^{2+} (reproduced with permission from Figure 3.25 of S.A. Cotton, *Lanthanides and Actinides*, Macmillan, 1991).

that repulsion (antibonding overlap) between oxygen p orbitals and occupied uranium 6d and 5f orbitals may destabilize the bonding d molecular orbitals, causing them to be higher in energy than the π bonding orbitals, as shown. Addition of further electrons puts them in the essentially non-bonding δ_u and ϕ_u orbitals, accounting for the existence of the rather less stable MO_2^{2+} (M = Np, Pu, Am) ions.

The uranium(V) species UO_2^+ exists, but is less stable than UO_2^{2+}, possibly owing to weaker overlap; it readily decomposes by disproportionation:

$$2\ UO_2^+(aq) + 4\ H^+(aq) \rightarrow UO_2^{2+}(aq) + U^{4+}(aq) + 2\ H_2O(l)$$

In contrast to the *trans* geometry of UO_2^{2+}, isoelectronic ($6d^0\ 5f^0$) ThO_2 molecules in the gas phase or in low temperature matrices are bent ($\angle$O–Th–O $\sim 122°$); this is thought to be due to the thorium 5f orbitals being much higher in energy (in uranium, the 5f orbitals are well below 6d), reducing possibilities of 5f–p overlap, so that thorium resorts to using 6d orbitals in the π bonding, like transition metals, as in the case of $MoO_2^{2+}(5d^0)$ and WO_2^{2+} ($6d^0$) ions which also adopt bent, *cis* geometries ($\angle$O–M–O $\sim 110°$).

11.5.2 Coordination Numbers and Geometries in Uranyl Complexes

An extensive range of uranyl complexes has been prepared and had their structures determined. Their structure can be summarized as a uranyl ion surrounded by a 'girdle' of 4, 5, or 6 donor atoms round its waist (a rare example of 2 + 3 coordination is known for a complex of the uranyl ion with a calixarene ligand; another is the amide complex [K(thf)$_2$] [UO$_2$\{N(SiMe$_3$)$_2$\}$_3$]). If the ligands are monodentate donors, there are usually 4 of them, unless they are small, like F or NCS, when five can be accommodated. When bidentate ligands with small steric demands like NO$_3$, CH$_3$COO, and CO$_3$ can be accommodated, six donor atoms can surround the uranyl group. In general if there are 4 or 5 donor atoms round the waist, they are reasonably coplanar, but puckering sometimes occurs when there are six. Table 11.3 shows examples of uranium complexes for 5-, 6-, 7-, and 8-coordination.

Table 11.3   Uranyl complexes

5-coordinate (2 + 3)	6-coordinate (2 + 4)	7-coordinate (2 + 5)	8-coordinate (2 + 6)
[UO$_2$\{N(SiMe$_3$)$_2$\}$_3$]$^-$	Cs$_2$[UO$_2$Cl$_4$]	UO$_2$Cl$_2$	UO$_2$F$_2$
	(Me$_4$N)$_2$[UO$_2$Br$_4$]	UO$_2$(superphthalocyanine)	UO$_2$CO$_3$
	MgUO$_4$	[UO$_2$(NO$_3$)$_2$(Ph$_3$PO)]	[UO$_2$(NO$_3$)$_2$(H$_2$O)$_2$]
	BaUO$_4$	[UO$_2$(L)$_5$]$^{2+}$	CaUO$_4$
		(L, e.g., H$_2$O, DMSO, urea)	SrUO$_4$
			Rb[UO$_2$(NO$_3$)$_3$]

The uranium(VI) aqua ion is now firmly established as the pentagonal bipyramidal [UO$_2$(OH$_2$)$_5$]$^{2+}$; it has been found in crystals of the salt [UO$_2$(OH$_2$)$_5$] (ClO$_4$)$_2$.2H$_2$O (U=O 1.702 Å, U–OH$_2$ 2.421 Å) and also found in solutions by X-ray diffraction studies. Similar [UO$_2$(L)$_5$]$^{2+}$ ions (L = urea, Me$_2$SO, HCONMe$_2$) also exist. Uranyl nitrate forms complexes with phosphine oxides of the type [UO$_2$(NO$_3$)$_2$(R$_3$PO)$_2$]; similar phosphate complexes [UO$_2$(NO$_3$)$_2$\{(RO)$_3$PO\}$_2$] are important in the extraction of uranium in nuclear waste processing (Figure 11.3 and Section 11.5.5). A number of structures of these complexes are known; for example, when R=isobutyl, U=O 1.757 Å; U–O(P) 2.372 Å; U–O(N) 2.509 Å.

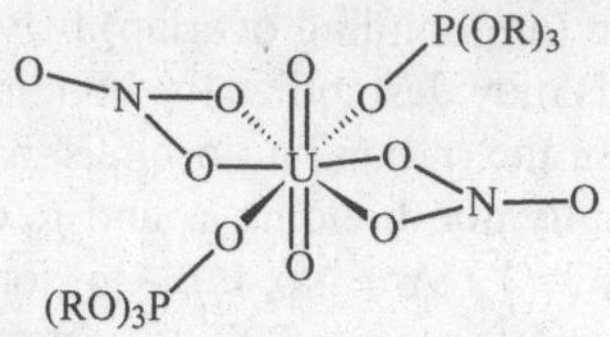

Figure 11.3
Structure of $UO_2(NO_3)_2[(RO)_3PO]_2$.

11.5.3   Some Other Complexes

Uranyl carbonate complexes have attracted considerable interest in recent years as they are intermediates in the processing of mixed oxide reactor fuels and in extraction of uranium from certain ores using carbonate leaching; more topically they can be formed when uranyl ores react with carbonate or bicarbonate ions underground, and can be present in relatively high amounts in groundwaters. The main complex formed in carbonate leaching of uranyl ores is 8 coordinate $[UO_2(CO_3)_3]^{3-}$, but around pH 6 a cyclic trimer $[(UO_2)_3(CO_3)_6]^{6-}$ has been identified.

UO_3 dissolves in acetic acid to form yellow uranyl acetate, $UO_2(CH_3COO)_2.2H_2O$. It formerly found use in analysis since, in the presence of M^{2+} (M = Mg or Zn), it precipitates sodium ions as NaM $[UO_2(CH_3COO)_3]_3.6H_2O$.

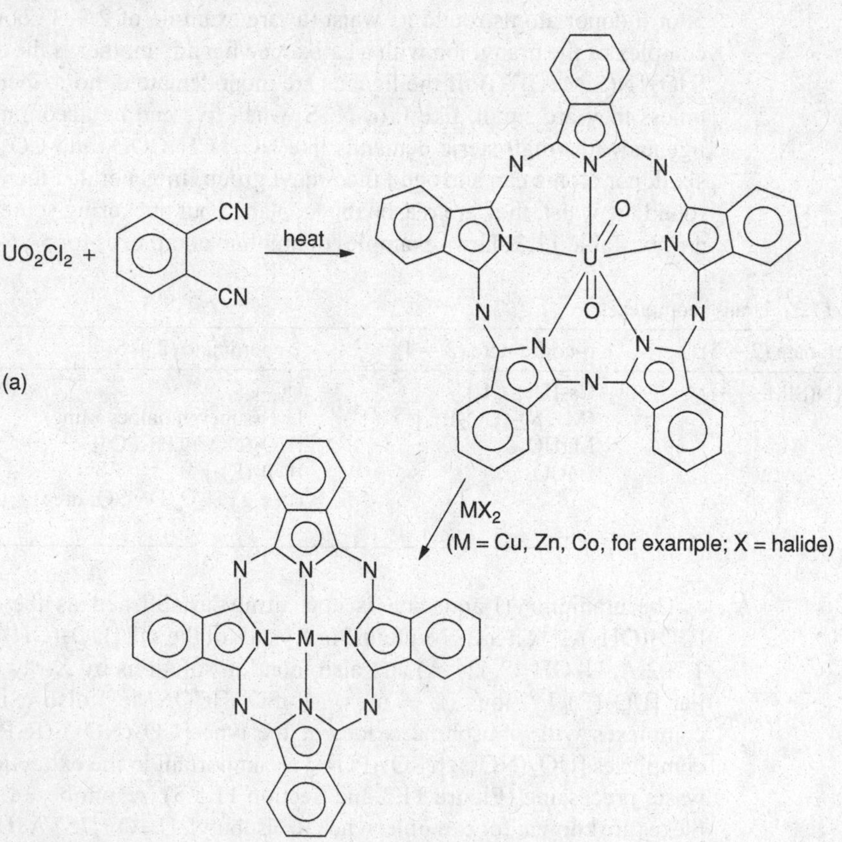

Figure 11.4
Structure and reaction of "uranyl superphthalocyanine".

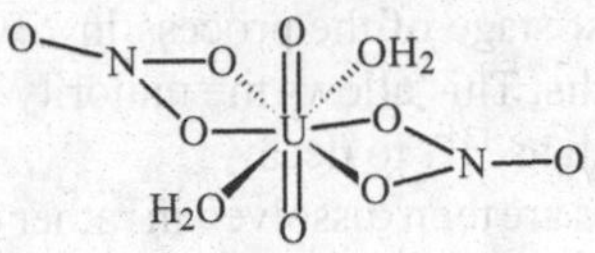

Figure 11.5
Structure of $UO_2(NO_3)_2(H_2O)_2$.

Uranyl chloride, UO_2Cl_2, reacts on heating with *o*-phthalodinitrile to form a so-called 'superphthalocyanine' complex with 2 + 5 coordination (Figure 11.4); other metals (lanthanides, Co, Ni, Cu) react with this, forming a conventional phthalocyanine, so that the uranyl ion has an important role in sustaining this unusual structure.

11.5.4 Uranyl Nitrate and its Complexes; their Role in Processing Nuclear Waste

Reaction of uranium oxide with nitric acid results in the formation of nitrates $UO_2(NO_3)_2.x$ H_2O ($x = 2, 3, 6$); the value of x depends upon the acid concentration. All contain $[UO_2(NO_3)_2(H_2O)_2]$ molecules; the nitrate groups are bidentate, so that uranium is 8 coordinate (Figure 11.5). Its most important property lies in its high solubility in a range of organic solvents in addition to water (Table 11.4), which is an important factor in the processing of nuclear waste.

Table 11.4 Solubility of uranyl nitrate in various solvents

Solvent	Solubility (g per g solvent)
Water	0.540
Diethyl ether	0.491
Acetone	0.617
Ethanol	0.675

By adding metal nitrates as 'salting out' agents, the solubility of uranyl nitrate in water can be decreased to favour its extraction from aqueous solution into the organic layer. Tributyl phosphate [TBP; $(C_4H_9O)_3P{=}O$] acts as a complexing agent (in the manner discussed in this chapter for a number of phosphine oxide ligands) and also as solvent, with no salting-out agent being needed. In practice, a solution of TBP in kerosene is used to give better separation, as pure TBP is too viscous and also has a rather similar density to that of water.

11.5.5 Nuclear Waste Processing

Nuclear fuel rods consist of uranium oxide pellets contained in zirconium alloy or steel tubes. As the fission process proceeds, uranium is used up and fission products accumulate. A lot of these fission products are good neutron absorbers and reduce the efficiency of the fission process (by absorbing neutrons before they reach uranium atoms) so that the rods are removed for reprocessing before all the ^{235}U content has undergone fission. Fission of a ^{235}U atom produces two lighter atoms of approximate relative atomic masses around 90–100 and 130–140, with the main fission products being the intensely radioactive and short lived ^{131}I ($t_{1/2} \sim 8$ d), ^{140}La, ^{141}Ce, ^{144}Pr, ^{95}Zr, ^{103}Ru, and ^{95}Nb, and longer-lived ^{137}Cs, ^{90}Sr, and ^{91}Y. These essentially useless and toxic products have to be separated from unchanged uranium and also from plutonium (the product of neutron absorption by ^{238}U), both of which can be used again as fuels.

1. The first stage of the process involves immersing the fuel rods in ponds of water for up to 3 months. This allows the majority of the short-lived and intensely radioactive fission products such as ^{131}I to decay.

2. The rods are then dissolved in rather concentrated (7M) nitric acid, producing a mixture of $UO_2(NO_3)_2$, $Pu(NO_3)_4$, and other metal nitrates.

3. The mixture is extracted with a counter-current of a solution of TBP in kerosene. Uranium and plutonium are extracted into kerosene as the complexes $[UO_2(NO_3)_2(TBP)_2]$ and $[Pu(NO_3)_4(TBP)_2]$, but the other nitrates, of metals such as the lanthanides and actinides beyond Pu, as well as fission products, do not form strong complexes with TBP and stay in the aqueous layer.

4. The mixture of uranium and plutonium is treated with a suitable reducing agent [iron(II) sulfamate, hydrazine, or hydroxylamine nitrate]; under these conditions, U^{VI} is not reduced and stays in the kerosene layer, but Pu^{IV} is reduced to Pu^{3+}, which is only weakly complexed by TBP and so migrates into the aqueous phase.

$$UO_2^{2+}(aq) + 4\,H^+(aq) + 2\,e^- \rightarrow U^{4+}(aq) + 2\,H_2O\,(l) \quad E = +0.27\,V$$
$$Pu^{4+}(aq) + e^- \rightarrow Pu^{3+}(aq) \quad E = +1.00\,V$$
$$Fe^{3+}(aq) + e^- \rightarrow Fe^{2+}(aq) \quad E = +0.77\,V$$

The uranyl nitrate is extracted back into the aqueous phase and crystallized as the hydrate $UO_2(NO_3)_2.xH_2O$; thermal decomposition to UO_3 is followed by hydrogen reduction to re-form UO_2. The plutonium is reoxidized to Pu^{4+}, precipitated as the oxalate $Pu(C_2O_4)_2.6H_2O$, which undergoes thermal decomposition to PuO_2.

The fission products remain to be dealt with. The preferred solution at the moment is to evaporate their solution and pyrolyse the product to convert it into a mixture of oxides; on fusion with silica and borax an inert borosilicate glass is formed, which encapsulates the radioactive materials. At present the main problem seems to be choosing suitably stable geological areas where these materials will not be disturbed by earthquakes nor dissolved by underground water.

11.6 Complexes of the Actinide(IV) Nitrates and Halides

11.6.1 Thorium Nitrate Complexes

A number of thorium nitrate complexes have been synthesized and studied. Hydrated thorium nitrate, $Th(NO_3)_4.5H_2O$, contains $[Th(NO_3)_4.(H_2O)_3]$ molecules and was one of the first 11-coordinate compounds to be recognized (Figure 11.6).

Reaction of thorium nitrate with tertiary phosphine oxides in solvents like acetone or ethanol has afforded a number of complexes. In particular, when thorium nitrate reacts with Me_3PO, a wide range of complexes is obtained depending upon the stoichiometry and solvent used, compounds with formulae $Th(NO_3)_4.5Me_3PO$, $Th(NO_3)_4.4Me_3PO$, $Th(NO_3)_4.3.67Me_3PO$, $Th(NO_3)_4.3Me_3PO$, $Th(NO_3)_4.2.67Me_3PO$, and $Th(NO_3)_4.2.33Me_3PO$ having been obtained. Thorium nitrate complexes display a fascinating range of structure (Table 11.5).

11.6.2 Uranium(IV) Nitrate Complexes

Uranium(VI) nitrate complexes have been discussed in Section 11.5.4, but uranium forms complexes in the +4 state that are generally similar to those of thorium.

$$Cs_2U(NO_3)_6 + 2\,Ph_3PO \rightarrow U(NO_3)_4(OPPh_3)_2 + 2\,CsNO_3$$

Table 11.5 Thorium nitrate complexes

Formula	Thorium species present	Coord. No.
$MTh(NO_3)_6$ (M = Mg, Ca)	$[Th(NO_3)_6]^{2-}$	12
$Ph_4P^+ [Th(NO_3)_5(OPMe_3)_2]^-$	$[Th(NO_3)_5(OPMe_3)_2]^-$	12
$Th(NO_3)_4.5H_2O$	$[Th(NO_3)_4(H_2O)_3]$	11
$Th(NO_3)_4.2.67Me_3PO$	$\{[Th(NO_3)_3(Me_3PO)_4]^+\}_2 [Th(NO_3)_6]^{2-}$	10, 12
$Th(NO_3)_4.2Ph_3PO$	$[Th(NO_3)_4(OPPh_3)_2]$	10
$Th(NO_3)_4.5Me_3PO$	$[Th(NO_3)_2(OPMe_3)_5]^{2+}$	9

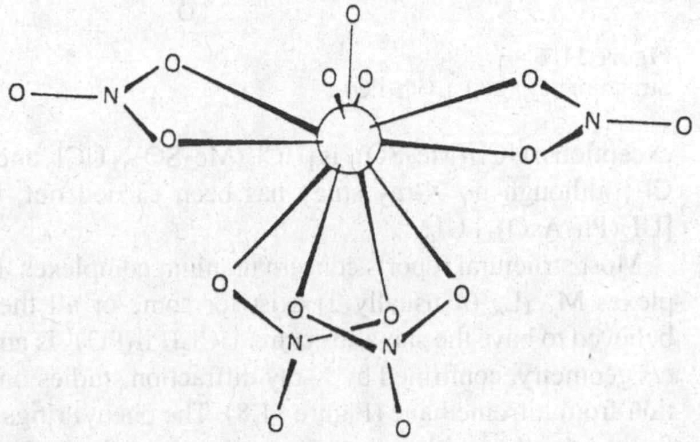

Figure 11.6
Structure of $Th(NO_3)_4(H_2O)_3$.

The reaction is carried out in a non-polar solvent like propanone; precipitated caesium nitrate is filtered off and green (the most characteristic colour of U^{IV} complexes) crystals of the nitrate complex are obtained on concentrating the solution. $U(NO_3)_4(OPPh_3)_2$ has a 10-coordinate structure (Figure 11.7) with phosphine oxide ligands *trans* to each other, and bidentate nitrates.

11.6.3 Complexes of the Actinide(IV) Halides

A large number of these have been synthesized, usually by reaction of the halides with the ligand in a non-polar solvent like MeCN or acetone, which will form a labile complex such as $[UCl_4(MeCN)_4]$ that will undergo ready substitution by a stronger donor:

$$ThCl_4 + 5 Me_2SO \rightarrow ThCl_4(Me_2SO)_5 \text{ (in MeCN)}$$

$$UCl_4 + 2 (Me_2N)_3PO \rightarrow UCl_4[(Me_2N)_3PO]_2 \text{ (in Me}_2CO)$$

Although sometimes direct reaction with a liquid ligand is possible:

$$UCl_4 + 3 THF \rightarrow UCl_4(THF)_3$$

$$ThCl_4 + 3 Me_3N \rightarrow ThCl_4(Me_3N)_3; UCl_4 + 2 Me_3N \rightarrow UCl_4(Me_3N)_2$$

Relatively few UI_4 complexes have been made. They can often be synthesized by reaction in a solvent like MeCN and, whilst relatively stable thermally, undergo ready oxidation to uranyl complexes in (moist) air. The structures of many of these compounds have been determined, such as UCl_4L_2 [L = Ph_3PO, $(Me_2N)_3PO$, Et_3AsO, $(Me_2N)_2PhPO$] and UBr_4L_2 [L = Ph_3PO, $(Me_2N)_3PO$] and $UX_4[(Me_2N)_2CO]_2(X = Cl$, Br, I). Both the 2:1 stoichiometry and *trans*-UX_4L_2 geometry are very common, but there are

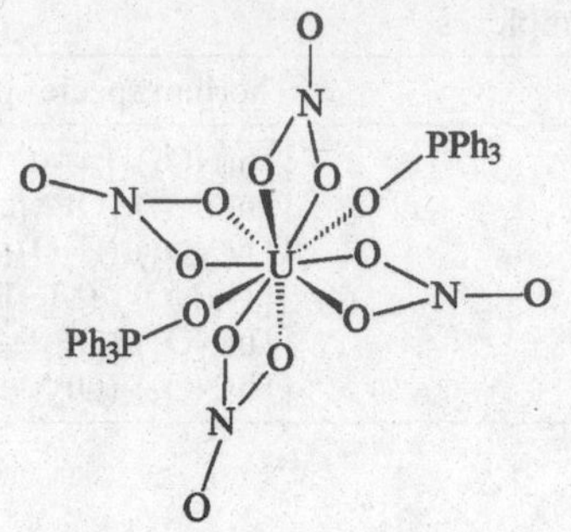

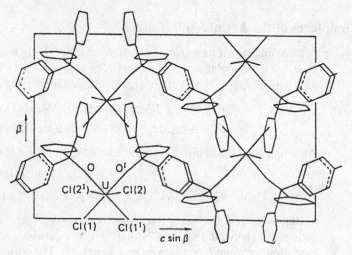

Figure 11.7
Structure of $U(NO_3)_4(Ph_3PO)_2$.

exceptions. $UCl_4(Me_2SO)_3$ is $[UCl_2(Me_2SO)_6]\,UCl_6$ and $UCl_4(Me_3PO)_6$ is $[UCl(Me_3PO)_6]\,Cl_3$; although no X-ray study has been carried out, $UI_4(Ph_3AsO)_2$ is almost certainly $[UI_2(Ph_3AsO)_4]\,UI_6$.

Most structural reports concern uranium complexes. Many cases are known where complexes MX_4L_n (n usually 2) exist for some or all the series Th–Pu; they are generally believed to have the same structure. $UCl_4(Ph_3PO)_2$ is an exceptional compound in having a *cis* geometry, confirmed by X-ray diffraction studies on crystals obtained on recrystallization from nitromethane (Figure 11.8). The phenyl rings in neighbouring Ph_3PO molecules face each other with a ring–ring separation of $\sim$3.52 Å, an early recognized example of π–π stacking in a coordination compound. Similar compounds can be obtained for Th, Pa, Np, and Pu.

It was subsequently noted that crystals obtained immediately from the reaction mixture had an IR spectrum different to that of authentic *cis*-$UCl_4(Ph_3PO)_2$ but strongly resembling those of samples of *trans*-$UBr_4(Ph_3PO)_2$. It seems likely that a less soluble *trans*-$UCl_4(Ph_3PO)_2$ crystallizes first but on recrystallization or contact with the mother liquor it isomerizes to the thermodynamically more stable *cis*-$UCl_4(Ph_3PO)_2$.

Figure 11.8
Reproduced with permission from JCS Dalton, (1975) 1875. G. Bombieri et al. Copyright (1975) RSC.

Table 11.6 Bond lengths in UX₄L₂ complexes

	UCl₄(tmu)₂	UBr₄(tmu)₂	UI₄(tmu)₂	UCl₄(hmpa)₂	UBr₄(hmpa)₂
Average U–O (Å)	2.209	2.197	2.185	2.23	2.18
Average U–X (Å)	2.62	2.78	3.01	2.615	2.781

Abbreviations: tmu = $(Me_2N)_2C=O$; hmpa = $(Me_2N)_3P=O$.

The only complete family of UX₄ complexes (X = Cl, Br, I) that has been examined crystallograpically is that with tetramethylurea, $UX_4[(Me_2N)_2CO]_2$. Table 11.6 lists structural data for these compounds and also for $UX_4[(Me_2N)_3PO]_2(X = Cl, Br)$ {though the compound $UI_4[(Me_2N)_3PO]_2$ has been made, its structure is not known}. Table 11.7 shows the formulae of ThCl₄ and UCl₄ complexes with the same ligands. The ionic radius of $U^{4+} = 1.00$ Å and that of $Th^{4+} = 1.05$ Å (values given for eight coordination, the pattern is similar for six coordination). Since the ions are of similar size, complexes usually have similar stoichiometry (and indeed geometry). In a few cases (Me_3N, Ph_3PO, THF) the slightly greater size of thorium allows one more ligand to be attached to the metal ion.

Table 11.7 A comparison of thorium(IV) and uranium(IV) complexes isolated with the same ligand

Ligand	Thorium	Uranium
MeCN	ThCl₄(MeCN)₄	UCl₄(MeCN)₄
Ph₃PO	ThCl₄(Ph₃PO)₃; ThCl₄(Ph₃PO)₂	UCl₄(Ph₃PO)₂
(Me₂N)₃PO	ThCl₄[(Me₂N)₃PO]₃; ThCl₄[(Me₂N)₃PO]₂	UCl₄[(Me₂N)₃PO]₂
(Me₂N)₂CO	ThCl₄[(Me₂N)₂CO]₃	UCl₄[(Me₂N)₂CO]₂
Ph₂SO	ThCl₄(Ph₂SO)₄	UCl₄(Ph₂SO)₄; UCl₄(Ph₂SO)₃
THF	ThCl₄(THF)₃(H₂O)	UCl₄(THF)₃
Mc₃N	ThCl₄(Me₃N)₃	UCl₄(Me₃N)₂
Me₂N(CH₂)₂NMe₂	ThCl₄[Me₂N(CH₂)₂NMe₂]₂	UCl₄[Me₂N(CH₂)₂NMe₂]₂
Me₂P(CH₂)₂PMe₂	ThCl₄[Me₂P(CH₂)₂PMe₂]₂	UCl₄[Me₂P(CH₂)₂PMe₂]₂

11.7 Thiocyanates

As the anhydrous actinide thiocyanates are not known {only hydrated $[M(NCS)_4(H_2O)_4]$}, thiocyanate complexes are prepared metathetically.

$$ThCl_4(Ph_3PO)_2 + 4\ KNCS + 2\ Ph_3PO \rightarrow Th(NCS)_4(Ph_3PO)_4 + 4\ KCl\ (in\ Me_2CO)$$

The precipitate of insoluble KCl is filtered off and the solution concentrated to obtain the actinide complex. Several structures have been reported; $Th(NCS)_4(L)_4$ [L = Ph_3PO, $(Me_2N)_3PO$]; $U(NCS)_4(L)_4$[L = Ph_3PO, $(Me_2N)_3PO$, Me_3PO] are all square antiprismatic; $Th(NCS)_4[(Me_2N)_2CO]_4$ is dodecahedral.

Some complexes are known where thiocyanate is the only ligand bound to uranium. The geometry of $(Et_4N)_4\ [U(NCS)_8]$ is cubic, whilst $Cs_4\ [U(NCS)_8]$ is square antiprismatic. Compared with the d-block transition metals, there is not much evidence for directional character in bonding in lanthanide and actinide compounds. Because of this, the size and geometry of the cation affects the packing arrangements in the lattice and energetically this factor must be more important than any crystal-field effects favouring a particular shape of the anion.

11.8   Amides, Alkoxides and Thiolates

Compounds $U(NR_2)_x$ and $U(OR)_x$ have been synthesized in oxidation states $+3$ to $+6$ (though the hexavalent amides are not well characterized). A limited number of thiolates $U(SR)_4$ are also known. These largely molecular species occupy a borderline between classical coordination chemistry and organometallic chemistry.

11.8.1   Amide Chemistry

Although they have been much less studied than the range of alkoxides in the $+4$, $+5$, and $+6$ oxidation states, a number of amides of uranium(IV) have been made by 'salt-elimination' reactions in solvents such as diethyl ether or THF, examples being:

$$UCl_4 + 4\,LiNEt_2 \rightarrow U(NEt_2)_4 + 4\,LiCl$$

$$UCl_4 + 4\,LiNPh_2 \rightarrow U(NPh_2)_4 + 4\,LiCl$$

$$UCl_4 + \text{excess of } LiN(SiMe_3)_2 \rightarrow UCl\{N(SiMe_3)_2\}_3 + 3\,LiCl$$

The steric demands of the ligands can be quantified in terms of the steric coordination number, CN_S [see J. Marçalo and A. Pires de Matos, *Polyhedron*, 1989, **8**, 2431; $CN_S =$ the ratio of the solid angle comprising the Van der Waals' spheres of the atoms of the ligand and that of the chloride ligand, when placed at a typical distance from the metal, values for these amide groups are: NEt_2^- 1.67; NPh_2^- 1.79; $N(SiMe_3)_2^-$ 2.17].

$U(NPh_2)_4$ is a monomer with a tetrahedral geometry, whilst in the solid state $U(NEt_2)_4$ is actually a dimer, $[U_2(NEt_2)_{10}]$, which contains five-coordinate uranium, with two bridging amides (Figure 11.9). As $N(SiMe_3)_2$ is a much bulkier ligand with larger solid cone angle, it seems likely that there is not room for a fourth amide group round uranium, hence the non-isolation of $U\{N(SiMe_3)_2\}_4$, with $MCl\{N(SiMe_3)_2\}_3$ (M = Th, U) being the most substituted species obtained; the halogen can be replaced by other groups such as methyl or tetrahydroborate. The compound $U(NMe_2)_4$ is harder to prepare. In the solid state it has a chain trimeric structure; each uranium is 6 coordinate, NMe_2 being less bulky than the other amides and has a smaller steric coordination number than even NEt_2.

So far, the amides described have all been in the $+4$ state, but one rare three coordinate U^{III} compound is $U\{N(SiMe_3)_2\}_3$ (pyramidal, like the lanthanide analogues). Unusual amides

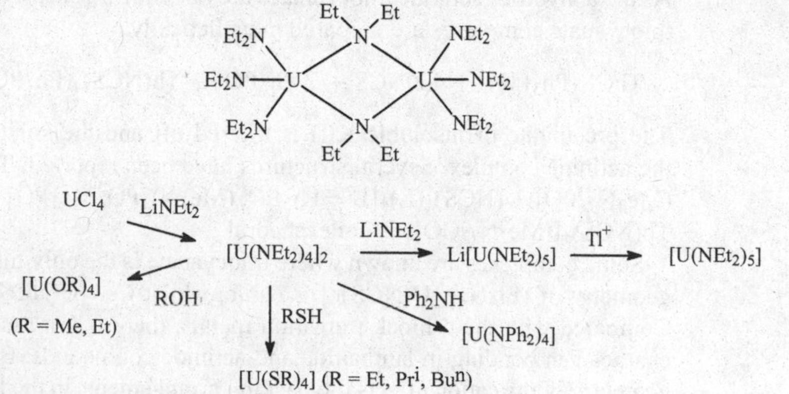

Figure 11.9
Synthesis, structure and reaction of $U(NEt_2)_4$.

have also been prepared with uranium in the +5 and +6 oxidation states. The first step in the syntheses involved preparing anionic complexes with saturated coordination spheres:

$$UCl_4 + 6 LiNMe_2 \rightarrow [Li(THF)]_2[U(NMe_2)_6];$$

$$UCl_4 + 5 LiNEt_2 \rightarrow [Li(THF)][U(NEt_2)_5]$$

The anions can then be oxidized to neutral molecular species using TlBPh$_4$ or, better, AgI:

$$[U(NMe_2)_6]^{2-} \rightarrow [U(NMe_2)_6]^- \rightarrow [U(NMe_2)_6]; \; [U(NEt_2)_5]^- \rightarrow [U(NEt_2)_5]$$

[U(NEt$_2$)$_5$] (which can be obtained as dark red crystals) is a monomer in benzene solution whilst [U(NMe$_2$)$_6$] is only known in solution. [U(NMe$_2$)$_6$] would be expected to be octahedral and [U(NEt$_2$)$_5$] trigonal bipyramidal, but no structures are known. The most versatile of all these compounds is U(NEt$_2$)$_4$. This is a low-melting (36 °C) and thermally stable substance (but like all alkoxides and alkylamides, immediately attacked by air or water). It can be distilled at ~40 °C at 10^{-4} mmHg pressure and on account of this volatility was investigated for some time as a possible material for isotopic separation of uranium by gaseous diffusion. U(NEt$_2$)$_4$ is a useful starting material for the synthesis of other amides, alkoxides, and thiolates (Figure 11.9)

Its utility stems from the fact that alcohols and thiols are better proton donors (more acidic) than amines, thus:

$$U(NEt_2)_4 + 4 \; EtOH \rightarrow U(OEt)_4 + 4 \; Et_2NH$$

11.8.2 Alkoxides and Aryloxides

U(OR)$_n$ compounds (R = alkyl or aryl) exist in oxidation states between +3 and +6; in addition, there are some uranyl alkoxides. An unusual feature is the number of thermally stable compounds in the otherwise unstable and rare +5 state.

A few uranyl alkoxides have been made, such as golden yellow [UO$_2$(OCHPh$_2$)$_2$(thf)$_2$], but more important are the large number of octahedral U(OR)$_6$ compounds that can be made, examples being U(OMe)$_6$, U(OPr^i)$_6$, U(OBu^t)$_6$, U(OCF$_2$CF$_3$)$_6$, and U(OCH$_2$Bu^t)$_6$ (U–O 2.001–2.002 Å).

$$UF_6 + 6 \; NaOMe \rightarrow U(OMe)_6 + 6 \; NaF$$

These compounds are volatile *in vacuo*, U(OCF$_2$CF$_3$)$_6$ remarkably boiling at 25 °C at 10 mmHg pressure. U(OMe)$_6$, which sublimes at 30 °C at 10^{-5} mmHg pressure, was investigated as a candidate for IR laser photochemistry leading to uranium isotopic enrichment.

Unlike the monomeric U(OR)$_6$, U(OR)$_5$ are usually associated; U(OPr^i)$_5$ is dimeric with two alkoxide bridges giving six coordination. U(OEt)$_5$ is the easiest compound to synthesize and can be converted into others by alcohol exchange (Figure 11.10).

Uranium(IV) compounds tend to be non-volatile, highly associated solids, though the aryloxide U[O(2,6-Bu$_2^t$C$_6$H$_3$)]$_4$ is made of monomeric tetrahedral molecules, doubtless owing to its very bulky ligand. A number of U^{III} aryloxide species have been synthesized, examples including U[O(2,6-Bu$_2^t$C$_6$H$_3$)]$_3$ being shown in Figure 11.11.

Since four of the same aryloxide ligands can bind to uranium in the uranium(IV) compound, a three coordinate U[O(2,6-Bu$_2^t$C$_6$H$_3$)]$_3$ is presumably coordinatively unsaturated, and so forms a uranium-ring bond; coordinative saturation can also be achieved by forming an adduct with a Lewis base.

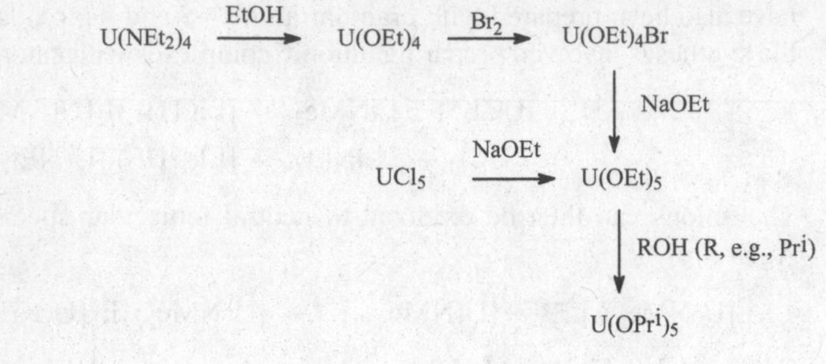

$$U(NEt_2)_4 \xrightarrow{EtOH} U(OEt)_4 \xrightarrow{Br_2} U(OEt)_4Br$$

$$\downarrow NaOEt$$

$$UCl_5 \xrightarrow{NaOEt} U(OEt)_5$$

$$\downarrow ROH \text{ (R, e.g., Pr}^i)$$

$$U(OPr^i)_5$$

Figure 11.10
Synthesis of uranium alkoxides.

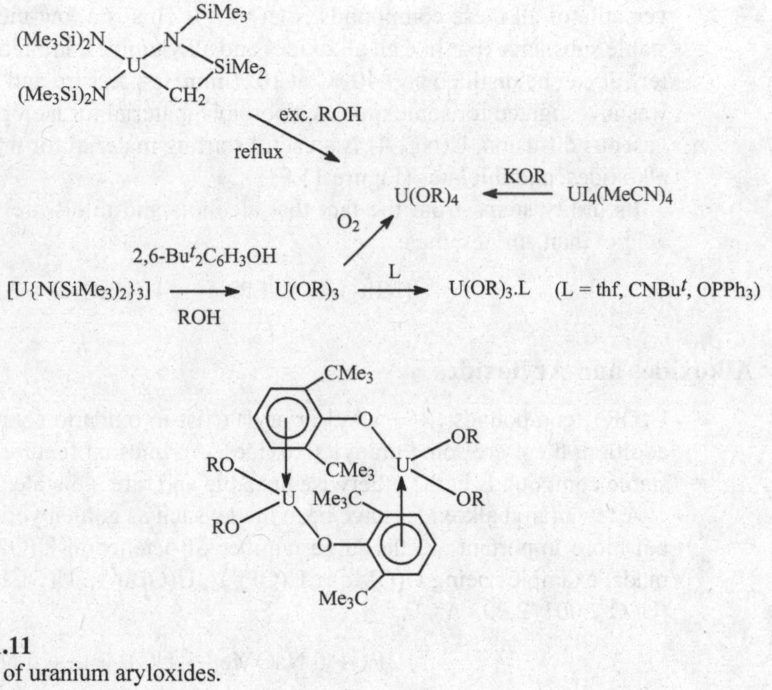

Figure 11.11
Synthesis of uranium aryloxides.

11.9   Chemistry of Actinium

Little is known about the chemistry of actinium. It strongly resembles the lanthanides, especially lanthanum, with identical reduction potentials ($Ac^{3+} + 3e^- \rightarrow Ac$; $E = -2.62$ V) and similar ionic radii ($La^{3+} = 106$ pm, $Ac^{3+} = 111$ pm). Its chemistry is dominated by the (+3) oxidation state; as expected for a f^0 ion, its compounds are colourless, with no absorption in the UV–visible region between 400 and 1000 nm. ^{227}Ac is strongly radioactive ($t_{1/2} = 21.77$ y) and so are its decay products. Most of the work carried out has been on the microgram scale, examining binary compounds such as the oxides and halides, and little is known about its complexes. Actinium metal itself is a silvery solid, obtained by reduction of the oxide, fluoride, or chloride with Group I metals; it is oxidized rapidly in moist air. Like lanthanum, it forms an insoluble fluoride (coprecipitating quantitatively with lanthanum on the tracer scale) and oxalate $Ac_2(C_2O_4)_3 \cdot 10H_2O$.

Since the chemistry of actinium is confined to the Ac^{3+} ion, it can readily be separated from thorium (and the lanthanides, for that matter) by processes like solvent extraction with thenoyltrifluoroacetone (TTFA) and by cation-exchange chromatography. The latter is an excellent means of purification, as the Ac^{3+} ion is much more strongly bound by the resin than its decay products.

11.10 Chemistry of Protactinium

Although its chemistry is not greatly studied at present, quite a lot of protactinium chemistry has been reported. The main isotope is the alpha-emitter ^{231}Pa, which has a long half-life ($t_{1/2} = 3.28 \times 10^4$ y) so few problems arise, once appropriate precautions are taken with the α-emission, the only other point of note being the ready hydrolysis of Pa^V in solution. Protactinium metal itself, formed by reduction (Ba) of PaF_4, or thermal decomposition of PaI_5 on a tungsten filament, is a high-melting (1565 °C), dense (15.37 g cm^{-3}), ductile silvery solid, which readily reacts on heating with many non-metals.

The aqueous chemistry is dominated by the readily hydrolysed Pa^{5+} ion, which in the absence of complexing ligands (e.g., fluoride) tends to precipitate as hydrated $Pa_2O_5.nH_2O$. A PaO_2^+ ion would be isoelectronic with the uranyl ion, UO_2^{2+}, but there is no evidence for it. However, solutions of Pa^V in fuming HNO_3 yield crystals of $PaO(NO_3)_3.x\ H_2O$ ($x = 1$–4), which may have Pa=O bonds; the sulfate $PaO(HSO_4)_3$ also exists. Reduction of Pa^V with zinc or Cr^{2+} gives Pa^{4+}(aq), only stable in strongly acidic solution, as, at higher pH, ions like $Pa(OH)_2^{2+}$, PaO^{2+}, and $Pa(OH)_3^+$ are believed to exist.

Many of the complexes of Pa that have been studied are halide complexes, falling into two types; anionic species with all-halide coordination, and neutral Lewis base adducts, usually of the chloride $PaCl_4$.

Historically, the fluoride complexes are most important. Aristid Von Grosse (1934) used K_2PaF_7 in his determination of the atomic mass of Pa in 1931. Colourless crystals of this and other complexes $MPaF_6$, M_2PaF_7, and M_3PaF_8 (M = K, Rb, Cs) can be obtained by changing the stoichiometry, e.g.:

$$2\ KF + PaF_5 \rightarrow K_2PaF_7 \quad \text{(in 17M HF, precipitate with propanone)}$$

This complex has tricapped trigonal prismatic nine coordination of Pa, whilst $MPaF_6$ has dodecahedral 8 coordination and Na_3PaF_8 the very rare cubic 8 coordination. Other halide complexes can be made by appropriate methods:

$$PaX_4 + 2\ R_4NX \rightarrow (R_4N)_2PaX_6 \quad (X = Cl,\ Br;\ R = Me,\ Et;\ \text{in MeCN})$$
$$PaCl_4 + 2\ CsCl \rightarrow Cs_2PaCl_6 \quad (\text{in } SOCl_2/ICl)$$
$$PaI_4 + 2\ MePh_3AsI \rightarrow (MePh_3As)_2PaI_6 \quad (\text{in MeCN})$$

Six-coordinate $[PaX_6]^{2-}$ ions have been confirmed by X-ray diffraction for $(Me_4N)_2PaX_6$ (X = Cl, Br) and Cs_2PaCl_6.

Many adducts of the tetrahalides of ligands like phosphine oxides have been synthesized by direct interaction of the tetrahalides with the ligands in solution in solvents like acetonitrile or propanone. These have similar stoichiometries and structures to analogous complexes of Th, U, Np, and Pu, examples being *trans*-$PaX_4[(Me_2N)_3PO]_2$ (X = Cl, Br); *cis*-$PaCl_4(Ph_3PO)_2$; $PaCl_4(MeCN)_4$; $PaCl_4(Me_2SO)_5$ {possibly $[PaCl_3(Me_2SO)_5]^+\ Cl^-$}

and $PaCl_4(Me_2SO)_3\{$possibly $[PaCl_2(Me_2SO)_6]^{2+}$ $[PaCl_6]^{2-}\}$. In addition, a few adducts of the pentahalides, $PaX_5(Ph_3PO)_n (n = 1,2)$ exist.

In addition to these, a number of other complexes have been isolated, such as the β-diketonates $[Pa(PhCOCHCOPh)_4]$ and antiprismatic $[Pa(MeCOCHCOMe)_4]$, which, like the TTFA complex $[Pa(C_4H_3SCOCHCOCF_3)_4]$, are soluble in solvents like benzene, and suitable for purification by solvent extraction. Pa forms a volatile borohydride $[Pa(BH_4)_4]$. These compounds are typical of those formed by other early actinides, such as Th and U.

11.11   Chemistry of Neptunium

The metal itself is a dense (19.5 g cm^{-3}) silvery solid, which readily undergoes oxidation in air. Chemical study uses one isotope, the long-lived ^{237}Np ($t_{1/2} = 2.14 \times 10^6$ y). The chemistry of neptunium shows interesting points of comparison between U and Pu, which flank it (and like them it shows a wide range of oxidation states). Thus, whilst the ($+6$) oxidation state is found in the $[NpO_2]^{2+}$ ion as well as in the halide NpF_6 (note that there is no $NpCl_6$, unlike U), the ($+6$) state is less stable than in the case of uranium, though it is more stable than Pu^{VI}. However, the ($+5$) state is more stable for neptunium than for uranium, with the $[NpO_2]^+$ ion showing no signs of disproportionation, though easily undergoing reduction by Fe^{2+} to Np^{4+}.

Np^{4+} is in many ways the most important oxidation state. It is formed by reduction of the higher oxidation states, and by aerial oxidation of Np^{3+}. Strong oxidizing agents like Ce^{4+} oxidize it back to $[NpO_2]^{2+}$, whilst electrolytic reduction of Np^{4+} affords Np^{3+}, which is stable in the absence of air (unlike U).

As neptunium has one more outer-shell electron than uranium, it has the possibility of a ($+7$) oxidation state, a possibility realized in alkaline solution, when ozone will oxidize Np^{VI} to Np^{VII}, an oxidation also achieved by XeO_3 or IO_4^- at higher temperatures. The potential for this is estimated as -1.24V (1M alkali). The relevant standard potentials for neptunium (1M acid) are:

$$Np^{3+} + 3 e^- \rightarrow Np \qquad\qquad\qquad\qquad\quad E = -1.79 \text{ V}$$
$$Np^{4+} + e^- \rightarrow Np^{3+} \qquad\qquad\qquad\qquad\quad E = -0.15 \text{ V}$$
$$NpO_2^+ + 4 H^+ + e^- \rightarrow Np^{4+} + 2 H_2O \qquad E = +0.67 \text{ V}$$
$$NpO_2^{2+} + e^- \rightarrow NpO_2^+ \qquad\qquad\qquad\quad\ E = +1.24 \text{ V}$$

11.11.1   Complexes of Neptunium

A significant amount of structural information has been gathered on neptunium complexes. For the main, they resemble corresponding U and Pu complexes. However, an EXAFS study of alkaline solutions of Np^{VII} found evidence for a *trans* dioxo ion of the type $[NpO_2(OH)_4(OH_2)]^{1-}$.

In the VI state, the 8 coordinate $Na[NpO_2(OAc)_3]$, $[NpO_2(NO_3)_2(H_2O)_2].4H_2O$, $K_4[NpO_2(CO_3)_3]$, and $Na_4[NpO_2(O_2)_3].9H_2O$ [where $(O_2)_3$ indicates three peroxide di-anion ligands] are isostructural with the U, Pu, and Am analogues. They demonstrate a contraction of ~ 0.01 Å in the M=O distance per unit increase in atomic number. Eight coordination is also found in $[NpO_2(NO_3)_2(bipy)]$. The aqua ion appears to be seven coordinate $[NpO_2(H_2O)_5]^{2+}$. Like the uranyl ion, the neptunyl ion complexes with expanded porphyrins like hexaphyrin, amethyrin, pentaphyrin, and alaskaphyrin.

In the Np^V state, structural comparison of $Ba[Np^VO_2(OAc)_3]$ with $Na[Np^{VI}O_2(OAc)_3]$ indicates a lengthening of 0.14 Å in the neptunium–oxygen bond length on going from Np^{VI} to Np^V, consistent with an electron added to an antibonding orbital. As with the VI state, the aqua ion is believed to be seven coordinate $[NpO_2(H_2O)_5]^+$; in the related $[NpO_2(urea)_5](NO_3)$ the neptunium atom has a typical pentagonal-bipyramidal environment with five oxygen atoms in the equatorial plane. A similar geometry is found in the acetamide complex of neptunium(V) nitrate, $[(NpO_2)(NO_3)(CH_3CONH_2)_2]$, where the pentagonal bipyramidal coordination of Np is completed (in the solid state) using oxygens of neighbouring NpO_2 groups; similarly in $NpO_2ClO_4.4H_2O$, the neptunyl(V) ion has four waters and a distant neptunyl oxygen occupying the equatorial positions. Eight and six coordination are, respectively, found in the crown ether complex, $[NpO_2(18\text{-crown-}6)]ClO_4$, and in $[NpO_2(OPPh_3)_4]ClO_4$. Carbonate complexes have been investigated as they could represent a means of leaching transactinides from underground deposits. EXAFS studies of Np^V carbonate complexes indicate the existence of $[NpO_2(H_2O)_3(CO_3)]^-$, $[NpO_2(H_2O)_2(CO_3)_2]^{3-}$ and $[NpO_2(CO_3)_3]^{5-}$; similar Np^{VI} species like $[NpO_2(CO_3)_3]^{4-}$ are indicated.

Among neptunium(IV) complexes, $[Np(S_2CNEt_2)_4]$ and $[Np(acac)_4]$ have dodecahedral and antiprismatic coordination of neptunium. A reminder of the small differences in energy between different geometries is that $[Me_4N]_4$ $[Np(NCS)_8]$ has tetragonal antiprismatic coordination, whilst $[Et_4N]_4$ $[Np(NCS)_8]$ has cubic coordination of Np. Neptunium(IV) is also eight coordinate in $[Np(urea)_8]SiW_{12}O_{40}.2Urea.11H_2O$.

As with Th, Pa, U, and Pu, a variety of neutral complexes are formed between the tetrahalides (and nitrate) and ligands like phosphine oxides, such as *trans*-$[NpX_4\{(Me_2N)_3PO\}_2]$ (X = Cl, Br); *cis*-$[NpCl_4(Ph_3PO)_2]$; $NpCl_4(Me_2SO)_3$ (possibly $[NpCl_2(Me_2SO)_6]^{2+}$ $[NpCl_6]^{2-}$); $[Np(NCS)_4(R_3PO)_2]$ (R = Ph, Me, Me_2N), and $[Np(NO_3)_4(Me_2SO)_3]$.

Only a few complexes of Np^{III} have been characterized, notably the dithiocarbamate Et_4N $[Np(S_2CNEt_2)_4]$, which has distorted dodecahedral coordination like the Pu and lanthanide analogues. $[NpI_3(thf)_4]$ and the silylamide $[Np\{N(SiMe_3)_2\}_3]$ are analogous to those of U and the lanthanides, the former potentially being a useful starting material. The amide reacts with a bulky phenol:

$$[Np\{N(SiMe_3)_2\}_3] + 3\ ArOH \rightarrow [Np(OAr)_3] + 3\ HN(SiMe_3)_2$$
$$(Ar = 2,6\text{-}Bu^t_2C_6H_3)$$

11.12 Chemistry of Plutonium

Plutonium presents particular problems in its study. One reason is that, since ^{239}Pu is a strong α-emitter ($t_{1/2} = 24,100$ years) and also tends to accumulate in bone and liver, it is a severe radiological poison and must be handled with extreme care. A further problem is that the accidental formation of a critical mass must be avoided.

11.12.1 Aqueous Chemistry

Because of the complicated redox chemistry, not all the oxidation states $+3$ to $+7$ are observed under all pH conditions. Blue Pu^{3+}(aq) resembles the corresponding Ln^{3+} ions whilst the brown Pu^{4+} ion requires strongly acidic conditions (6M) to prevent oligomerization and disproportionation. The purple-pink PuO_2^+ ion readily disproportionates, whilst the orange-yellow PuO_2^{2+} ion resembles the uranyl ion, but is less stable, tending to be reduced, not least by its own α-decay. Blue Pu^{VII}, possibly PuO_5^{3-}(aq) or some oxo/hydroxy species like $[PuO_2(OH)_4(OH_2)]^{3-}$, exists only at very high pH; it is formed by ozonolysis of Pu^{VI}, and some Pu^{VIII} may be generated.

11.12.2 The Stability of the Oxidation States of Plutonium

In comparison with uranium, the $(+3)$ state has become much more stable (Table 9.5), witness the standard reduction potential of $+0.98$ V for $Pu^{4+} + e \rightarrow Pu^{3+}$, the comparative value for uranium being -0.63 V. This means that quite strong oxidizing agents such as manganate(VII) are needed to effect this oxidation in the case of plutonium.

The relevant standard potentials for plutonium are:

$$Pu^{3+} + 3\,e^- \rightarrow Pu \qquad\qquad\qquad E = -2.00 \text{ V}$$
$$Pu^{4+} + e^- \rightarrow Pu^{3+} \qquad\qquad\qquad E = +0.98 \text{ V}$$
$$PuO_2^+ + 4\,H^+ + e^- \rightarrow Pu^{4+} + 2\,H_2O \qquad E = +1.04 \text{ V}$$
$$PuO_2^{2+} + e^- \rightarrow PuO_2^+ \qquad\qquad\qquad E = +0.94 \text{ V}$$

The Pu^{III}–Pu^{IV} and the Pu^V–Pu^{VI} couples are both reversible, but not the Pu^{IV}–Pu^V, as the latter involves the making and breaking of Pu=O bonds and significant changes in geometry (cf. uranium). Reactions involving the making and breaking of Pu=O bonds are also kinetically slow, so that it is possible for ions in all four oxidation states between $(+3)$ and $(+6)$ to coexist in aqueous solution under certain conditions. The even spacing of the potentials linking the four oxidation states means that disproportionation and reproportionation reactions are feasible. For example, a possible route for disproportion of Pu^{4+} in aqueous solution can be written:

$$2\,Pu^{4+} + 2\,H_2O \rightarrow Pu^{3+} + PuO_2^+ + 4\,H^+$$

This is composed of two redox processes:

$$Pu^{4+} + e^- \rightarrow Pu^{3+} \quad \text{and} \quad Pu^{4+} + 2\,H_2O \rightarrow PuO_2^+ + 4\,H^+ + e^-$$

Because the potentials are of roughly equal magnitude (and of opposite sign in this case), the free energy change is small (<20 kJmol^{-1}). The following process:

$$2\,PuO_2^+ + 4\,H^+ \rightarrow PuO_2^{2+} + Pu^{4+} + 2\,H_2O$$

is also favoured; overall, the two can be combined as:

$$3\,Pu^{4+} + 2\,H_2O \rightarrow 2\,Pu^{3+} + PuO_2^{2+} + 4\,H^+.$$

Another feasible reproportionation reaction is of course:

$$PuO_2^+ + Pu^{3+} + 4\,H^+ \rightarrow 2\,Pu^{4+} + 2\,H_2O$$

A further problem is that compounds in the $+5$ and $+6$ oxidation states tend to be reduced (autoradiolysis), as ^{239}Pu is a strong α-emitter (1 mg ^{239}Pu emits over a million α-particles a second) decomposing water molecules into $^\bullet$H, $^\bullet$OH and $^\bullet$O radicals which, in turn, participate in redox reactions. In acidic solution, decomposition of the solvent to H_2O_2 and the acid can occur. In nitric acid, for example, both HNO_2 and nitrogen oxides are formed, so that, starting with plutonium(VI), ions in the lower oxidation states are generated

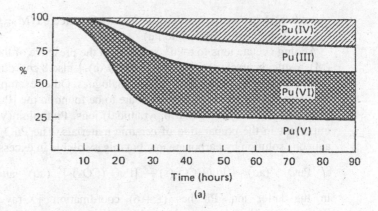

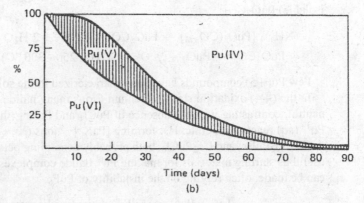

Figure 11.12
(a) Disproportionation of Pu^V in 0.1M HNO_3 + 0.2 M $NaNO_3$. (b) Self-reduction of Pu^{VI} due to its own α-emission, in 5M $NaNO_3$ (after P.I. Artiukhin, V.I. Medicedovskii, and A.D. Gel'man, *Radiokhimiya*, 1959, **1** 131; *Zh. Neorg. Khim.*, 1959, **4**, 1324) (permission applied for reproduction).

by both the reduction and disproportionation/reproportionation reactions described above. Figure 11.12 shows the effects of these processes. In Figure 11.12(a), the disproportionation of Pu^V into all the (+3) to (+6) states can be seen, whilst in Figure 11.12(b), over a longer time scale, the self-reduction of Pu^{VI} to Pu^{IV} is observed.

In perchloric acid, the (+4) state is less abundant, and the (+3) state more abundant, a reflection of the higher oxidizing power of nitric acid.

11.12.3 Coordination Chemistry of Plutonium

In its coordination chemistry, plutonium(VI) resembles uranium, where corresponding complexes exist. Thus crystallization of solutions of Pu^{VI} in concentrated nitric acid affords red-brown to purple crystals of the hexahydrate $PuO_2(NO_3)_2.6H_2O$, which is isostructural with the U analogue, and contains 8-coordinate $[PuO_2(NO_3)_2(H_2O)_2]$ molecules. An X-ray absorption spectroscopy study of tributyl phosphate complexes of $[AnO_2(NO_3)_2]$ (An = U, Np, Pu) in solution, probably present as $[AnO_2(NO_3)_2(tbp)_2]$, has indicated major changes in the actinide's coordination sphere on reduction to the An^{IV} state, but no significant changes across the series UO_2^{2+}, NpO_2^{2+}, PuO_2^{2+}. An actinide contraction of about 0.02 Å between successive actinides was detected (M = O changes overall from 1.79 to 1.75 Å on passing from U to Pu).

Adding acetate ions to a Pu^{VI} solution, in the presence of the appropriate Group I metal (M), results in precipitation of M [PuO$_2$(OAc)$_3$], also 8 coordinate, with bidentate acetates, isostructural with the U, Np, and Am analogues. Other examples of polyhedra resembling the corresponding uranyl complexes are to be found in the [PuO$_2$Cl$_4$]$^{2-}$(*trans*-octahedral) and [PuO$_2$F$_5$]$^{3-}$ (pentagonal bipyramidal) ions. Plutonium(VI) carbonate complexes are important in the preparation of ceramic materials. The PuO$_2{}^{2+}$ ion is precipitated from aqueous solution by carbonate ion, but this is soluble in excess of carbonate.

$$PuO_2{}^{2+}(aq) \rightarrow PuO_2CO_3(s) \rightarrow [PuO_2(CO_3)_2]^{2-}(aq) \quad and \quad [PuO_2(CO_3)_3]^{4-}(aq)$$

In the latter ion, Pu has (2 + 6) coordination {X-ray diffraction of [C(NH$_2$)$_4$] [PuO$_2$(CO$_3$)$_3$]}. The ammonium salt undergoes two-stage thermal decomposition, via (the isolable) PuO$_2$CO$_3$:

$$(NH_4)_4 [PuO_2(CO_3)_3] \rightarrow PuO_2CO_3 + 4 NH_3 + 2 H_2O + 2 CO_2 \text{ (at 110–150 °C)}$$
$$PuO_2CO_3 \rightarrow PuO_2 + \tfrac{1}{2} O_2 + CO_2 \text{ (at 250–300 °C)}$$

Few Pu(+5) compounds have been characterized in the solid state.

In the (+4) oxidation state, plutonium forms many halide complexes, both anionic and neutral, contrasting with the absence of PuCl$_4$ and PuBr$_4$ (though no iodides are known). Pu^{4+}(aq) reacts with conc. HX forming [PaX$_6$]$^{2-}$ ions (X = Cl, Br) isolable as salts such as (Et$_4$N)$_2$PuBr$_6$ and Cs$_2$PuCl$_6$, both probably containing octahedrally coordinated Pu, and useful as starting materials for making Pu^{IV} halide complexes. Various fluoride complexes can be made, often relying on the instability of PuF$_6$:

$$LiF + PuF_6 \rightarrow LiPuF_5 \text{ (simplified equation)} \quad \text{(at 300 °C)}$$
$$LiPuF_5 \rightarrow Li_4PuF_8 \quad \text{(at 400 °C)}$$
$$CsF + PuF_6 \rightarrow Cs_2PuF_6 \quad \text{(at 300 °C)}$$
$$NH_4F + PuF_6 \rightarrow NH_4PuF_5 + (NH_4)_2PuF_6 \quad \text{(at 70–100 °C)}$$

High coordination numbers are usual in fluoride complexes; (NH$_4$)$_4$PuF$_8$ has nine-coordinate Pu.

Neutral halide complexes are usually got starting from [PuX$_6$]$^{2-}$ salts (X = Cl, Br), though bromides can also be made by bromine oxidation of Pu^{III} species.

$$Cs_2PuCl_6 + 2 Ph_3PO \rightarrow cis\text{-}PuCl_4(Ph_3PO)_2 + 2 CsCl \quad \text{(in MeCN)}$$
$$PuBr_3 + \tfrac{1}{2} Br_2 + 2 Ph_3PO \rightarrow trans\text{-}PuBr_4(Ph_3PO)_2 \quad \text{(in MeCN)}$$

A wide range of these complexes, also involving nitrate and thiocyanate ligands, has been made:

$$Cs_2Pu(NO_3)_6 + 2 Ph_3PO \rightarrow Pu(NO_3)_4(Ph_3O)_2 + 2 CsNO_3$$
$$PuCl_4(R_3PO)_2 + 2 R_3PO + 4 KNCS \rightarrow Pu(NCS)_4(R_3PO)_4 + 4 KCl$$
$$(R, \text{ e.g., Ph, Me}_2N)$$

Others include *trans*-PuX$_4$[(Me$_2$N)$_3$PO]$_2$ (X = Cl, Br); PuCl$_4$(Ph$_2$SO)$_n$ (n = 3, 4); PuCl$_4$(Me$_2$SO)$_n$ (n = 3, 7); PuCl$_4$(Me$_3$PO)$_6$ {probably [PuCl(Me$_3$PO)$_6$]$^{3+}$ (Cl$^-$)$_3$}. Structurally, they resemble the corresponding complexes of Th, Pa, U, and Np. The tridentate ligand 2,6-[Ph$_2$P(O)CH$_2$)]$_2$C$_5$H$_3$NO combines two phosphine oxide and one N-oxide functional groups; it forms a 2:1 Pu^{IV} complex which has the same structure, [PuL$_2$(NO$_3$)$_2$] (NO$_3$)$_2$, in both the solid state and solution. The 2:1 complex of the bidentate ligand 2-[Ph$_2$P(O)CH$_2$]C$_5$H$_4$NO with plutonium nitrate also has an ionic structure [PuL$_2$(NO$_3$)$_3$]$_2$ [Pu(NO$_3$)$_6$]. As is the case with uranium, nitrate complexes are important in the separation chemistry of plutonium. Solutions of Pu^{4+}(aq) in conc. HNO$_3$ deposit dark green crystals of Pu(NO$_3$)$_4$.5H$_2$O, which contain 11-coordinate [Pu(NO$_3$)$_4$(H$_2$O)$_3$] molecules (as is the case with Th). [Pu(NO$_3$)$_6$]$^{2-}$ ions can be fished out of very concentrated (10–14M) nitric acid solutions as salts R$_2$Pu(NO$_3$)$_6$(R = Cs, Et$_4$N). [Pu(NO$_3$)$_6$]$^{2-}$ is strongly adsorbed by anion-exchange resins, and this is made use of in the purification of Pu commercially.

Plutonium carbonate complexes are important as they contribute to an understanding of what happens to plutonium in the environment, Pu(+4) being the most stable oxidation state under normal conditions. [Pu(CO$_3$)$_5$]$^{6-}$ has been identified by EXAFS as the Pu^{IV} species in solution at high carbonate concentration; the structure of crystalline [Na$_6$Pu(CO$_3$)$_5$]$_2$.Na$_2$CO$_3$.33H$_2$O has also been determined; it features 10-coordinate plutonium. The sulfate complex K$_4$Pu(SO$_4$)$_4$.2H$_2$O contains dimeric [(SO$_4$)$_3$Pu(μ-SO$_4$)$_2$Pu(SO$_4$)$_3$]$^{8-}$ anions with nine-coordinate Pu.

In many aspects of its coordination chemistry, Pu^{IV} compounds resemble their uranium analogues; thus it forms eight-coordinate complexes with diketonate ligands, like [Pu(acac)$_4$], and similar complexes with 8-hydroxyquinolinate and tropolonate. There is a limited alkoxide chemistry, [Pu(OBu^t)$_4$] being volatile at 112 °C (0.05 mmHg pressure) and very likely being a monomer. The borohydride [Pu(BH$_4$)$_4$] is a blue-black volatile liquid (mp 15 °C) and has a 12 coordinate molecular structure, with tridentate borohydride ligands, like the neptunium analogue, but unlike 14 coordinate [U(BH$_4$)$_4$].

Little coordination chemistry is as yet known in the (+3) state, but these compounds are interesting. Reaction of Pu with iodine in THF affords off-white [PuI$_3$(thf)$_4$], a useful starting material; [PuI$_3$(dmso)$_4$] and [PuI$_3$(py)$_4$] have also been reported. The σ-alkyl [Pu{CH(SiMe$_3$)$_2$}$_3$] and the silylamide [Pu{N(SiMe$_3$)$_2$}$_3$] are analogous to those of U and the lanthanides, the latter potentially being a useful starting material (though neither has been completely characterized). The amide reacts with a bulky phenol forming the aryloxide [Pu(OAr)$_3$] (Ar = 2,6-Bu$^t_2$C$_6$H$_3$), which is probably three coordinate. Plutonium forms insoluble oxalates Pu$_2$(C$_2$O$_4$)$_3$.x H$_2$O ($x = 10$, 11) similar to those of the lanthanides.

Reaction of plutonium metal with triflic acid affords blue Pu(CF$_3$SO$_3$)$_3$.9H$_2$O, isostructural with the lanthanide triflates and which contains tricapped trigonal prismatic [Pu(H$_2$O)$_9$]$^{3+}$ ions; similarly, Pu reacts with AgPF$_6$ (or TlPF$_6$) in MeCN suspension, dissolving to form [Pu(MeCN)$_9$] (PF$_6$)$_3$, in which the coordination polyhedron is rather more distorted.

11.12.4 Plutonium in the Environment

Actinides are potentially present in the environment from a number of sources, not just the 'natural' thorium and uranium. Nuclear power plants and uses in nuclear weapons are obvious areas for concern, raised in recent times by possibilities that terrorist groups could gain control of such weapons, possibly 'dirty' bombs. Plutonium is an obvious source of concern. Any plutonium in the environment is believed to be largely present as rather insoluble Pu^{IV} species. Pu^{IV} ions are hydrolysed easily, unless in very acidic solution, forming light green colloidal species, which age with time, their solubility decreasing. The possibility of colloid-facilitated transport of plutonium occurring was raised by the observation of migration of plutonium in ground water by rather more than a mile from the location of underground weapons testing in Nevada, USA, over a period greater than 20 years. It is therefore necessary to consider the interaction of plutonium, not just with inorganic ions like carbonate (see Section 11.12.3), phosphate, silicate and sulfate, but with natural organic substances like humic acid.

Disposal of nuclear waste materials – especially the high-level waste from the cores of reactors and nuclear weapons – so that they do not escape into the environment is a pressing problem. Currently the preferred solution in the USA and most of Europe is immobilization in a suitable geological repository. The waste materials are evaporated if necessary, then heated with silica and borax to form a borosilicate glass, encased in an inert (e.g., lead) material, then buried in a stable zone, such as rock salt or clay. The conditions surrounding such sites are important. Yucca Mountain, Nevada, is the leading candidate among possible American sites. Solubility studies have shown that Np is more than 1000 times more soluble in the Yucca Mountain waters than is Pu, because under the ambient conditions Pu tends to adopt the (+4) state, associated with insolubility, whilst Np adopts the (+5) state, where compounds are much more soluble. However, at the WIPP site in New Mexico, which is built on a deep salt formation, the very salty brines favour formation of soluble Pu(+6) complexes such as $[PuO_2Cl_4]^{2-}$, which will present much more of a problem unless reducing conditions can be generated.

As a potential environmental hazard, plutonium is particularly toxic because of the remarkable similarity of Pu^{IV} and Fe^{III} – which means that Pu can be taken up in the biological iron transport and storage system of mammals. This similarity is being used in a biomimetic approach to Pu-specific ligands for both *in-vivo* actinide decorporation and nuclear waste remediation agents. It has also been suggested that plutonium could be solubilized by microbial siderophores. These are ligands used by bacteria to bind Fe^{3+} (an ion which, like Pu^{4+}, is essentially 'insoluble' at near-neutral pH) and carry it into cells. In fact, such siderophores bind plutonium strongly (log $\beta = 30.8$ for the desferrioxamine B complex), only taking it up as Pu^{IV} with plutonium in other oxidation states being oxidized or reduced appropriately. They do not solubilize plutonium rapidly, however; they are, in fact, less effective in solubilizing plutonium than are simple complexing agents like EDTA and citrate, leading to the suggestion that the siderophores passivate the surface of the plutonium hydroxide. (The overall formation constant for [Pu-EDTA] in acidic solution is log $\beta = 26.44$; the strength of the complexation means that the possibility of EDTA, widely used in former nuclear weapons programmes, solubilizing and transporting plutonium is a present concern).

Unlike the exclusively 6-coordinate iron(III) siderophore complexes, higher coordination numbers are possible with plutonium; the complex of Pu^{IV} with desferrioxamine E (DFE; a hexadentate iron-binding siderophore ligand) has shown it to contain 9-coordinate $[Pu(dfe)(H_2O)_3]^+$ ions with a tricapped trigonal prismatic geometry (Figure 11.13).

The microbe *Mycena flavescens* has been found to take up Pu (but not U) in the form of a siderophore complex though at a much slower rate than it takes up iron; the Fe and Pu complexes inhibit each other, indicating competition for the same binding site on the microbe. It has similarly been found that pyoverdin, the main siderophore in iron-gathering capacity produced by *Pseudomonas aeruginosa*, will also transport plutonium across cell membranes. It has been suggested that using phytosiderophores such as desferrioxamine and mugineic acids could solubilize actinides and promote their uptake by plants and consequent removal from the soil.

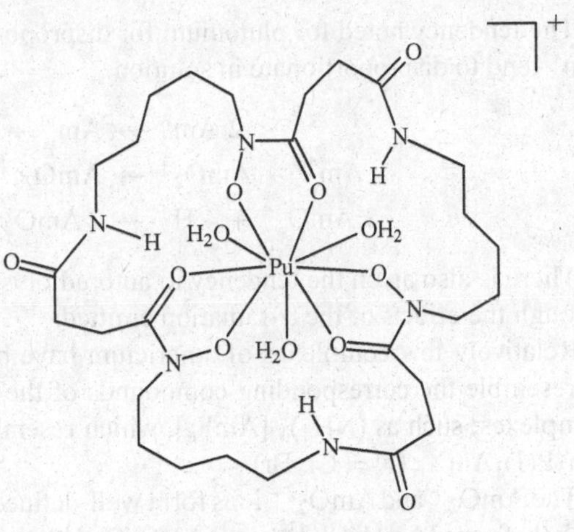

Figure 11.13
Structure of a Pu^{IV} complex of a siderophore.

11.13 Chemistry of Americium and Subsequent Actinides

Two isotopes, ^{241}Am ($t_{1/2} = 433$ y), a decay product of ^{241}Pu, and ^{243}Am ($t_{1/2} = 7380$ y), have half-lives suitable for their use in chemical reactions. The metal itself is a silvery, ductile and malleable solid, which is obtained by metallothermic (Ba, Li) reduction of the trifluoride; alternatively by heating the dioxide with lanthanum and using the difference in boiling point between La (bp 3457 °C) and americium (2607 °C) to displace the equilibrium to the right as americium distils.

$$3 \, AmO_2 + 4 \, La \rightleftharpoons 2 \, La_2O_3 + 3 \, Am$$

Americium undergoes slow oxidation in air and dissolves in dil. HCl, as expected from its favourable potential [$E(Am/Am^{3+}) = -2.07$ V].

11.13.1 Potentials

The relevant standard potentials for americium (1M acid) are:

$$
\begin{array}{ll}
Am^{3+} + 3e^- \rightarrow Am & E = -2.07 \, V \\
Am^{4+} + e^- \rightarrow Am^{3+} & E = +2.3 \, V \\
AmO_2{}^+ + 4H^+ + e^- \rightarrow Am^{4+} + 2 H_2O & E = +0.82 \, V \\
AmO_2{}^{2+} + e^- \rightarrow AmO_2^+ & E = +1.60 \, V
\end{array}
$$

The tendency for the (+3) state to become more stable in the sequence U < Np < Pu continues with americium (and dihalides, AmX$_2$, make their appearance for the first time). The pink Am^{3+} ion is thus the most important species in aqueous solution, Am^{IV} being unstable in the absence of complexing agents. Hypochlorite oxidizes Am^{3+} in alkaline solution to what may be Am(OH)$_4$, soluble in NH$_4$F solution, very likely as the fluoride complex [AmF$_8$]$^{4-}$. The (+5) state is accessible, again through oxidation in alkaline solution (e.g., with O$_3$ or peroxydisulfate, S$_2$O$_8{}^{2-}$), as the AmO$_2{}^+$ ion, though this tends to disproportionate to Am^{III} and Am^{V}; the (+6) state, in the form of the AmO$_2{}^{2+}$ ion, is obtained by oxidation (Ag^{2+} or S$_2$O$_8{}^{2-}$) of lower oxidation states in acid solution.

The tendency noted for plutonium for disproportionation occurs here, as both Am^{IV} and Am^V tend to disproportionate in solution:

$$2\,Am^{IV} \rightarrow Am^{III} + Am^V$$
$$Am^{4+} + AmO_2^+ \rightarrow AmO_2^{2+} + Am^{3+}$$
$$3\,AmO_2^+ + 4\,H^+ \rightarrow 2\,AmO_2^{2+} + Am^{3+} + 2\,H_2O$$

There is also again the tendency to autoradiolysis, the reduction of high oxidation states through the effects of the α-radiation emitted.

Relatively few complexes of americium have been characterized; those that have tend to resemble the corresponding compounds of the three previous metals. Many are halide complexes, such as $(NH_4)_4\,[AmF_8]$, which resembles the U analogue, $Cs_2NaAmCl_6$, and $(Ph_3PH)_3AmX_6$ ($X = Cl, Br$).

The AmO_2^+ and AmO_2^{2+} ions form well-defined complexes; thus in HCl solution, where it likely forms $[AmO_2Cl_4]^{3-}$ and $[AmO_2Cl_4]^{2-}$, the symmetric $Am=O$ stretching vibrations can be detected at 730 (AmO_2^+) and 796 cm^{-1} (AmO_2^+), respectively. $Na[AmO_2(OAc)_3]$ has eight coordination with bidentate acetates, just like the U–Pu analogues, the same coordination number also being found in AmO_2F_2, $M\,[AmO_2F_2]$ ($M = Rb, K$), and $M\,[AmO_2CO_3]$ ($M = Rb, K, Cs$). Six coordination occurs in $Cs_2[AmO_2Cl_4]$ and in $NH_4[AmO_2PO_4]$.

A significant chemistry occurs in the (+3) state, similar to that of Ln^{3+} ions, with studies often made in the context of separating Am from lanthanide fission products. Amide ligands like N, N, N', N'-tetraethylmalonamide (TEMA) have been investigated as possible extractants. Solution studies are assisted by some of the newer spectroscopic techniques; thus EXAFS (Extended X-ray Absorption Fine Structure) studies on solutions of $[Am(TEMA)_2(NO_3)_3]$ indicate a similar geometry to $[Nd(TEMA)_2(NO_3)_3]$. Carboxylate-derived calix[4]arenes show high selectivity for Am^{3+}, whilst complexation of Am^{3+} by crown ethers and diazacrown ethers has also been studied. The resemblance extends to complexes isolated. The sulfate $Am_2(SO_4)_3.8H_2O$ is isomorphous with the 8-coordinate lanthanide analogies (Pr–Sm) and insoluble oxalates $Am_2(C_2O_4)_3.x\,H_2O$ ($x = 7, 11$) have been characterized. The resemblance extends to diketonate complexes such as $[\{Am(Me_3CCOCHCOCMe_3)_3\}_2]$, a dimer with seven-coordinate Am, like the Pr analogue. The adduct $[Am(CF_3CCOCHCOCCF_3)_3\{(BuO)_3PO\}_2]$ is volatile at 175 °C and potentially could be used in separations. In the iodate complex $K_3Am_3(IO_3)_{12}\cdot HIO_3$, the $[AmO_8]$ polyhedra are made of eight $[IO_3]$ oxygen atoms in a distorted bicapped trigonal prismatic array. There is additionally one very long Am–O contact to complete a distorted tricapped trigonal prismatic Am coordination sphere.

11.14 Chemistry of the Later Actinides

The succeeding actinides (Cm, Bk, Cf, Es, Fm, Md, No, Lr) mark the point where the list of isolated compounds tends to involve binary compounds (oxides, halides and halide complexes, chalcogenides, and pnictides) rather than complexes. Those studies of complexes that have been made are usually carried out in solution and, from Fm, onwards, have been tracer studies.

Fermium coprecipitates with lanthanide fluorides and hydroxides, showing it to form lanthanide-like Fm^{3+} ions. These elute from cation-exchange resins slightly before Es^{3+}, whilst its chloride and thiocyanate complexes are eluted from anionic exchange resins just *after* Es^{3+}. This shows that Fm^{3+} forms slightly stronger complexes than does Es^{3+}, as expected purely on electrostatic considerations for a slightly smaller ion. In the case of nobelium, slightly different results have been obtained. Nobelium ions coprecipitate with BaF_2, rather than with LaF_3, suggesting that they are present as No^{2+} ions, but, after adding Ce^{4+} oxidant to the initial solution, nobelium precipitated with LaF_3, indicating that the cerium had oxidized the nobelium to No^{3+} ions. Complexing studies using Cl^- ions also indicated alkaline-earth-like behaviour.

Question 11.1 Use the reduction potentials to show why the $UO_2^+(aq)$ tends to disproportionate.

$$UO_2^+(aq) + 4 H^+(aq) + e^- \rightarrow U^{4+}(aq) + 2 H_2O(l) \quad E = +0.62\,V$$
$$UO_2^{2+}(aq) + e^- \rightarrow UO_2^+(aq) \quad E = +0.06\,V$$

Answer 11.1 Combining the potentials shows that for

$$2\,UO_2^+(aq) + 4 H^+(aq) \rightarrow U^{4+}(aq) + UO_2^{2+}(aq) + 2 H_2O(l) \quad E = +0.56\,V$$

Since E is positive, the reaction is energetically feasible; although this only predicts that the UO_2^+ (aq) ion is *thermodynamically unstable* with respect to disproportion and says nothing about *kinetic stability*, the fact is that the uranium(V) aqua ion is very short-lived.

Question 11.2 Write the expression for K_1 and K_2 for complex formation between Th^{4+} and F^- ions.
Answer 11.2

$$K_1 = [ThF^{3+}(aq)]/[Th^{4+}(aq)][F^-(aq)]; K_2 = [ThF_2^{2+}(aq)]/[ThF^{3+}(aq)][F^-(aq)]$$

Question 11.3 The highest occupied orbitals in the uranyl ion have the electronic arrangement $\sigma_u^2\, \sigma_g^2\, \pi_u^4\, \pi_g^4$. By referring to Figure 11.2, suggest the electronic arrangements similarly of UO_2^+ and PuO_2^{2+}.
Answer 11.3 UO_2^+ is $\sigma_u^2\, \sigma_g^2\, \pi_u^4\, \pi_g^4\, (\delta_u,\ \phi_u)^1$ and PuO_2^{2+} is $\sigma_u^2\, \sigma_g^2\, \pi_u^4\, \pi_g^4$ $(\delta_u,\ \phi_u)^2$ (The relative positions of δ_u and ϕ_u are uncertain).

Question 11.4 Suggest reasons why actinides from Cm onwards do not appear to form similar ions like CmO_2^{2+}.
Answer 11.4 Heavier actinides such as Cm do not display high oxidation states as the f orbitals appear to be more contracted and their electrons are not available for bonding, (and could be required for π bonding, for example). A further factor is that the additional electrons would need to be placed in antibonding orbitals such as σ_u^* and π_u^*, which would destabilize such an ion.

Question 11.5 Suggest coordination numbers for each of the following complexes.
(a) $[UO_2F_5]^{3-}$; (b) $[UO_2Br_4]^{2-}$; (c) $[UO_2(NO_3)_3]^-$; (d) $[UO_2(NCS)_5]^{3-}$; (e) $[UO_2(CH_3COO)_3]^-$; (f) $[UO_2(NO_3)_2(H_2O)_2]$; (g) $[UO_2(NO_3)_2(Ph_3PO)_2]$; (h) $[UO_2(CH_3COO)_2(Ph_3PO)]$.
Answer 11.5 (a) 7; (b) 6; (c) 8 (nitrate is bidentate) (d) 7; (e) 8 (acetate is bidentate) (f) 8; (g) 8; (h) 7.

Question 11.6 Explain why you can get $2 + 6$ coordination in $[UO_2(OH_2)_2(O_2NO)_2]$, $[UO_2(O_2NO)_3]^-$, etc. rather than the $2 + 5$ coordination in the uranyl aqua ion.

Answer 11.6 Although bidentate in most complexes, nitrate groups take up little space (they are said to have a small 'bite angle').

Question 11.7 Read section 11.5.5 and explain why Pu^{4+} can be reduced by Fe^{2+}, but UO_2^{2+} can't be so reduced.

Answer 11.7 $Fe^{3+}(aq)$ has a more negative reduction potential than $Pu^{4+}(aq)$, making Fe^{2+} a better reducing agent, one that will reduce $Pu^4(aq)$. Thus for:

$$Pu^{4+}(aq) + Fe^{2+}(aq) \rightarrow Pu^{3+}(aq) + Fe^{3+}(aq) \qquad E = +0.23 \text{ V}$$

and the process is energetically feasible. However, for:

$$UO_2^{2+}(aq) + 4H^+(aq) + 2Fe^{2+}(aq) \rightarrow U^{4+}(aq) + 2H_2O\ (l) + 2Fe^{3+}(aq)$$
$$E = -0.50 \text{ V}$$

and the reduction is not energetically feasible under these conditions.

Question 11.8 The infrared spectrum of $Th(NO_3)_4.4\,Me_3PO$ displays bands due to bidentate nitrate and ionic nitrate groups. Suggest a possible structure affording a reasonable coordination number.

Answer 11.8 Phosphine oxide ligands and bidentate nitrate groups take up different amounts of space round thorium. Assuming that all the phosphine oxide groups are bound to thorium, reference to Table 11.5 suggests that a coordination number of 10 is likely; four Me_3PO groups bound would leave room for three bidentate nitrates. The suggested structure is $[Th(NO_3)_3(Me_3PO)_4]^+NO_3^-$.

Question 11.9 (Refer to section 11.6.2 in connection with this) Hexamethylphosphoramide (HMPA), $OP(NMe_2)_3$, is a good σ-donor, behaving like a typical monodentate phosphine oxide ligand such as Ph_3PO. A number of complexes of HMPA with uranium(IV) nitrate have been prepared. Green crystals of $U(NO_3)_4(hmpa)_2$ are made thus:

$$Cs_2(NO_3)_6 + 2\ HMPA\ (acetone) \rightarrow 2\ CsNO_3 + U(NO_3)_4(hmpa)_2$$

The IR spectrum of this compound shows only one type of nitrate group; all are coordinated.

Question A How are the nitrates likely to be bound in $U(NO_3)_4(hmpa)_2$? Suggest a coordination number and coordination geometry for uranium.

Answer A By analogy with $U(NO_3)_4(OPPh_3)_2$, a 10-coordinate compound with a *trans* geometry and bidentate nitrates is probable (*this is known to be the case*).

Using excess of HMPA, a green compound $U(NO_3)_4(hmpa)_4$ is obtained. Its IR spectrum shows two different modes of nitrate coordination, and no ionic nitrate.

Question B Suggest a credible structure for the compound $U(NO_3)_4(hmpa)_4$, indicating the coordination number of uranium.

Answer B Various structures are possible. It is unlikely that all nitrates are bidentate, as this product would be 12 coordinate. If one nitrate is monodentate, the compound would be 11 coordinate $U(O_2NO)_3(ONO_2)(hmpa)_4$.

If $U(NO_3)_4(hmpa)_4$ is dissolved in non-polar solvents such as CH_3CN, a non-conducting solution is obtained, but with a simpler IR spectrum possibly indicating just one type of nitrate coordination.

Question C Suggest a structure for the compound present in solution.
Answer C One alternative is an 11-coordinate complex $U(NO_3)_4(hmpa)_3$ formed by dissociation of one hmpa ligand.

When $U(NO_3)_4(hmpa)_4$ is treated with $NaBPh_4$, a crystalline solid is obtained which has the analysis $U(NO_3)_3(hmpa)_4BPh_4$. This is a 1:1 electrolyte in solution. Its IR spectrum shows only one type of nitrate group, and no ionic nitrate.

Question D Suggest a credible structure for this compound, indicating the coordination number of uranium.
Answer D Assuming all nitrates are bidentate and that the BPh_4^- group remains ionic, a structure $[U(NO_3)_3(hmpa)_4]^+(BPh_4)^-$ would contain 10-coordinate uranium.

Question 11.10 Comment on the patterns in bond length in Table 11.6.
Answer 11.10 The U–X distances increase with increasing radius of halogen, as expected. There appears to be a slight decrease in U–O distance as the σ-donor power of the halogen decreases, possibly because this means that the ligands are competing for metal orbitals.

Question 11.11 Complete these equations:

$$UCl\{N(SiMe_3)_2\}_3 + LiCH_3 \rightarrow$$
$$UCl\{N(SiMe_3)_2\}_3 + NaBH_4 \rightarrow$$

Answer 11.11

$$UCl\{N(SiMe_3)_2\}_3 + LiCH_3 \rightarrow U(CH_3)\{N(SiMe_3)_2\}_3 + LiCl$$
$$UCl\{N(SiMe_3)_2\}_3 + NaBH_4 \rightarrow U(BH_4)\{N(SiMe_3)_2\}_3 + NaCl.$$

12 Electronic and Magnetic Properties of the Actinides

By the end of this chapter you should be able to:

- understand that the Russell–Saunders coupling scheme is not a good approximation here;
- recognize that the electronic spectra of compounds in the +3 and +4 states are dominated by f–f transitions;
- know that transitions are more sensitive to ligand than with the 4f metals;
- appreciate that interpretation of spectroscopic and magnetic properties is more difficult than for the lanthanides.

12.1 Introduction

In general, it is more difficult to interpret the spectra and magnetic behaviour of actinide compounds than those of lanthanide compounds, so this chapter will provide a qualitative and somewhat superficial discussion of the phenomena rather than a qualitative one, concentrating largely upon uranium compounds. The reasons for this is that spin–orbit coupling plays a more important part in actinide chemistry as the 5f orbitals and their electrons are not so 'core-like' as the 4f, particularly in the early part of the actinide series. This means that the Russell–Saunders (RS) coupling scheme, which treats spin–orbit coupling as being much weaker than interelectronic repulsion terms, is not applicable in most cases. Neither, however, can one usually apply the other extreme, the j–j coupling scheme, which relies on spin–orbit coupling being strong compared with electrostatic repulsion. Thus the 'intermediate' coupling scheme (intermediate between RS and jj) is used. It will be recalled that the f–f electronic transitions in the spectra of lanthanide complexes are relatively weak in comparison with those of transition metal complexes. However, in the case of the actinides, the 5f orbitals are larger than lanthanide 4f orbitals, so that they interact more with ligand orbitals, causing much higher extinction coefficients and also, because covalency is greater, to create greater nephelauxetic effects in actinide spectra. This means that there is more variation in both position and intensity of absorption bands than in lanthanide compounds. The 'forbidden' electronic dipole transitions are allowed in the presence of an asymmetric ligand field, which can arise by either a permanent distortion or by temporary coupling with an asymmetric metal–ligand vibration (vibronic coupling). Apart from the f–f transitions, there are two more types of absorption bands to note in actinide spectra. In general, the parity-allowed 5f–6d transitions occur above $20\,000\ cm^{-1}$, since the 6d levels are considerably above 5f for most actinides; these are more intense (and broader) than the f–f transitions. In the case of the free U^{3+} ion, the $5f^2 6d^1$ level is over 30 000 cm^{-1} above the $5f^3$ ground state, while in U^{3+}(aq) the charge-transfer transitions start around 24 000 cm^{-1}; solvation thus has a very significant effect upon the relative energies of the 5f and 6d electrons. Metal–ligand charge-transfer transitions have their maxima out in the ultraviolet, but, as with transition metals, the tail of these broad and often very intense absorption bands runs into the visible region of the spectrum, and is responsible for the red, brown, or yellow colours often noted for actinide complexes with polarizable ligands like Br or I.

12.2   Absorption Spectra

12.2.1   Uranium(VI) – UO_2^{2+} – f^0

The ground state of the uranyl ion has a closed-shell electron configuration. There is a characteristic absorption $\sim$25 000 cm^{-1} (400 nm) which frequently gives uranyl compounds a yellow colour (though other colours like orange and red are not infrequent). This absorption band often exhibits fine structure due to progressions in symmetric O=U=O vibrations in the excited state, sometimes very well resolved, sometimes not (Figures 12.1 and 12.2).

It should also be remarked that uranyl complexes tend to emit a bright green fluorescence under UV irradiation, from the first excited state. This is used by geologists both to identify and to assay uranium-bearing minerals in deposits of uranium ores.

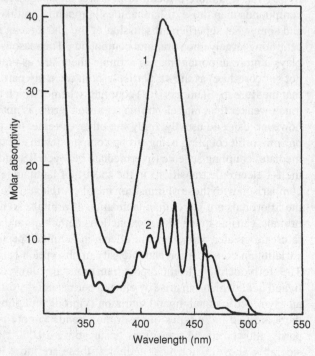

Figure 12.1
The absorption spectrum of (1) $[UO_2(OAc)_4]^{2-}$ in liquid Et$_4$NOAc.H$_2$O, showing the lack of vibronic structure, due to hydrogen bonding; (2) $[UO_2(OAc)_3]^-$ in MeCN solution, showing the progression due to the O=U=O stretching vibration (from J.L. Ryan and W.E. Keder, *Adv. Chem. Ser.*, 1967, **71**, 335 and reproduced by permission of the American Chemical Society).

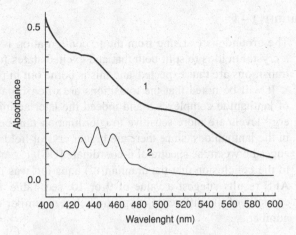

Figure 12.2
Absorption spectra of THF solutions of 1 [$UO_2Cl\{\eta^3\text{-}CH(Ph_2PNSiMe_3)_2\}(thf)$] and and 2 [$UO_2Cl\{\eta^3\text{-}N(Ph_2PNSiMe_3)_2\}(thf)$] (reproduced with permission of the Royal Society of Chemistry from M.J. Sarsfield, H. Steele, M. Helliwell, and S.J. Teat, *Dalton Trans.*, 2004, 3443).

12.2.2 Uranium(v) – f^1

The ^2F ground state is split into two levels, $^2F_{7/2}$ and $^2F_{5/2}$, by spin–orbit coupling in the free ion. These are split further under the influence of a crystal field; the effect on the energy levels of increasing the crystal field up to the strong field limit is shown in Figure 12.3.

Four transitions are thus expected in the electronic spectrum and generally, in practice, four groups of lines are seen, between the near-IR and the visible, bearing out this prediction. The ground state is a Kramers' doublet (Γ_7), so U^V compounds are EPR active.

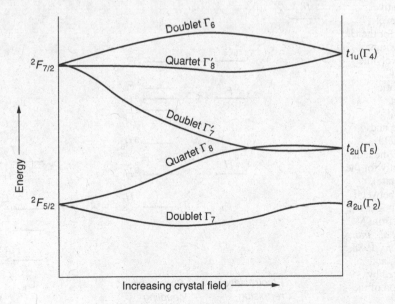

Figure 12.3
The effect of increasing crystal field upon the energy of an electron in an f^1 system such as U^v (reproduced with permission from S.A. Cotton. Lanthanides and Actinides, Macmillan, 1991, p. 109).

12.2.3   Uranium(IV) – f^2

The ground state arising from the f^2 configuration is $^3H_4$ (Figure 12.4) and the effect of a crystal field is to split both that and excited states further. A large number of electronic transitions are thus expected, and this is borne out in practice (Figures 12.5 and 12.6).

It will be noted that the transitions are often broader than those found in the spectra of lanthanide complexes – and indeed the later actinides, see Section 12.2.4. The 5f energy levels are more sensitive to coordination number than are the corresponding levels in the lanthanides; since there are bigger crystal-field effects, one sees pronounced differences between the spectra of 6-coordinate $[UCl_6]^{2-}$ and of U^{4+}(aq) (Figure 12.5), leading to the conclusion that the uranium(IV) aqua ion was not six coordinate (most recent EXAFS results suggest a value of 9 or 10, see Table 11.1). Figure 12.6 displays another example of the difference in spectra between similar complexes of different coordination number.

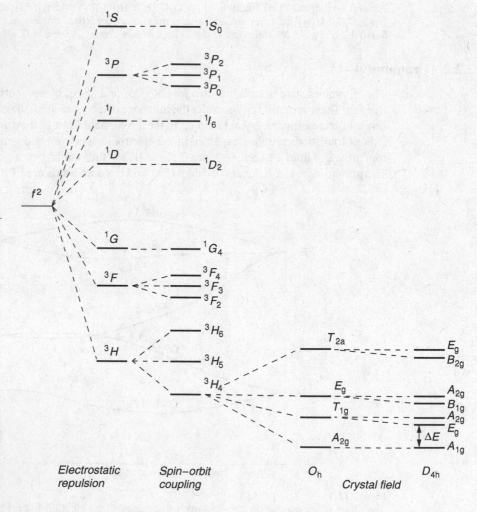

Figure 12.4 A qualitative energy-level diagram for the U^{4+} ion, showing successively the effects of electrostatic repulsion, spin–orbit coupling, and crystal-field splitting (the latter shown only for the ground state). Overlap between levels is neglected. Adapted from M. Hirose *et al.*, *Inorg. Chim. Acta*, 1988, **150**, L93, and reproduced by permission of the Editor.

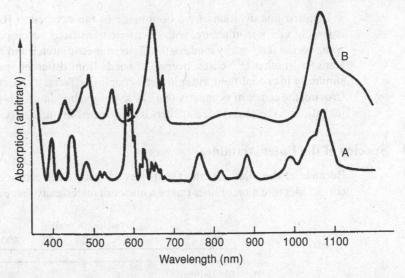

Figure 12.5
The absorption spectra of octahedral $[UCl_6]^{2-}$ (A) and 9–10-coordinate U^{4+}(aq) (B) (redrawn from D.M. Gruen and R.L. Macbeth, *J. Inorg. Nucl. Chem.*, 1959, **9**, 297 and reproduced by permission of Elsevier Science Publishers).

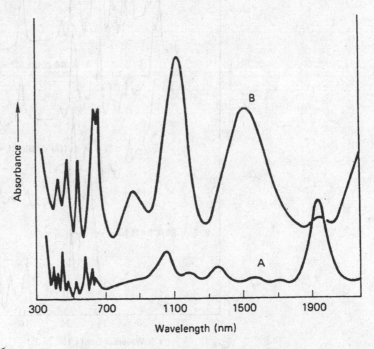

Figure 12.6
Solid-state absorption spectra of octahedral $[UCl_4(Bu^t_2SO)_2]$ (A) and 8-coordinate $[U(Me_2SO)_2]\,I_4$ (B) (redrawn from J.G.H. DuPreez and B. Zeelie, *Inorg. Chim. Acta*, 1989, **161**, 187 and reproduced by permission of Elsevier Science Publishers).

Detailed investigations have been made of the octahedral $[UCl_6]^{2-}$ ion. Its spectrum is largely vibronic in nature, with electronic transitions accompanied by vibrations of the complex ion (odd-parity modes – the T_{1u} asymmetric stretch and T_{1u} and T_{2u} deformations). Here, as in other U^{IV} cases, overlap of bands from different states occurs because of the similarity in crystal-field and spin–orbit coupling effects. Its spectrum can be altered by destroying the centre of symmetry (e.g., by hydrogen bonding), which enables pure electronic transitions to be observed, and alters band patterns in multiplets.

12.2.4 Spectra of the Later Actinides

Because of the relatively short half-lives of many later actinides, purity of samples and correct identification of lines can be a matter of uncertainty, but Figure 12.7 shows how this

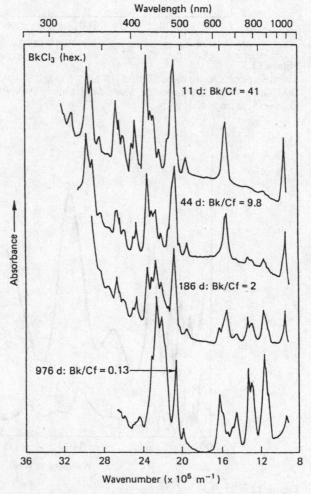

Figure 12.7
The absorption spectrum of the hexagonal form of $BkCl_3$ as a function of time. The changes in the spectrum are due to the formation of $CfCl_3$. Note the sharp 'lanthanide-like' transitions characteristic of the later actinide (+3) state. (from J.R. Peterson *et al.*, *Inorg. Chem.*, 1986, **25**, 3779 reproduced by permission of the American Chemical Society).

can be turned to advantage. It shows spectra obtained over a period of time from a sample of $^{249}BkCl_3$ beginning 11 days after synthesis. Now, ^{249}Bk is a β-emitter with a half-life of 320 days, and the spectrum obtained over a 976 day period (three half-lives, during which time the berkelium decays to some 12.5 % of its original amount) shows the loss of the characteristic absorptions due to the $^{249}BkCl_3$ and their replacement by a spectrum due to $^{249}CfCl_3$. These spectra are very reminiscent of the sharp, line-like absorptions · obtained from the lanthanides. This reflects the fact that chemically the heavy actinides are lanthanide-like, suggesting that with increasing atomic number the 5f orbitals are now more core-like and thus less readily influenced by environment.

After the emission of a β-particle from the Bk nucleus the californium ion regains an electron to maintain the (+ 3) oxidation state:

$$^{249}_{97}Bk^{3+} \rightarrow \, ^{249}_{98}Cf^{4+} + e^-$$

$$Cf^{4+} + e^- \rightarrow Cf^{3+}$$

It may also be noted that crystal type is retained; X-ray diffraction confirms that the $CfCl_3$ retains the hexagonal structure of the original $BkCl_3$ rather than adopting the orthorhombic modification.

12.3 Magnetic Properties

Uranium(VI) compounds are expected to be diamagnetic, with their $^1S_0$ (f^0) ground state. However, compounds like UF_6 and uranyl complexes in fact exhibit temperature-independent paramagnetism, explained by a coupling of paramagnetic excited states with the ground state.

Uranium(V) compounds are, as expected for an f^1 system, paramagnetic, usually exhibiting Curie–Weiss behaviour, with large Weiss constants; g-values, expected to be 6/7, are modified by the mixing in of higher states and by orbital-reduction effects (covalency), experimental g-values including values of 1.2 in Na_3UF_8 and 0.71 in $CsUF_6$.

Matters are more complicated for uranium(IV); this f^2 system has a $^3H_4$ ground state, the energy level diagram has already been given (Figure 12.4). In a regular octahedral geometry, there is no contribution to the paramagnetic susceptibility from the first-order Zeeman term. Species like the $[UCl_6]^{2-}$ ion (and isoelectronic PuF_6) display temperature-independent paramagnetism, caused by the second-order Zeeman term mixing the $^3T_{1g}$ excited state into the ground state. In lower symmetry, such as a D_{4h} *trans*-UX_4L_2 complex, both the first- and second-order Zeeman effects contribute to the susceptibility. If there is a small distortion from a regular octahedron, the splitting of the $^3T_{1g}$ excited state is small, so that ΔE in Figure 12.4 is large. There is thus little thermal population of these excited states, so the first-order Zeeman effect is small; the paramagnetic susceptibility shows little or no temperature dependence. As the distortion becomes larger, in complexes like $UBr_4(Et_3AsO)_2$ and $UI_4[(Me_2N)_3PO)_2]$, thermal population of a component of the $^3T_{1g}$ excited state becomes more feasible, and thus the susceptibility shows a greater temperature dependence. In the case of a *cis*-UX_4L_2 complex, with D_{4h} symmetry, there is no first-order Zeeman term, so that the second-order Zeeman effect causes temperature-independent paramagnetism.

Figure 12.8 shows variable-temperature susceptibility data for some U^{IV} complexes of these types, which is in keeping with these explanations.

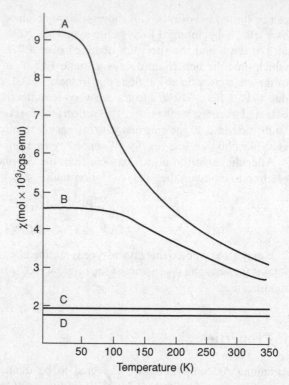

Figure 12.8
Temperature dependence of the magnetic susceptibility of some uranium(IV) complexes: (A) *trans*-[UBr$_4$(Et$_3$AsO)$_2$]; (B) *trans*-[UCl$_4$(Et$_3$AsO)$_2$]; (C) (Ph$_4$P)$_2$ [UCl$_6$]; (D) *cis*-[UCl$_4$(Ph$_3$PO)$_2$] (redrawn from B.C. Lane and L.M. Venanzi, *Inorg. Chim. Acta*, 1969, **3**, 239 and reproduced by permission of the editor).

Few data are available for uranium(III) compounds, but a number of compounds with the f^3 configuration, like Cs$_2$NaUCl$_6$, have $\mu \sim 3.2$ μ_B, which is largely temperature independent.

13 Organometallic Chemistry of the Actinides

By the end of this chapter you should be able to:

- recall that these compounds are very air- and water-sensitive;
- recall that the bonding has a significant polar character, especially in the +3 state, but that +4 compounds, especially those of uranium, have more stability;
- recall that these compounds usually have small molecular structures with appropriate volatility and solubility properties;
- recall appropriate bonding modes for the organic groups;
- suggest suitable synthetic routes;
- suggest structures for suitable examples;
- appreciate the contribution of the C_5Me_5 ligand in uranium chemistry;
- recall the unique properties of uranocene;
- appreciate the unusual compounds that can be made using imide and CO as ligands.

13.1 Introduction

Although lacking π-bonded compounds in low oxidation states that characterize the d-block elements, the actinides have a rich organometallic chemistry. Their compounds frequently exhibit considerable thermal stability, but like the lanthanide compounds are usually intensely air- and moisture-sensitive. They are often soluble in aromatic hydrocarbons such as toluene and in ethers (e.g., THF) but are generally destroyed by water. Sometimes they are pyrophoric on exposure to air. Most of the synthetic work has been carried out with Th and U; this is partly due to the ready availability of MCl_4 (M = Th, U) and also because of the precautions that have to be taken in handling compounds of other metals, especially Pu and Np.

13.2 Simple σ-Bonded Organometallics

During the Manhattan project, a wide range of compounds was screened in the search for volatile uranium compounds that could be used for separation of isotopes of uranium, since UF_6 was not an easy compound to handle. The leading organometallic chemist of the day, Henry Gilman, investigated the synthesis of simple alkyls and aryls, concluding that such compounds did not exist, or at least were unstable. Even today, just one neutral homoleptic

---

alkyl of known structure exists, $[U\{CH(SiMe_3)_2\}_3]$, made from an aryloxide by a route ensuring no contamination with lithium halide:

$$3\,LiCH(SiMe_3)_2 + U(O\text{-}2,6\text{-}Bu_2^tC_6H_3)_3 \rightarrow [U\{CH(SiMe_3)_2\}_3]$$
$$+\,3\,Li(O\text{-}2,6\text{-}Bu_2^tC_6H_3)$$

If $LiCH(SiMe_3)_2$ is treated directly with $UCl_3(THF)_3$ in a salt-elimination reaction the alkylate salt $[(THF)_3Li\text{-}Cl\text{-}U\{CH(SiMe_3)_2\}_3]$ is obtained.

$[U\{CH(SiMe_3)_2\}_3]$ has a trigonal pyramidal structure, like the corresponding lanthanide alkyls (Section 6.2.1) and also like the isoelectronic amides $[M\{N(SiMe_3)_2\}_3]$ (M = U, lanthanides). There are some agostic U . . . . H–C interactions. Similar Np and Pu compounds $[M\{CH(SiMe_3)_2\}_3]$ have been made. *Ab initio* MO calculations for hypothetical $M(CH_3)_3$ (M = U, Np, Pu) molecules also predict pyramidal structures as a result of better 6d involvement (*not 5f*) in the M–C bonding orbitals in pyramidal geometries.

The benzyl $[Th(CH_2Ph)_4]$ has been synthesized and the pale yellow dimethylbenzyl compound $Th(1,3,5\text{-}CH_2C_6H_3Me_2)_4$ has also been reported, but their structures are not known (and may have multihapto-coordination of benzyl to thorium). The heptamethylthorate anion $[ThMe_7]^{3-}$ is found in the salt $[Li(tmeda)]_3[ThMe_7]$ (tmeda = $Me_2NCH_2CH_2NMe_2$). The latter contains thorium in monocapped trigonal prismatic coordination; one methyl is terminal (Th–C 2.571 Å) whilst the other six methyl groups pairwise bridge the lithium and thorium (Th–C 2.655–2.765 Å). The methyl groups are, however, equivalent in solution (NMR).

Certain simple alkyls are stabilized by the presence of either phosphine or amide ligands, thus:

$$UCl\{N(SiMe_3)_2\}_3 + CH_3Li \rightarrow U(CH_3)\{N(SiMe_3)_2\}_3 + LiCl$$

$$2\,ThCl\{N(SiMe_3)_2\}_3 + Mg(CH_3)_2 \rightarrow 2\,Th(CH_3)\{N(SiMe_3)_2\}_3 + MgCl_2$$

These are pentane-soluble, volatile tetrahedral molecules. They have an extensive chemistry involving metallacycles which will be dealt with later (Section 13.6). Although MMe_4 (M = Th, U) have not been isolated, this unit can be stabilized as eight-coordinate phosphine adducts, where the role of phosphines is to saturate the coordination sphere. The coordination of a 'soft' phosphine ligand to a 'hard' metal ion like uranium(IV) is noteworthy.

$$MCl_4(dmpe)_2 + 4\,CH_3Li \rightarrow M(CH_3)_4(dmpe)_2 + 4\,LiCl$$

These react with phenol forming phenoxides:

$$Th(CH_3)_4(dmpe)_2 + 4\,C_6H_5OH \rightarrow Th(OC_6H_5)_4(dmpe)_2 + 4\,CH_4$$

Similar compounds $[M(CH_2Ph)_4(dmpe)_2]$ (stable to 85 °C in the absence of air) and $[M(CH_2Ph)_3Me(dmpe)_2]$ have been isolated; structures of $[Th(CH_2Ph)_4(dmpe)_2]$ and $[U(CH_2Ph)_3Me(dmpe)_2]$ confirm their identity, they have short M–C contacts. Compounds of alkyl groups like CH_2CH_3 that are capable of undergoing elimination have not been isolated.

Compounds containing one or more cyclopentadienyl-type ligands are much more robust. (Throughout this section, C_5H_5 is abbreviated to Cp; C_5Me_5 is abbreviated to Cp*; $C_5H_4.CH_3$ is MeCp.)

Cyclopentadienyls

These are regarded as derivatives of the $C_5H_5^-$ ligand; substituted rings also occur. Some of these were the first organoactinide compounds to be reported, shortly after the discovery of ferrocene. The most important compounds are in the +4 state but there is a significant chemistry of U(+3) as well as examples in oxidation states (+5) and (+6). One factor that should be noted is that replacing one or more hydrogens in the cyclopentadienyl ring can have a significant effect upon the stability and isolability of complexes made from them, through both steric and electronic effects. Thus three or four Cp ligands can be accommodated around uranium(IV), whereas with Cp* two is the norm. In another example, [UCp₃] gives no sign of reacting with carbon monoxide; [U{C₅H₃(SiMe₃)₂}₃] forms a compound that can be detected in solution; and [U(C₅Me₄H)₃] forms [U(C₅Me₄H)₃(CO)], which can be isolated in the solid state.

13.3.1 Oxidation State (VI)

High oxidation states are not normally associated with organometallic chemistry, but a remarkable uranium(VI) imide has been made (Figure 13.1):

$$UCp^*_2(CH_3)Cl + Li[PhNN(H)Ph] \rightarrow UCp^*_2(= NPh)_2 + CH_4 + LiCl.$$

The reaction may go via an intermediate [UCp*₂(η^2-PhN–NPh)]. There are no f–f transitions in the visible spectrum, confirming its U^{VI}-f^0 character, whilst the structure of this remarkable compound shows the expected pseudotetrahedral geometry; the short U–N bond length (1.952 Å) confirms its multiple-bond character and the U=N–Ph linkage is virtually linear.

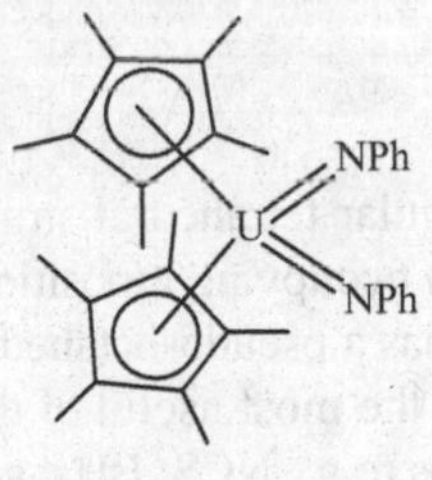

Figure 13.1
A uranium(VI) imide.

13.3.2 Oxidation State (V)

Rare organoimides U(MeCp)₃(=NR) have been made by oxidation of U(MeCp)₃(thf) with RN₃ (R = Me₃Si and Ph). The U–N bond length of 2.109 Å is regarded as having multiple-bond character (cf. the U=O bond in uranyl compounds).

$$U(MeCp)_3(THF) + RN_3 \rightarrow U(MeCp)_3(= NR) + THF + N_2$$

13.3.3 Oxidation State (IV)

There are three families of compounds whose structures can be regarded as derived from a tetrahedral UX₄ molecule with one or more halogens replaced by one or more cyclopentadienyl rings; [Cp$_4$An], [Cp$_3$AnX]; [CpAnX$_3$] (X = halogen, most usually Cl). Since halogens have less steric influence than a cyclopentadienyl group, a CpAnX$_3$ molecule would be coordinatively unsaturated, so CpAnX$_3$ actually exist as solvates like CpAnX$_3$(THF)$_2$. Cp$_2$AnX$_2$ are unknown for X = halogen, attempts to prepare them resulting in mixtures of [Cp$_3$AnX] and [CpAnX$_3$], but they are known for X = NEt$_2$ and BH$_4$.

Syntheses

$$UCl_4 + 4\,KCp \rightarrow [UCp_4] + 4\,KCl \quad (C_6H_6)$$

$$UCl_4 + 3\,TlCp \rightarrow [UCp_3Cl] + 3\,TlCl \quad (1,2\text{-dimethoxyethane})$$

$$UCl_4 + TlCp \rightarrow [UCpCl_3(THF)_2] + TlCl \quad (THF)$$

Some similar compounds can be made with other actinides, namely [MCp$_4$] (M = Th, Pa, Np), [MCp$_3$Cl] (M = Th, Np), and [ThCpCl$_3$].

Sometimes other starting materials are used:

$$U(NEt_2)_4 + 3\,CpH \rightarrow [UCp_3(NEt_2)] + 3\,HNEt_2$$

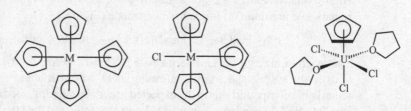

Figure 13.2
Structures of MCp$_4$, MCp$_3$Cl and MCpCl$_3$ (THF)$_2$.

Structures

[MCp$_4$] have regular tetrahedral structures whilst [MCp$_3$Cl] are analogous (Figure 13.2) with the halogen occupying a position corresponding to the centroid of a fourth ring and [UCpCl$_3$(thf)$_2$] has a pseudo-octahedral structure.

[MCp$_3$Cl] are the most useful of these compounds as the halogen can be replaced by a number of groups (e.g., NCS, BH$_4$, acac, Me, Ph) (Fig. 13.3).

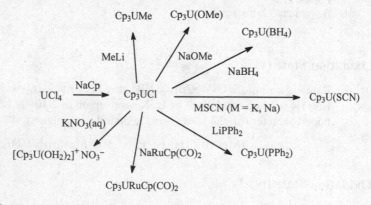

Figure 13.3
Synthesis and reactions of Cp$_3$UCl.

There is evidence for a significant covalent contribution to the bonding in some of these actinide(IV) compounds; UCp_3Cl does not react with $FeCl_2$ forming ferrocene, whereas MCp_3 do (M, e.g., Ln, U). It appears to undergo ionization in aqueous solution as a stable green $[UCp_3(OH_2)_n]^+$ ($n \sim 2$) cation. There is also some evidence from Mössbauer spectra of neptunium-(III) and -(IV) cyclopentadienyl compounds that in a Np^{IV} compound such as $NpCp_4$ (but not Np^{III} compounds) there is a greater shielding of the 6s shell than in an ionic compound like $NpCl_4$, leading to enhanced covalence in the bond. Bonds are reasonably strong; the mean bond dissociation energy in UCp_4 has been determined to be $247 \, kJ \, mol^{-1}$, compared with a value of $297 \, kJ \, mol^{-1}$ for $FeCp_2$.

The most interesting $[MCp_3X]$ compounds are the alkyls and aryls, $[MCp_3R]$ being made by reaction with either organolithium compounds or Grignard reagents.

$$[MCp_3Cl] + RLi \rightarrow [MCp_3R] + LiCl; [MCp_3Cl] + RMgX \rightarrow [MCp_3R] + MgXCl$$

$$(M = Th, U; R, e.g., CH_3, Pr^n, Pr^i, Bu^n, Bu^t, C_6H_5, C_6F_5, C\equiv CPh, CH_2Ph, etc.)$$

They were independently synthesized by three independent research groups at the start of the 1970s and represent the first compounds to have well-defined actinide-carbon σ bonds (structures for M = U; R, e.g., Bu, $CH_2C_6H_4CH_3$, etc.). (See, e.g., the papers by T.J. Marks group in *J. Am. Chem. Soc.*, 1973, **95**, 5529; 1976, **98**, 703.) Thermally stable *in vacuo*, they react with methanol, forming $MCp_3(OMe)$ and RH. A low-temperature NMR study of $[UCp_3Pr^i]$ indicates restricted rotation about the U–C σ bond at low temperatures.

Bond disruption energies of 315–375 kJ/mol for the Th–C bonds in $[ThCp_3R]$ (R, e.g., CH_3, CH_2Ph) indicate the thermodynamic stability of these compounds, comparable with transition metal alkyls (the bond energies are 10–30 kJ/mol less for the uranium compounds). Decomposition does not occur by the β-elimination route; instead intermolecular abstraction of a hydrogen from a Cp ligand occurs (Figure 13.4) forming RH. A stable compound $[Cp_2Th(\mu\text{-}C_5H_4)_2ThCp_2]$ is the other product (Figure 13.5).

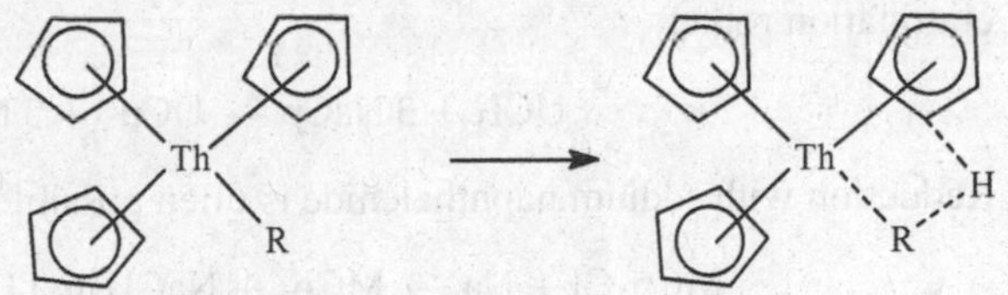

Figure 13.4
Decomposition of $ThCp_3R$.

These compounds also undergo insertion reactions of CO, CO_2, and R'NC into the M–C sigma bond, forming $[Cp_3M(\eta^2\text{-}COR)]$, $[Cp_3M(\eta^2\text{-}O_2CR)]$, and $[Cp_3M(\eta^2\text{-}C(R)NR')]$, respectively (Figure 13.6).

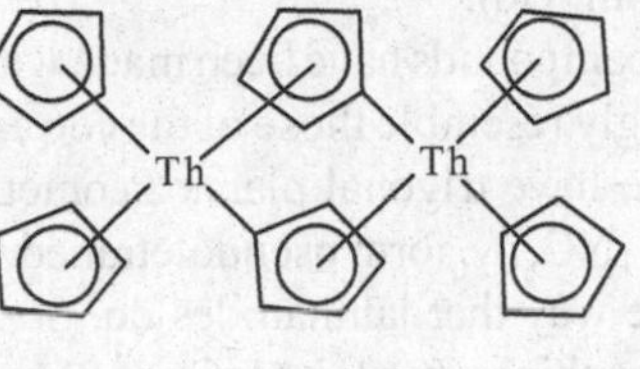

Figure 13.5
Structure of $[Cp_2Th(\mu\text{-}C_5H_4)_2ThCp_2]$.

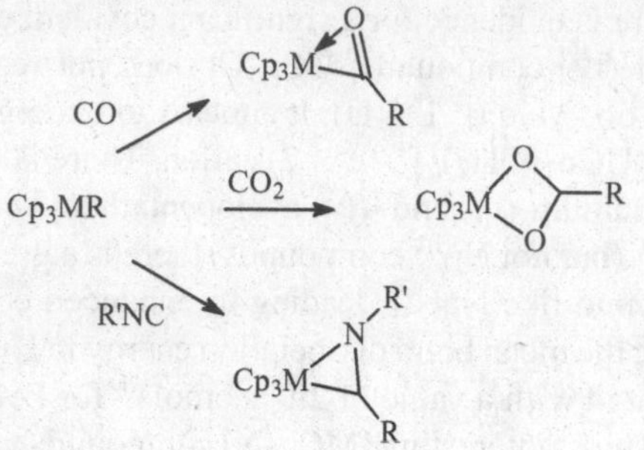

Figure 13.6
Insertion reactions of Cp_3MR.

As already mentioned, there is no UCp_2Cl_2 – attempts to make it giving mixtures of $[UCp_3Cl]$ and $[UCpCl_3]$, but some UCp_2X_2 can be made for example:

$$[U(NEt_2)_4] + 2\,C_5H_6 \rightarrow UCp_2(NEt_2)_2 + 2\,HNEt_2$$

$$UCl_4 + 2\,NaBH_4 \rightarrow UCl_2(BH_4)_2 + 2\,NaCl \quad \text{followed by}$$

$$UCl_2(BH_4)_2 + 2\,TlCp \rightarrow UCp_2(BH_4)_2 + 2\,TlCl$$

$UCp_2(NEt_2)_2$ reacts with ROH [R = bulky group, e.g., $[(Bu^t)_3C; 2,6\text{-}Me_2C_6H_3]$:

$$UCl_2(NEt_2)_2 + 2\,ROH \rightarrow UCp_2(OR)_2 + 2\,HNEt_2$$

13.3.4 Oxidation State (III)

A variety of syntheses are possible to these compounds, including the expected salt-elimination route:

$$UCl_3 + 3\,NaCp \rightarrow UCp_3 + 3\,NaCl\,(C_6H_6)$$

Reduction with sodium naphthalenide is often possible:

$$MCp_3Cl + Na \rightarrow MCp_3 + NaCl\,(in\ THF)\,(M = U, Th, Np)$$

An unusual route used for several of the transuranium elements is a microscale reaction of the metal trichlorides with molten $BeCp_2$ at 65 °C.

$$2\,MCl_3 + 3\,BeCp_2 \rightarrow 2\,MCp_3 + 3\,BeCl_2\,(M = Pu, Am, Cm, Bk, Cf)$$

The cyclopentadienyl products were sublimed out of the reaction mixture and characterized by powder diffraction data, showing they were isostructural with the corresponding $LnCp_3$ (Ln = Pr, Sm, Gd).

Similar compounds have been made with substituted cyclopentadienyl groups; their properties strongly resemble those of the corresponding $LnCp_3$. The unsolvated $U(C_5H_4SiMe_3)_3$ is known to have trigonal planar geometry, as is $U(C_5Me_4H)_3$ (X-ray). Lewis bases, e.g. THF, $C_6H_{11}NC$, py, form pseudotetrahedral adducts $UCp_3.L$ {X-ray, e.g. for $[UCp_3(THF)]$} in the same way that lanthanides do. Some adducts are known with rather soft bases like tertiary phosphines, e.g $[U(MeCp)_3(PMe_3)]$, that might not have been expected for a fairly 'hard' metal such as U^{III}.

Evidence from UV–visible spectra (Cm and Am compounds) and Mossbauer spectra (Np compounds) indicates that these compounds have probably rather more covalent contributions to their bonding than do the corresponding lanthanide compounds, but are largely ionic compared with cyclopentadienyls in the +4 state.

Analogous compounds $AnCp_3$ exist for other actinides, e.g. Th, Np (?), Pu, Am, Bk, Cf, Cm. The thorium(III) compound is especially noteworthy, on account of the rarity of this oxidation state; the structure of the similar $Th\{C_5H_3(SiMe_3)_2\}_3$ is known to have trigonal planar geometry around thorium.

$$3\,Th\{C_5H_3(SiMe_3)_2\}_2Cl_2 + Na/K(alloy) \rightarrow Th + 2\,Th\{C_5H_3(SiMe_3)_2\}_3 + 6\,Na/KCl$$

These Th^{III} compounds are EPR-active; spectra indicate a $6d^1$ ground state for $Th\{C_5H_3(SiMe_3)_2\}_3$, despite the fact that the free Th^{3+} ion has a $5f^1$ configuration.

13.4 Compounds of the Pentamethylcyclopentadienyl Ligand ($C_5Me_5 = Cp^*$)

13.4.1 Oxidation State(IV)

Compounds of the type $[MCp^*_2X_2]$ (M = Th, U) are rather important. They owe their importance to the fact that although three Cp^* rings can be accommodated round uranium in UCp^*_3Cl, this compound cannot be made by direct synthesis (nor can MCp^*_3), partly owing to the bulk of the C_5Me_5 ligands, and the $[MCp^*_2X_2]$ systems are in practice more readily obtained.

$$MCl_4 + 2\,Cp^*MgBr \rightarrow MCp^*_2Cl_2 + 2\,MgBrCl \quad (M = Th, U)\,in\,toluene$$

Halides can be replaced by suitable ligands, thus reactions with NaOMe and $LiNMe_2$ afford products like $ThCp^*_2(OMe)_2$, $ThCp^*_2Cl(NMe_2)$, and $ThCp^*_2(NMe_2)_2$. Li_2S_5 reacts forming $[ThCp^*_2(S_5)]$. Importantly, the chloride compounds can be alkylated and arylated. Either one or two chlorines can be replaced.

$$MCp^*_2Cl_2 + LiR \rightarrow MCp^*_2(R)Cl + LiCl$$
$$MCp^*_2(R)Cl + LiR \rightarrow MCp^*_2R_2 + LiCl$$
$$(M = Th, U; R, e.g., Me, Ph, CH_2CMe_3, CH_2SiMe_3)$$

They have the usual pseudotetrahedral coordination; the structure of $Cp^*_2Th(CH_2CMe_3)_2$ reveals agostic Th....H–C interactions with the α-C–H bonds. The M–C σ bonds seem to have a significant ionic contribution; thus $ThCp^*_2Me_2$ reacts with acetone to afford a *tert*-butoxy compound.

Reactions with R'OH, R'SH, and R^2_2NH afford products such as $ThCp^*_2(R)(OR')$, $ThCp^*_2(OR')_2$, $ThCp^*_2(SR')_2$, and $ThCp^*_2(NR^2_2)_2$ (R', e.g., Pr; R^2, e.g., Et).

The compounds $Cp^*_2Th(CH_2CMe_3)_2$ and $Cp^*_2Th(CH_2SiMe_3)_2$ undergo most remarkable elimination reactions to form metallacyclic complexes on heating to only 50 °C in toluene solution:

$$Cp^*_2Th(CH_2SiMe_3)_2 \rightarrow Cp^*_2\overline{Th(CH_2CMe_2\text{-}CH_2)} + CMe_4$$

$$Cp^*_2Th(CH_2SiMe_3)_2 \rightarrow Cp^*_2\overline{Th(CH_2SiMe_2-CH_2)} + CMe_4$$

Reactions of $Cp^*_2\overline{Th(CH_2CMe_2\text{-}CH_2)}$ are shown in Figure 13.7.

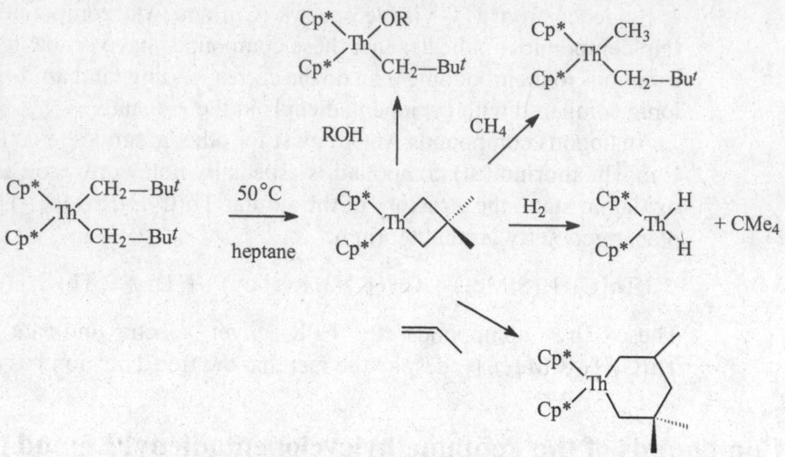

Figure 13.7
Reactions of Cp*$_2$Th(CH$_2$CMe$_2$-CH$_2$).

Some mono(Cp*) complexes are known, reaction of MCl$_4$ (M = U, Th) with MeMgCl.THF yielding [Cp*MCl$_3$(thf)$_2$], with structures analogous to the [CpLnCl$_2$(thf)$_3$] compounds. The halogens can be replaced with the isolation of [Cp*Th(CH$_2$Ph)$_3$]; this has a piano-stool structure with Th–C σ-bonds of 2.578–2.581 Å. There is some interaction between Th and the benzene rings, so this is not a classical η^1-alkyl.

13.4.2 Cationic Species and Catalysts

Cationic alkyls have been studied with profit. Thus, in benzene solution,

$$Cp^*_2Th(CH_3)_2 + [HNBu^n_3]_3{}^+[B(C_6F_5)_4]^- \rightarrow [Cp^*_2Th(CH_3)]^+[B(C_6F_5)_4]^-$$
$$+ NBu^n_3 + CH_4$$

[Cp*$_2$Th(CH$_3$)]$^+$ [B(C$_6$F$_5$)$_4$]$^-$ in fact has a very weakly bound anion [Th–F 2.757, 2.675 Å; Th–C (ring) 2.754 Å, Th–CH$_3$ 2.399 Å], so that it is an active catalyst for ethene polymerization and hex-1-ene hydrogenation (Figure 13.8).

If this type of abstraction reaction is carried out in THF, [Cp*$_2$Th(CH$_3$)(THF)$_2$]$^+$ [BPh$_4$]$^-$ is formed [Th–C (ring) 2.801 Å , Th–CH$_3$ 2.491 Å].

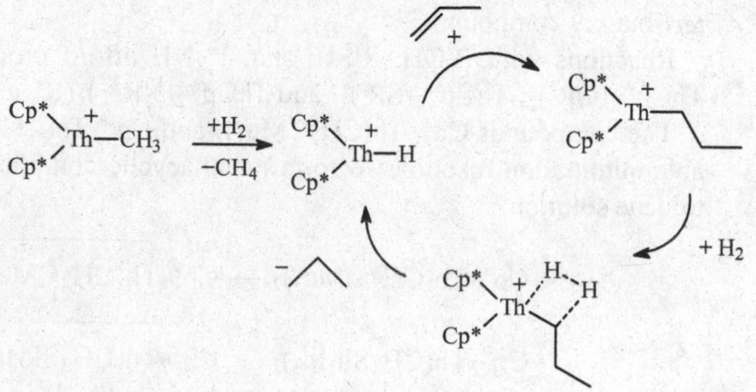

Figure 13.8
[Cp*$_2$ThCH$_3$]$^+$ as a catalyst for hydrogenation.

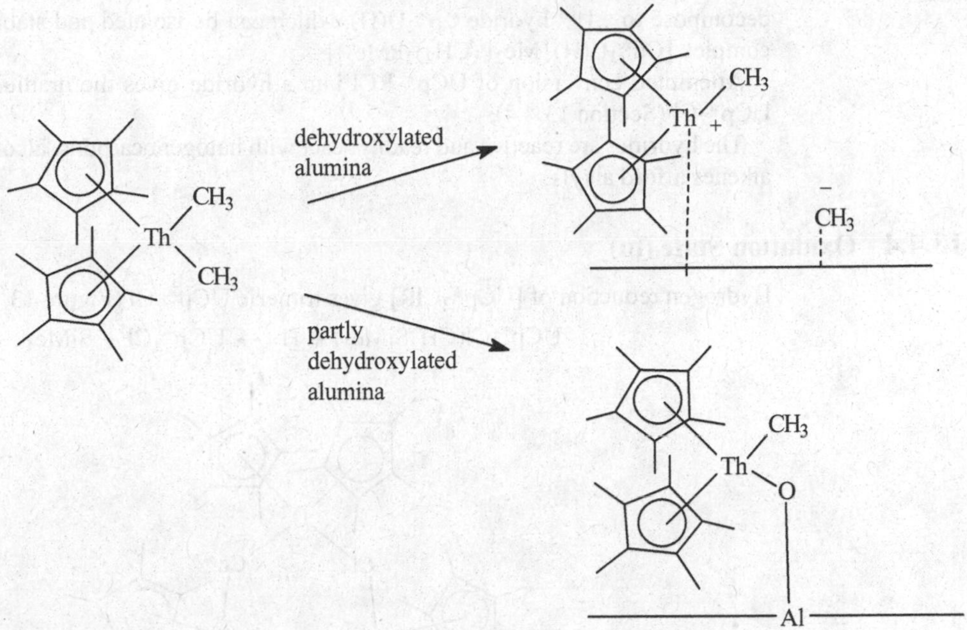

Figure 13.9
Surface bound cationic thorium alkyls.

On alumina surfaces cationic alkyls exhibit catalytic activity, due to the formation of coordinatively unsaturated species (Figure 13.9).

13.4.3 Hydrides

Striking hydride derivatives are formed, which are efficient catalysts for the homogeneous hydrogenation of hex-1-ene (Figure 13.10):

$$2\,ThCp^*_2R_2 + 4\,H_2 \rightarrow Cp^*_2Th(H)(\mu\text{-}H)_2Th(H)Cp^*_2 + 4\,RH \quad R, \text{e.g., Me}$$

This remarkable thorium compound is stable to 80 °C. In solution it has rapid hydrogen exchange with dihydrogen gas. Solution ^{1}H NMR shows the bridge and terminal hydrogens are equivalent even at −90 °C. In the solid state, Th–H is 2.03 Å (terminal) and 2.29 Å (bridge); a Th–Th distance of 4.007 Å indicates minimal metal–metal bonding. In the IR, ν Th–H (terminal) vibrations occur at 1404 and 1336 cm^{-1}; bridging vibrations occur at 1215, 1114, 844, and 650 cm^{-1}. $Cp^*_2U(H)(\mu\text{-}H)_2U(H)Cp^*_2$ can be made too but tends to

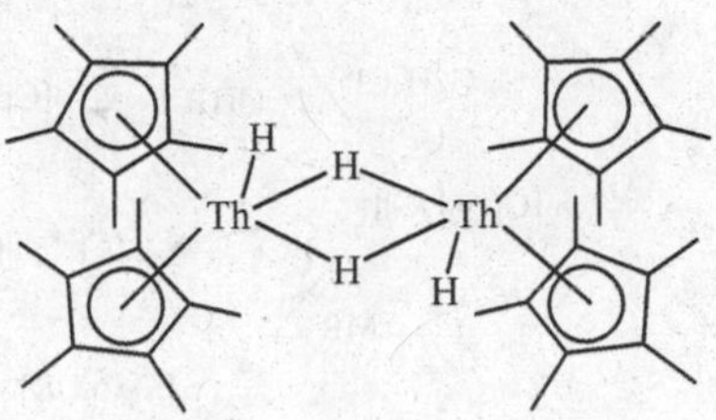

Figure 13.10
Structure of $[Cp^*_2Th(H)(\mu\text{-}H)_2Th(H)Cp^*_2]$.

decompose to a U^{III} hydride $Cp*_2U(H)$, which can be isolated and stabilized as a DMPE complex $[Cp*_2U(H)\{Me_2P(CH_2)_2PMe_2\}]$.

Attempted conversion of $UCp*_2RCl$ into a hydride gives the uranium (III) compound $UCp*_2Cl$ (Section 13.4.4).

The hydrides are reactive and readily react with halogenocarbons, alcohols, and ketones; alkenes afford alkyls.

13.4.4   Oxidation State (III)

Hydrogen reduction of $[UCp*_2ClR]$ gives trimeric $UCp*_2Cl$ (Figure 13.11).

$$UCp*_2Cl(CH_2SiMe_3) + H_2 \rightarrow UCp*_2Cl + SiMe_4$$

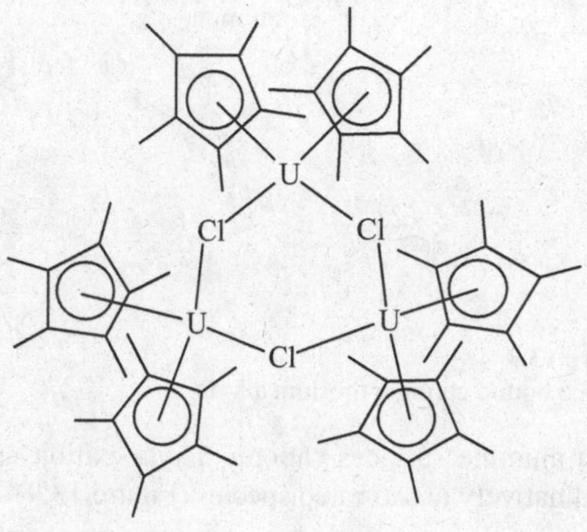

Figure 13.11
The structure of trimeric $[Cp*_2UCl]_3$.

In the similar reaction with R = methyl, the organic product is CH_4. If the reaction is carried out with the analogous thorium starting material, $ThCp*_2Cl(R)$ forms the thorium(IV) compound $[Cp*_2Th(\mu\text{-H})Cl]_2$.

The uranium compound has a cyclic trimeric structure. Each uranium is individually in pseudotetrahedral coordination. It undergoes various reactions (Figure 13.12).

These may be classed as straightforward substitutions using a bulky anionic ligand; adduct formation with Lewis bases retaining pseudotetrahedral coordination; and redox. The photoelectron spectrum of the alkyl $[UCp*_2(CH_2SiMe_3)]$ indicates a $5f^3$ ground state but also that the $5f^26d^1$ state is only slightly higher in energy.

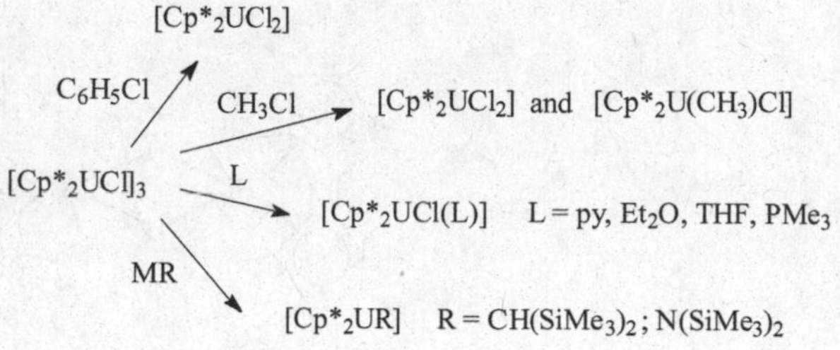

Figure 13.12
Reactions of $[Cp*_2UCl]_3$.

13.5 Tris(pentamethylcyclopentadienyl) Systems

For many years it was believed that [LnCp*$_3$] systems could not be made. After the success-ful isolation of [SmCp*$_3$] and similar compounds (Section 6.3.2), attention turned to the possibility of isolating the uranium analogue. The hydride Cp*$_2$UH(dmpe) proved to be a useful starting point, reacting with either tetramethylfulvene or PbCp*$_2$ to form [UCp*$_3$], but routes analogous to those used for lanthanides have also worked (Figure 13.13).

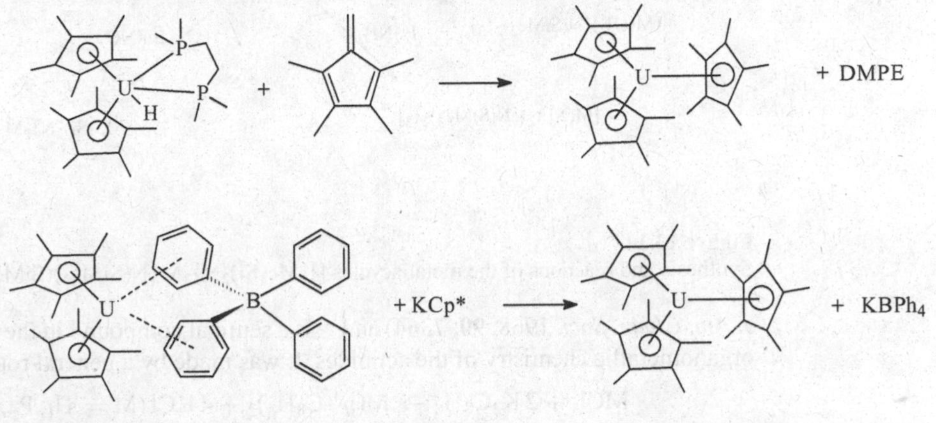

Figure 13.13
Synthesis of [UCp*$_3$].

With the isolation of [UCp*$_3$] accomplished, the next target was what promised to be even harder, the synthesis of the uranium(IV) species [UCp*$_3$Cl] and related compounds. [UCp*$_3$] underwent oxidation with HgF$_2$, forming [UCp*$_3$F], and reaction with a stoichio-metric amount of PhCl, forming [UCp*$_3$Cl]; reaction of [UCp*$_3$] with 2 molecules of PhCl afforded [UCp*$_2$Cl$_2$].

The structure of [UCp*$_3$] has trigonal planar coordination of U; similarly in [UCp*$_3$Cl]; the three ring centroids are coplanar with U, and with a very long U–Cl bond at 2.90 Å – compare 2.637 Å in [U(C$_5$Me$_4$H)$_3$Cl].

13.6 Other Metallacycles

Cp*$_2$Th(CH$_2$CMe$_2$-CH$_2$) has already been mentioned. Another kind of metallacycle has been synthesized from the alkyls [M(CH$_3$){N(SiMe$_3$)$_2$}$_3$] (M = Th, U, see Section 11.8.1). They undergo thermolysis in benzene solution to afford [{(Me$_3$Si)$_2$N}$_2$M{N(SiMe$_3$)(SiMe$_2$)}]. These in turn have a rich chemistry (Figure 13.14), undergoing insertion re-actions with CO, carbonyl compounds, nitriles, and isocyanides. In the reactions involving compounds with acidic hydrocarbons, such as tertiary alcohols, phenols, and thiols, the metallacyclic ring is broken to re-form the amide ligand.

13.7 Cyclooctatetraene Dianion Compounds

These are regarded as derivatives of the C$_8$H$_8^{2-}$ ligand; substituted rings also occur. Ura-nocene, U(η^8-C$_8$H$_8$)$_2$, was reported in 1968 (A. Streitweiser, Jr., and V. Mueller-Westerhoff,

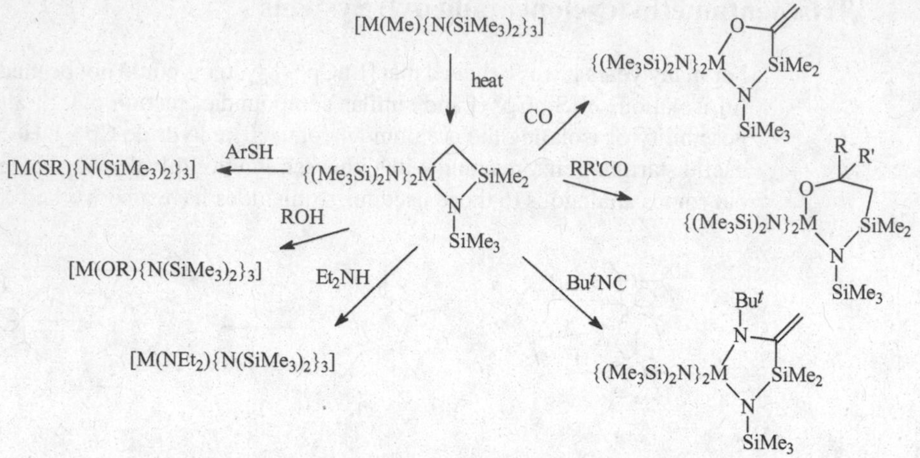

$[M(Me)\{N(SiMe_3)_2\}_3]$

$(Me_3Si)_2N\}_2M$ ... SiMe_2 / N / SiMe_3 (with O=)

heat ↓

CO ↗

ArSH ←

$[M(SR)\{N(SiMe_3)_2\}_3]$ ← $\{(Me_3Si)_2N\}_2M$ ... SiMe_2 / N / SiMe_3

RR'CO →

$(Me_3Si)_2N\}_2M$ ... SiMe_2 (with O, R, R')

ROH ↙

$[M(OR)\{N(SiMe_3)_2\}_3]$

Et_2NH ↙

$[M(NEt_2)\{N(SiMe_3)_2\}_3]$

Bu^{t}NC ↘

$(Me_3Si)_2N\}_2M$ ... SiMe_2 (with N–Bu^t) / N / SiMe_3

Figure 13.14
Synthesis and reactions of the metallacycles $[\{(Me_3Si)_2N\}_2M\{N(SiMe_3)(SiMe_2)\}]$.

J. Am. Chem. Soc., 1968, **90**, 7364) and was a seminal compound in the development of the organometallic chemistry of the actinides. It was made by a general route:

$$MCl_4 + 2\,K_2C_8H_8 \rightarrow M(\eta^8\text{-}C_8H_8)_2 + 4\,KCl\,(M = Th, Pa, U, Np)$$

The Pu analogue was made from $(pyH)_2PuCl_6$, since plutonium does not form a tetrachloride. All five compounds are isomorphous and believed to have similar sandwich structures, with the two (planar) 8-membered rings in the eclipsed configuration (Figure 13.15).

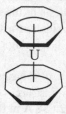

Figure 13.15
Structure of uranocene, $[U(C_8H_8)_2]$.

Uranocene forms green crystals that are pyrophoric in air, but it is otherwise rather unreactive, not readily hydrolysed by water or even acetic acid. It is stable to vacuum sublimation at 200 °C. The other $M(\eta^8\text{-}C_8H_8)_2$ are more reactive and easily hydrolysed. A mean bond-dissociation energy of 347 kJ mol^{-1} has been measured for $U(\eta^8\text{-}C_8H_8)_2$; it is substantially greater than the value of 247 kJ mol^{-1} for $U(C_5H_5)_4$. Of substituted uranocenes, two stand out. $[U(1,3,5,7\text{-}Me_4C_8H_4)_2]$ exists in the crystal as a mixture of the 'eclipsed' and 'staggered' rotamers, whilst $[U(1,3,5,7\text{-}Ph_4C_8H_4)_2]$ (present in the crystal with phenyls staggered in the two different rings) sublimes at 400 °C, 10^{-5} mmHg, and is totally air stable, doubtless owing to the phenyl groups hindering access of oxygen molecules to the uranium atom.

Some sandwich compounds $[K(\text{solvent})]\,[M(\eta^8\text{-}RC_8H_7)_2$ $(M = U, Np, Pu, Am; R, e.g., H, Me)$ may also be made in the +3 state by analogous reactions starting from:

$$MCl_3 + 2\,K_2(RC_8H_7) \xrightarrow{\text{Diglyme}} [K(\text{diglyme})][M(\eta^8\text{-}RC_8H_7)_2] + 3\,KCl$$

Figure 13.16
Possible f-orbital involvement in the bonding in uranocene.

Bonding in uranocene has been the subject of controversy for some 30 years. It was early on pointed out that orbital symmetry interactions, involving f orbitals, could be drawn analogous to those in ferrocene (Figure 13.16).

More sophisticated MO calculations, especially including relativistic corrections, suggest that whilst uranium 6d orbitals interact with the ligand orbitals, the 5f orbitals are relatively unperturbed. Photoelectron spectra do suggest that 5f interaction increases as more alkyl groups are introduced into the cyclooctatetraene rings.

Half-sandwich compounds have been made, in the case of thorium by redistribution:

$$Th(\eta^8\text{-}C_8H_8)_2 + ThCl_4 \rightarrow 2\,(\eta^8\text{-}C_8H_8)ThCl_2(THF)_2 \quad \text{(in THF)}$$

This compound has a piano-stool-type structure (Figure 13.17).

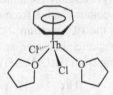

Figure 13.17
Structure of the 'half-sandwich' [Th(C$_8$H$_8$)Cl$_2$(THf)$_2$].

Uranium analogues can be made, not always by identical routes:

$$C_8H_8 + 2\,NaH + UCl_4 \rightarrow (\eta^8\text{-}C_8H_8)UCl_2(THF)_2 + 2\,NaCl + H_2 \quad \text{(in THF)}$$

$$U(\eta^8\text{-}C_8H_8)_2 + I_2 \rightarrow (\eta^8\text{-}C_8H_8)UI_2(THF)_2 + C_8H_8 \quad \text{(in THF)}$$

13.8 Arene Complexes

The reaction between UCl$_4$, Al, and AlCl$_3$ in refluxing benzene affords [(η^6-C$_6$H$_6$)U{(μ-Cl)$_2$AlCl$_4$}$_3$]. (Figure 13.18).

Certain other compounds of this type are known in the (+3) state, such as [(η^6-C$_6$H$_3$Me$_3$)U(BH$_4$)$_3$]; also dimeric [(C$_6$Me$_6$)$_2$U$_2$Cl$_7$]$^+$ (AlCl$_4$)$^-$, which is [(η^6-C$_6$Me$_6$)Cl$_2$U(μ-Cl)$_3$UCl$_2$(η^6-C$_6$Me$_6$)]$^+$ (AlCl$_4$)$^-$. These are considerably more reactive than uranocene.

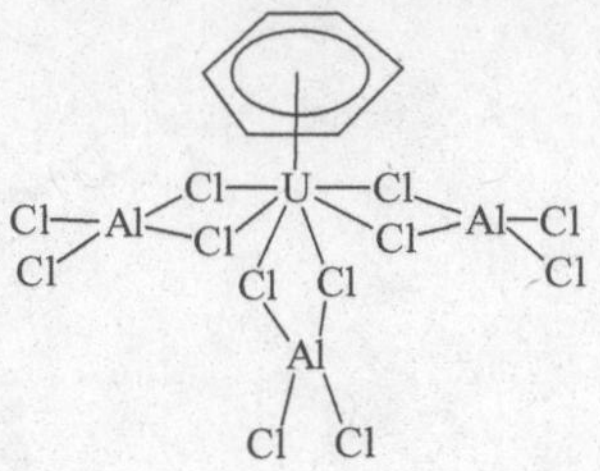

Figure 13.18

However, in contrast to the lanthanides, no arene complexes in a formal zero oxidation state have been isolated.

13.9 Carbonyls

The early transition metals are conspicuous in forming binary carbonyls like $M(CO)_6$ ($M =$ Cr, Mo, W); $M'(CO)_5$ ($M' =$ Fe, Ru, Os); and $Ni(CO)_4$, which have considerable thermal stability.

Co-condensation of uranium atoms with CO in an argon matrix at 4 K followed by annealing gives carbonyls $U(CO)_n$ detectable by IR, but which decompose above 20 K (as do those of the lanthanides). A complex identified as $U(CO)_6$ has ν_{CO} at 1961 cm^{-1}, very similar to that in $W(CO)_6$ (1987 cm^{-1}), a significant decrease from the value of 2145 cm^{-1} in CO itself, indicating a substantial amount of π-donation by uranium.

However, unlike the lanthanides, some stable carbonyl complexes in normal oxidation states have been identified for uranium. No adduct has been identified for [UCp$_3$] but $U\{C_5H_4(SiMe_3)\}_3$ reversibly combines with CO in both the solid state and hexane solution, the latter with a colour change from deep green to burgundy, and the appearance of ν_{CO} at 1976 cm^{-1} in the IR spectrum. $[U\{C_5H_3(SiMe_3)_2\}_3]$ forms a similarly detectable but non-isolable CO adduct; however, an adduct $[U(C_5Me_4H)_3(CO)]$ has been isolated in the solid state, starting from $[U(C_5Me_4H)_3]$:

$$[U(C_5Me_4H)_3] + CO \rightleftharpoons [U(C_5Me_4H)_3(CO)]$$

It has a linear U–CO linkage, with U–C 2.383 Å, and C–O slightly lengthened from 1.128 Å in free CO to 1.142 Å. ν_{CO} occurs at 1880 cm^{-1} in $[U(C_5Me_4H)_3(CO)]$, shifted to 1840 cm^{-1} in $[U(C_5Me_4H)_3(^{13}CO)]$ and 1793 cm^{-1} in $[U(C_5Me_4H)_3(C^{18}O)]$.

Question 13.1 Suggest why, at the time of the Manhattan project in the 1940s, Gilman attempted the synthesis of simple uranium alkyls and aryls as vehicles for isotope separation.
Answer 13.1 At that time, few d-block alkyls and aryls were known. Gilman would have reasoned that by analogy with compounds of the Group IV elements like carbon and silicon, uranium(IV) alkyls and aryls might have simple molecular structures and thus be volatile.

Question 13.2 Why are simple actinide alkyls rare in view of the number of lanthanide species like $[M\{CH(SiMe_3)_2\}_3]$, $[LnPh_3(thf)_3]$, and $[LnMe_6]^{3-}$ (see Chapter 5)?
Answer 13.2 Good question! There are several possible reasons. One is that most studies have involved uranium, where there may be redox problems (note that CH_3Li reduces Eu^{3+}, too). Another is that the later actinides, which might be expected on size grounds to form

compounds in the +3 state similar to those of the lanthanides, are very inaccessible owing to radioactivity and non-availability of starting materials.

Question 13.3 Predict the products of the reaction of [Cp$_3$UCl] with (a) LiNEt$_2$, (b) LiSiPh$_3$, (c) CH$_3$CH$_2$CH$_2$CH$_2$Li, (d) C$_6$H$_5$CH$_2$MgBr. Would you expect [Cp$_3$U(SCN)] to have U–S or a U–N bond? Why?
Answer 13.3 (a) [Cp$_3$U(NEt$_2$)]; (b) [Cp$_3$U(SiPh$_3$)]; (c) [Cp$_3$U(CH$_2$CH$_2$CH$_2$CH$_3$)]; (d) [Cp$_3$U(CH$_2$C$_6$H$_5$)]. [Cp$_3$U(SCN)] is N-bonded, owing to the preference of uranium for the 'hard' donor N.

Question 13.4 Attempts to make the *tert*-butyl compound ThCp*$_2$(CMe$_3$)$_2$ by the reaction of ThCp*$_2$Cl$_2$ with *tert*-butyllithium were unsuccessful, the hydride ThCp*$_2$H$_2$ {Cp*$_2$Th(H)(μ-H)$_2$Th(H)Cp*$_2$} being obtained. Suggest why.
Answer 13.4 The most likely answer is a steric one; attaching two bulky *tert*-butyl groups to the Cp*$_2$Th unit would result in a very congested environment.

Question 13.5 Read Section 13.4.1. Suggest a reason for the reaction of ThCp*$_2$(CH$_2$CMe$_3$)$_2$ in forming a cyclic compound, also a reason for the reactivity of the metallacycles.
Answer 13.5 The agostic Th....H–C interactions in the structure of ThCp*$_2$(CH$_2$CMe$_3$)$_2$ referred to earlier hint that these molecules may be congested. The cyclic compounds contain a 4-membered ring; presumably the reactions relieve the strain.

Question 13.6 How would you confirm IR assignments for the bridged hydride [Cp*$_2$Th(H)(μ-H)$_2$Th(H)Cp*$_2$]?
Answer 13.6 Since the hydride groups are labile, isolating the deuteriated compound by exchange under a D$_2$ atmosphere in solution should be straightforward.

Question 13.7 Suggest why the uranium compound [Cp*$_2$U(H)(μ-H)$_2$U(H)Cp*$_2$] is less stable than the thorium analogue.
Answer 13.7 The +3 oxidation state is more accessible for uranium and so facilitates a decomposition pathway.

Question 13.8 Of the adducts [UCp*$_2$(CH$_2$SiMe$_3$)L] (L = PMe$_3$, py, Et$_2$O, THF), the Lewis base is most easily removed from the PMe$_3$ adduct. Suggest why.
Answer 13.8 Tertiary phosphines are 'soft' bases and actinides are 'hard' acids. This is likely to be the weakest linkage thermodynamically.

Question 13.9 The U–C and Th–C bond lengths are 2.647 and 2.701 Å, respectively, in M(η^8-C$_8$H$_8$)$_2$(M = U, Th). In [K(diglyme)][U(η^8-MeC$_8$H$_7$)$_2$] the U–C bond length averages 2.719 Å. Explain these values.
Answer 13.9 The ionic radius of Th^{4+} is greater than that of U^{4+}, whilst the radius of U^{3+} is greater than that of U^{4+}.

Question 13.10 Comment upon the C–O bond length and IR data for [U(C$_5$Me$_4$H)$_3$(CO)] (hint, remember $E = h\nu$). The C–O distance is slightly lengthened from 1.128 Å in free CO to 1.142 Å. ν_{CO} occurs at 1880 cm^{-1} in [U(C$_5$Me$_4$H)$_3$(CO)], shifted to 1840 cm^{-1} in

$[U(C_5Me_4H)_3(^{13}CO)]$ and 1793 cm^{-1} in $[U(C_5Me_4H)_3(C^{18}O)]$. It occurs at 2145 cm^{-1} in CO itself.

Answer 13.10 The C–O bond lengthens upon coordination, showing that it is getting weaker. This can be explained in terms of uranium (5f) to CO (2π) back-bonding, with electrons introduced into ligand π* antibonding orbitals (thus weakening the bond). The CO stretching frequency moves to lower frequency, also indicating bond weakening. This frequency also decreases on introducing C or O atoms of higher mass, confirming that vibrations of these atoms were indeed involved.

14 Synthesis of the Transactinides and their Chemistry

By the end of this chapter you should be able to:

- recognize that the transactinide elements have very short half-lives;
- know that studying their chemistry presents serious practical difficulties;
- appreciate ways in which these elements may be synthesized and their chemistry studied;
- have some knowledge of the chemistry of elements 104–108 and relate them to their neighbours in the Periodic Table;
- appreciate how the principles of the Periodic system may be used to predict the properties of other elements yet to be made and studied.

14.1 Introduction

As noted in Section 9.2, the heaver actinides were made an 'atom at a time'. The last one, lawrencium, was synthesized in 1961 by bombarding a target of californium isotopes (masses 249–252) with ^{10}B and ^{11}B nuclei. The isotope ^{258}Lr has a half-life of 3.8 s.

$$^{249}Cf + ^{11}B \rightarrow ^{258}Lr + 2^{1}n$$

Chemical studies were not possible until 1970 with the synthesis of longer-lived ^{256}Lr ($t_{1/2} = 26$ s). The even more stable neutron-rich ^{262}Lr with a half-life of 3.6 hours is now known. At the time, though, there seemed few prospects of extending the Periodic Table into another series and in any case it was also not clear what chemical properties were expected for these elements, nor how they might be studied. Experience with the actinides had shown that half-lives for α-decay of each element were steadily growing shorter as the atomic number increased, whilst the drop in half-lives for spontaneous fission was even more dramatic. Extrapolation to elements with atomic numbers around 107–108 indicated that half-lives would be so short ($t_{1/2} = 10^{-3}$ s) that it would be impossible to isolate further elements.

Theoreticians thought that stable heavier elements might be in prospect. The stability of a nucleus (based on a model of nuclear stability analogous to that of the Rutherford–Bohr model of electronic structure) is determined by the inter-nucleon forces (nucleons are protons and neutrons), an attractive force between all nucleons and a Coulombic repulsion force between protons, the latter becoming proportionately more important as the number of protons increases. Extra stability is associated with filled shells of nucleons, 'magic numbers'; for neutrons they are 2, 8, 20, 28, 50, 82, 126, 184, and 196; and for protons they are 2, 8, 20, 28, 50, 82, 114, and 164.

Lanthanide and Actinide Chemistry S. Cotton
© 2006 John Wiley & Sons, Ltd.

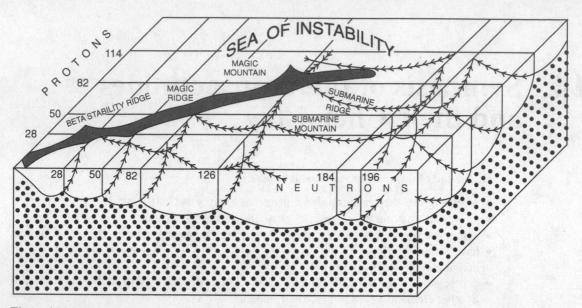

Figure 14.1
Known and predicted regions of nuclear stability (reproduced with permission from G.T. Seaborg, *J. Chem. Educ.*, 1969, **46**, 631).

Nuclei with closed shells of nucleons are stabilized, particularly against spontaneous fission. Nuclei where the numbers of protons and neutrons are both magic numbers ('double-magic' nuclei) are particularly stable, such as ^{4}He, ^{16}O, ^{40}Ca, and ^{208}Pb. After ^{208}Pb, the next 'double-magic' nucleus would be ^{298}114. The stability of nuclei can be represented on 3-D maps (Figure 14.1), with peaks corresponding to the stablest 'magic number' nuclei. Originally element 114 was represented by an 'island of stability' but new calculations suggest relatively stable nuclei may extend towards it. In the past thirty years, elements 104–116 and 118 have all been reported and some chemistry of elements 104–108 investigated. These investigations represent one of the greatest scientific achievements.

14.2 Finding New Elements

Unsuccessful attempts were made to find elements with $Z \sim 110$–114 in case they were stable. Searches for element 110 ('eka-platinum') in platinum metal ores led to negative results, corresponding to a limit of 10^{-11} g per g of stable element. Similar inconclusive results were obtained looking for element 114 ('eka-lead') in lead ores. Scientists therefore attempted to synthesize these new elements, using essentially the same methods used for the later actinides.

14.3 Synthesis of the Transactinides

Discovery of these elements has largely resulted from the researches of three groups: from Russian scientists at the Joint Institute for Nuclear Research (JINR), Dubna; German workers at the Gesellschaft für Schwerionenforschung (GSI), Darmstadt; and Americans,

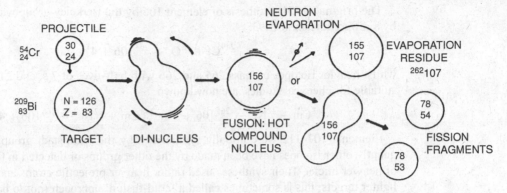

Figure 14.2
Synthesis of element 107 by nuclear fusion (reproduced with permission from G. Herrmann, *Angew. Chem., Int. Ed. Engl.*, 1988, **27**, 1421).

principally at the Lawrence Berkeley National Laboratory, Berkeley, California; recent contributions have come from Japanese workers at the RIKEN Linear Accelerator Facility. The approach used is to bombard a target nucleus (generally an actinide) with the positively charged nuclei of other elements, using a particle accelerator (cyclotron or linear accelerator) to supply the high energy needed to overcome the repulsion between the positively charged target nucleus and positive projectile nucleus, as already discussed in Section 9.2. In this method the atoms are produced one at a time, obviating 'normal' chemical experiments on bulk samples. The chance of a collision producing a combined nucleus is at best about one in a billion collisions.

Moreover, since the new atomic nucleus produced is in an excited nuclear state, it has an excess of energy to get rid of, whether by emission of neutrons and γ-rays, or by fission (Figure 14.2) (If the energy of the bombarding particles is too low, they do not possess enough energy to overcome the Coulombic repulsion between them.)

For elements 104 and 105, competing claims of first discovery have been made by the American and Russian teams. In 1969 workers from Berkeley reported the synthesis of $^{257-259}104$:

$$^{249}\text{Cf} + {}^{12}\text{C} \rightarrow {}^{257}104 + 4\,{}^{1}\text{n};\ {}^{249}\text{Cf} + {}^{13}\text{C} \rightarrow {}^{259}104 + 3\,{}^{1}\text{n}$$
$$^{248}\text{Cm} + {}^{16}\text{O} \rightarrow {}^{258}104 + 6\,{}^{1}\text{n}$$

Almost simultaneously an alternative route was reported by the Dubna workers:

$$^{242}\text{Pu} + {}^{22}\text{Ne} \rightarrow {}^{260}104 + 4\,{}^{1}\text{n};\ {}^{242}\text{Pu} + {}^{22}\text{Ne} \rightarrow {}^{259}104 + 3\,{}^{1}\text{n}$$

Since different isotopes of the transactinides have frequently been reported from different synthetic routes, it has been very difficult to decide who 'discovered' a new element.

Element 105 was reported by the Berkeley group in 1970 and the Dubna group in 1971. The Berkeley team used

$$^{249}\text{Cf} + {}^{15}\text{N} \rightarrow {}^{260}105 + 4\,{}^{1}\text{n}$$

whilst the Dubna group employed

$$^{243}\text{Am} + {}^{22}\text{Ne} \rightarrow {}^{260}105 + 5\,{}^{1}\text{n}$$

The original (1974) synthesis of element 106 by the Berkeley group was repeated there by a different team in 1993:

$$^{249}Cf + {}^{18}O \rightarrow {}^{263}106 + 4{}^{1}n$$

whilst heavier isotopes of mass 265 and 266 with half-lives of 7 s. and 21 s, respectively, suitable for chemical studies, are now known.

$$^{248}Cm + {}^{22}Ne \rightarrow {}^{265}106 + 5{}^{1}n; {}^{248}Cm + {}^{22}Ne \rightarrow {}^{266}106 + 4{}^{1}n$$

Elements 107–112 were initially synthesized by the Darmstadt group, though subsequently other isotopes have been made by the other groups or detected in the decay chains of heavier nuclei. Their syntheses used rather heavier projectile atoms and concomitantly lighter targets; this is sometimes called a 'cold-fusion' approach (not to be confused with the notorious room-temperature fusion reactions reported in 1989 as a route to 'limitless energy'). This method uses projectile particles with the minimum kinetic energy to just result in fusion so that the product nucleus has little excessive energy to get rid of, so that not only is the likelihood of fission minimized, but very few neutrons 'evaporate' from the 'hot' nucleus, so that the mass number of the resulting product is as high as possible. Another significant point is that the Pb and Bi targets have magic- or near-magic-number structures, as do the nickel projectile nuclei used in the syntheses of elements 110 and 111, so that a large part of the projectile energy is consumed in breaking into the filled shell and this also reduces the possibility of the new nucleus undergoing fission.

$$^{209}Bi + {}^{54}Cr \rightarrow {}^{262}107 + {}^{1}n; {}^{208}Pb + {}^{58}Fe \rightarrow {}^{265}108 + {}^{1}n$$
$$^{209}Bi + {}^{58}Fe \rightarrow {}^{265}109 + 2{}^{1}n; {}^{208}Pb + {}^{62}Ni \rightarrow {}^{269}110 + {}^{1}n$$
$$^{208}Pb + {}^{64}Ni \rightarrow {}^{271}110 + {}^{1}n; {}^{209}Bi + {}^{64}Ni \rightarrow {}^{272}111 + {}^{1}n$$
$$^{208}Pb + {}^{70}Zn \rightarrow {}^{277}112 + {}^{1}n$$

A different approach to element 107 has led to an isotope with a half-life of 14 s, used in chemical studies; similarly, synthesis of $^{269}108$ with a half-life of around 11s has permitted the first chemical investigations of element 108.

$$^{249}Bk + {}^{22}Ne \rightarrow {}^{267}107 + 4{}^{1}n; {}^{248}Cm + {}^{26}Mg \rightarrow {}^{269}108 + 5{}^{1}n$$

In 1999–2000, syntheses of elements 114, 116, and 118 were first reported, but the results for experiment 118 have proved to be irreproducible, so it cannot yet be claimed to have been made.

Bombardment of ^{244}Pu with neutron-rich ^{48}Ca for several weeks led to reports of one atom of element 114. The atom had a lifetime of about 30 s before decaying to form an atom of element 112:

$$^{244}Pu + {}^{48}Ca \rightarrow {}^{289}114 + 3{}^{1}n; {}^{289}114 \rightarrow {}^{285}112 + {}^{4}He$$

Particularly long-lived atoms detected in the decay chain were identified as isotopes of elements 112 and 108 with respective lifetimes of around 15 minutes. Other isotopes $^{287}114$ and $^{288}114$ have also been reported.

A similar route, bombardment of ^{248}Cm with ^{48}Ca, was used to make element 116 (Uuh); it alpha-decayed after 47 ms to a known isotope of element 114:

$$^{248}Cm + {}^{48}Ca \rightarrow {}^{292}116 + 4{}^{1}n; {}^{292}116 \rightarrow {}^{288}114 + {}^{4}He$$

Isotopes $^{290}116$ and $^{291}116$ have also been reported.

An alternative route was followed by workers at LBNL in California, who used a 'cold fusion' approach with a heavier projectile nucleus (krypton) and a lighter target (lead); bombardment over an eleven-day period afforded three atoms having a lifetime of less than a millisecond, believed to be element 118, though these claims must be regarded as unsubstantiated.

$$^{208}Pb + {}^{86}Kr \rightarrow {}^{293}118 + {}^{1}n$$

Most recently, in February 2004, experiments were reported giving serious evidence for elements 113 and 115, again synthesized using ^{48}Ca as the projectile:

$$^{243}Am + {}^{48}Ca \rightarrow {}^{287}115 + 4{}^{1}n \text{ (3 atoms)}$$
$$^{243}Am + {}^{48}Ca \rightarrow {}^{288}115 + 3{}^{1}n \text{ (1 atom)}$$

These atoms all decayed (with lifetimes in the range 19–280 milliseconds) by α-decay to afford what are believed to be atoms of element 113:

$$^{287}115 \rightarrow {}^{283}113 + {}^{4}He; {}^{288}115 \rightarrow {}^{284}113 + {}^{4}He$$

These atoms of element 113 have longer lifetimes, one in excess of a second.

An alternative route for synthesis of an atom of element 113 was reported by a team of Japanese (largely) and Chinese workers in October 2004:

$$^{209}Bi + {}^{70}Zn \rightarrow {}^{278}113 + {}^{1}n \text{ (1 atom)}$$

This decayed with a lifetime of 344 ms to an isotope of element 111.

$$^{278}113 \rightarrow {}^{274}111 + {}^{4}He; {}^{274}111 \rightarrow {}^{270}109 + {}^{4}He$$

14.4 Naming the Transactinides

The two groups who claim to have synthesized elements 104 and 105 gave them both a name. The International Union of Pure and Applied Chemistry therefore suggested a system of surrogate names for these elements until definite identification could be credited; this assigned each element a name based on its atomic number, using the 'code' 0 = nil, 1 = un, 2 = bi, 3 = tri, 4 = quad, 5 = pent, 6 = hex, 7 = sept, 8 = oct, and 9 = enn. Thus element 104 is unnilquadium.

In 1997, the International Union of Pure and Applied Chemistry decided names for elements 104–109, which have generally met with acceptance; they have recently decided names for 110 (2003) and for 111 (2004).

No decisions have yet been made about the heavier elements.

Table 14.1

Atomic number	Name	Symbol
104	Rutherfordium	Rf
105	Dubnium	Db
106	Seaborgium	Sg
107	Bohrium	Bh
108	Hassium	Hs
109	Meitnerium	Mt
110	Darmstadtium	Ds
111	Roentgenium	Rg

14.5 Predicting Electronic Arrangements

The 5f orbitals are gradually filled as the actinide series is crossed and there is general agreement that the 6d orbitals will be filled next, followed by 7p orbitals. Table 14.2 lists predicted electron configurations.

Table 14.2 Election configurations of elements 104–121[a]

Element	6d	7s	$7p_{1/2}$	$7p_{3/2}$	8s	7d (or $8p_{1/2}$)
104	2	2				
105	3	2				
106	4	2				
107	5	2				
108	6	2				
109	7	2				
110	8	2				
111	9	2				
112	10	2				
113	10	2	1			
114	10	2	2			
115	10	2	2	1		
116	10	2	2	2		
117	10	2	2	3		
118	10	2	2	4		
119	10	2	2	4	1	
120	10	2	2	4	2	
121	10	2	2	4	2	1

[a] Electron configurations are given outside the core [Rn] $5f^{14}$.

14.6 Identifying the Elements

Classical chemical routes are impractical with ultrasmall quantities of very radioactive, short-lived atoms. Even the most stable transactinides have such short half-lives (Table 14.3) that they had to be rapidly removed from the accelerator target for chemical study. This is achieved by the recoil of the new atom from the target and its transport by an aerosol in a current of helium to a collection site. Figure 14.3 shows the system used to study the chlorination of elements 104 and 105, using an MoO_3 aerosol to transport the atoms prior to halogenation. The volatile halides formed are transported by the helium along the latter part of the tube, which acts as an isothermal gas chromatography column; on leaving the quartz tube they are attached to new aerosols and detected either with a rotating wheel system or on a moving computer tape.

The millisecond half-lives of the first isotopes of elements 107–112 meant that there was no chance of carrying out chemical separations, so the Darmstadt group have used a 'velocity filter' which relies on electric and magnetic fields to separate the excess of fast-moving 'projectile' nuclei from slow-moving products recoiling from the target so that the reaction products can be detected (Figure 14.4). The beams of particles are focussed with magnetic lenses and then separated by electric and magnetic fields before colliding with solid-state detectors.

Table 14.3 Half-lives and decay modes for the isotopes of elements 104–116 and 118

Element	Half-life	Decay mode
104		
253	0.05 ms	SF
254	0.5 ms	SF
255	1.4 s	SF, α
256	7 ms	α, SF
257	4.8 s	α, SF
258	13 ms	SF, α?
259	3.0 s	α, SF
260	20 ms	SF
261	78 s	α, SF
262	47 ms	SF?
263	10 m	α, SF
105		
255	1.6 s	α, SF
256	2.6 s	EC, α
257	1.3 s	α, SF
258	20 s	EC, α
259		α
260	1.5 s	α, SF
261	1.8 s	α, SF
262	34 s	α, SF
263	27 s	SF, α
265	7.4 s	α
266	21 s	α
106		
258	2.9 ms	SF
259	0.9 s	α, SF
260	3.6 ms	α, SF
261	0.23 s	α, SF?
263	0.8 s	α, SF
265	16 s	α
266	21 s	α
107		
260		α
261	12 ms	α, SF?
262	8 ms/0.1 s	α
264	0.44 s	α
267	17 s	α
108		
263	<0.1s	α
264	0.08 ms	α, SF
265	1.8 ms	α
267	33 ms	α
269	11 s	α
270	3.6 s	α
273	1.2 s	α
277	11.4 m	α
109		
266	3.4 ms	α
268	70 ms	α

Continued

Table 14.3 *Continued*

Element	Half-life	Decay mode
110		
267	4 μs	α
269	0.17 ms	α
271	1.1 ms	α
273	0.1 s/0.1 ms	α
277		
280	7.4 s	SF
281	1.1 m	α
111		
272	1.5 ms	α
112		
277	242 μs	α
281		
283	3 m	SF
284	9.7 s	α
285	10.7 m	α
113		
283	~1 s	α
284	~0.4 s	α
114		
285	<1 ms	α
287	5 s	α
288	1.9 s	α
289	~21 s	α
115		
287	~46 ms	α
288	20–200 ms	α
116		
289	47 ms	α
118		
293		α

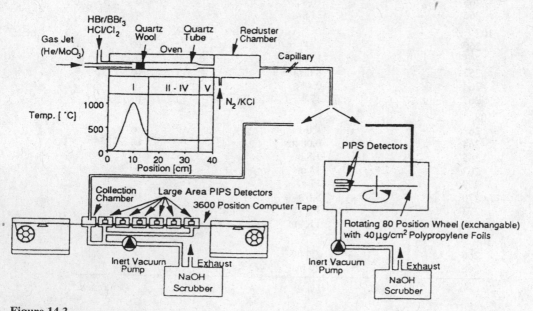

Figure 14.3
Apparatus for the study of the volatile halides of elements 104 and 105 (reproduced with permission from J.V. Kratz, *J. Alloys Compd.*, 1994, **213–214**, 22).

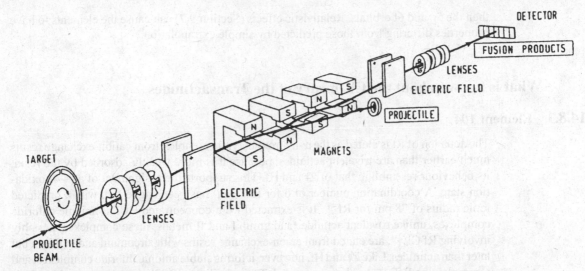

Figure 14.4
Velocity filter SHIP to separate transactinide fusion products from the projectile beam (reproduced with permission from G. Herrmann, *Angew. Chem., Int. Ed. Engl.*, 1988, **27**, 1422).

Detection of these elements relies on studying their radioactive decay, which is usually either α-emission or spontaneous fission. A time-correlation process is used; in this, a solid-state detector monitors both the time and position of arrival of fusion products. Subsequent decay events *at this position* give not just the decay information of the atom (half-life, α-particle energy) but also the corresponding information for its decay products, which are recognizable and thus 'known' nuclei. This therefore gives a history of stepwise decay of the initial product.

Thus reaction of ^{209}Bi with ^{54}Cr led to very short-lived nuclei ($t_{1/2} = 4.7$ ms) emitting α-particles with an energy of 10.38 MeV. This α-emitter correlated with a decay chain composed of three successive α-emitters, identified as ^{258}105 (4.4 s; α-particles of 9.17 MeV); ^{254}Lr (13 s; α-particles of 8.46 MeV); and ^{250}Md (52 s; α-particles of 7.75 MeV). The product of the fusion reaction was thus ^{262}107.

$$^{209}\text{Bi} + {}^{54}\text{Cr} \rightarrow {}^{262}\text{Bh} + {}^{1}\text{n}$$
$$^{262}107 \rightarrow {}^{258}\text{Db} + {}^{4}\text{He}$$
$$^{258}105 \rightarrow {}^{254}\text{Lr} + {}^{4}\text{He}$$
$$^{254}\text{Lr} \rightarrow {}^{250}\text{Md} + {}^{4}\text{He}$$

The limiting factor in detecting the reaction product is thus the time of flight of the product from the target to the detector, which is of the order of microseconds.

14.7 Predicting Chemistry of the Transactinides

Elements from 104 to 112 are believed to have electron configurations resulting from filling the 6d orbitals, so they are predicted to be transition metals. For the elements with $Z > 112$, it is assumed that the 7p orbitals are filled first; elements 119 and 120 involve filling the 8s orbital; whilst from 121 to 153, one electron is first placed in the 7d (or 8p) orbital then the 5g and 6f orbitals. Relativistic effects (Section 9.7) can cause the elements to have properties differing from those predicted by simple extrapolation.

14.8   What is known about the Chemistry of the Transactinides

14.8.1   Element 104

The aqua ion of Rf is eluted as the α-hydroxybutyrate complex from cation-exchange resins much earlier than are trivalent actinides, because it is more weakly adsorbed by the resin, its behaviour resembling that of Zr and Hf. This supports the assignment of a (+4) oxidation state. A coordination number of 6 for Rf^{4+}(aq) has been suggested, with a predicted ionic radius of 78 pm for Rf^{4+}. It is extracted from concentrated HCl as anionic chloride complexes, unlike trivalent actinides and group I and II metals; these complexes, possibly involving $RfCl_6^{2-}$, are eluted from anion-exchange resins with zirconium and hafnium but later than actinides. Like Zr and Hf, however, it forms stable anionic fluoride complexes, and a complex ion RfF_6^{2-} has been proposed. The volatility of $RfCl_4$ (condensing ~220 °C) is similar to that of $ZrCl_4$ but greater than that of $HfCl_4$; it is much greater than that of actinide tetrachlorides. $RfBr_4$ is more volatile than $HfBr_4$ (though the bromide is less volatile than the chloride).

14.8.2   Element 105

The aqua ion of Db is readily hydrolysed, even in strong HNO_3, and resembles those of Nb^V, Ta^V and Pa^V rather; the +4 ions of the Group IV elements don't do this, so this is an argument in favour of the +5 oxidation state in aqueous solution. Similarly, elution of Db as the α-hydroxybutyrate complex from cation-exchange resins shows it to be eluted rapidly, like Nb^V, Ta^V, and Pa^V but unlike Zr^{4+} and Eu^{3+} ions, which are strongly retained on the column. Study of complex formation in HCl indicates resemblances to Nb^V and Pa^V rather than Ta^V. It forms anionic fluoride complexes which, like those of Nb, Ta, and Pa, are strongly adsorbed on anion-exchange resins. Study of the volatility of the halides (assumed to be pentahalides) indicates that $DbBr_5$ is less volatile than $NbBr_5$ and $TaBr_5$ (in contradiction of recent theoretical predictions) and $RfBr_4$. The chloride appears to be more volatile than the bromide (at present, it still remains to be confirmed that it was the binary halides studied in these experiments rather than oxyhalides, as well as confirming the oxidation state of the element).

14.8.3   Element 106

A comparative study of the reaction of ^{263}Sg, Mo, and W with $SOCl_2$ in air suggests that $SgCl_2O_2$ is formed, with similar volatility to the Mo and W homologues. Investigations of Sg in aqueous solution indicate +6 to be the most stable oxidation state and that it forms neutral or ionic oxo and oxohalide compounds, species possibly including $[SgO_4]^{2-}$, $[SgO_3F]^-$, or $[SgO_2F_4]^{2-}$. Chromatographic study of solutions in very dilute HF indicates the formation of SgO_4^{2-}, analogous to MoO_4^{2-} and WO_4^{2-}. So far, 'seaborgium' behaves as a 6d transition metal, similar to Mo and W, with no evidence for deviations due to relativistic effects and little resemblance to uranium.

14.8.4   Element 107

In 2000, chemists synthesized atoms of ^{267}Bh ($t_{1/2} \sim 17$ s), using the reaction ^{22}Ne (^{249}Bk, 4n) ^{267}Bh. Atoms recoiling from the back of the berkelium target were absorbed onto carbon particles suspended in a flow of helium gas and transported to a quartz wool trap in an oven at 1000 °C. There the carbon and the nuclear reaction products were reacted with a mixture of HCl and O_2 gases. Some six atoms of bohrium were detected as a volatile oxychloride, believed to be BhO_3Cl, by analogy with Tc and Re.

14.8.5   Element 108

In the only studies so far reported (2005), hassium atoms were oxidized to a very volatile oxide (thought to be HsO_4 by analogy with Ru and Os – its absorption enthalpy is comparable with that of OsO_4), and deposited on an NaOH surface. Six correlated α-decay chains of Hs were observed and attributed to $Na_2[HsO_4(OH)_2]$, sodium hassate(VIII), by analogy with osmium, the 5d homologue of Hs.

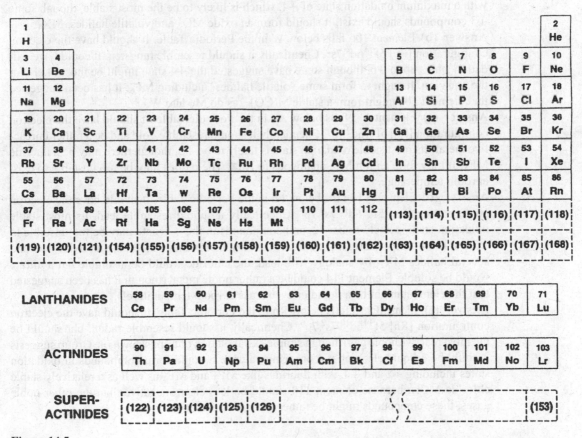

Figure 14.5
'Future' Periodic Table (reproduced with permission from G.T. Seaborg, *J. Chem. Soc., Dalton Trans.*, 1996, 3904).

14.9   And the Future?

Already many more elements have been synthesized than could be predicted in the early 1960s. Higher neutron-flux reactors could make the necessary amounts of the heavier transuranium elements needed for synthesis of the elements beyond 112. Much will depend upon whether the more neutron-rich isotopes with longer half-lives, essential for any study of chemical properties, can be made. The only safe prediction is the unpredictability of this area.

Question 14.1 Study the electron configurations in Table 14.2, and use your knowledge of the properties of lighter elements to predict properties of the following elements and their compounds:
(a) Element 104 and its compounds
(b) Element 106 and its compounds
(c) Element 111 and its compounds
(d) Element 114 and its compounds
(e) Element 118 and its compounds

Answer (a) Element 104 falls below Hf in the Periodic Table. It should have the electron configuration [Rn] $5f^{14}$ $6d^2$ $7s^2$. Chemically it should resemble zirconium and hafnium, with a maximum oxidation state of +4, which is likely to be the most stable, though some +3 compounds should exist. It should form an oxide MO_2 and volatile halides, MX_4.

Answer (b) Element 106 falls below W in the Periodic Table. It should have the electron configuration [Rn] $5f^{14}6d^4$ $7s^2$. Chemically it should resemble tungsten. It could have oxidation states up to +6, though some have suggested the +4 state might be most likely for the aqua ion. It ought to form some volatile halides, including MF_6. It has been suggested that element 106 might form a stable $M(CO)_6$, as do Mo and W.

Answer (c) Element 111 falls below Au in the Periodic Table. It should have the electron configuration [Rn] $5f^{14}6d^9$ $7s^2$. Chemically it should resemble gold and be a rather unreactive metal. Its most stable oxidation state is likely to be +3, though there may be a fluoride MF_5. It may form an ion M^-, analogous to Au^-. The oxides may be too unstable to exist at room temperature.

Answer (d) Element 114 falls below Pb in the Periodic Table. It should have the electron configuration [Rn] $5f^{14}6d^{10}$ $7s^2$ $7p^2$. Chemically it should resemble lead. Its most stable oxidation state is likely to be +2, and compounds in the +4 state may be too unstable to exist at room temperature. It should form several compounds, including an oxide MO and several halides MX_2; by analogy with lead, the chloride would be insoluble but a nitrate would be soluble. Element 114 could be a rather noble metal (though it has been suggested that the element might be a liquid or even a gas at room temperature!).

Answer (e) Element 118 falls below Rn in the Periodic Table. It should have the electron configuration [Rn] $5f^{14}6d^{10}$ $7s^2$ $7p^6$. Chemically it should resemble radon, but should be the most reactive noble gas (though simple extrapolation of trends down the Group suggests it could be a liquid element at room temperature). It could form compounds in oxidation states including +2 and +4, with fluorides like MF_2 and MF_4 as well as a relatively stable chloride MCl_2; because element 118 should be less electronegative than the other noble gases, these compounds might be rather ionic and involatile.

Bibliography

J. Emsley, *The Elements*, OUP, 1997; *Nature's Building Blocks*, OUP, 2001.

D.R. Lide (ed.), *CRC Handbook of Chemistry and Physics*, 85th edition, Central Rubber Company Press, 2004.

Acronyms for General Sources

AIP N.M. Edelstein (ed.), *Actinides in Perspective*, Pergamon, 1982.

ACE J.J. Katz, G.T. Seaborg and L.R. Morss (eds), *The Chemistry of the Actinide Elements*, 2nd edition, Chapman and Hall, 1986.

APU D. Hoffman (ed.), *Advances in Plutonium Chemistry 1967–2000*, American Nuclear Society, 2002.

CCC-1 G. Wilkinson, R.D. Gillard and J.A. McCleverty (eds), *Comprehensive Coordination Chemistry*, 1st edition, Pergamon, Oxford, 1987.

CCC-2 J.A. McCleverty and T.J. Meyer (eds), *Comprehensive Coordination Chemistry II*, Elsevier, Amsterdam, 2004.

CIC J.C. Bailar Jr., H.J. Emeleus, R.S. Nyholm and A.F. Trotman-Dickenson (eds), *Comprehensive Inorganic Chemistry*, Pergamon, Oxford, 1973.

COMC-1 G. Wilkinson, F.G.A. Stone and E. Abel (eds), *Comprehensive Organometallic Chemistry*, Pergamon, Oxford, 1982.

COMC-2 E.W. Abel, F.G.A. Stone and G. Wilkinson (eds), *Comprehensive Organometallic Chemistry-2*, Pergamon, Oxford, 1995.

EIC-1 R.B. King (ed.), *Encyclopedia of Inorganic Chemistry*, 1st edition, Wiley, Chichester, 1994.

EIC-2 R.B. King (ed.), *Encyclopedia of Inorganic Chemistry*, 2nd edition, Wiley, Chichester, 2005.

GMEL *Gmelins Handbuch der Anorganischer Chemie*.

HAC A.J. Freeman and G.H. Lander (eds), *Handbook on the Physics and Chemistry of the Actinides*, North Holland, 1984 onwards, Vols 1–5.

HRE K.A. Gschneider, LeRoy Eyring, *et al.* (eds), *Handbook on the Physics and Chemistry of Rare Earths*, North Holland, 1978 onwards, Vols 1–33 (at present).

IAR K.A. Gschneider (ed.), *Industrial Applications of Rare Earth Elements*, ACS Symposium Series, 1981.

LAS N.M. Edelstein (ed.) , *Lanthanide and Actinide Chemistry and Spectroscopy*, vol. 131. ACS Symposium Series, 1980.

LPL J.-C.G. Bunzli and G.R. Choppin (eds), *Lanthanide Probes in Life, Chemistry and Earth Sciences*, Elsevier, 1989.

MTP 1 and *MTP2* K.W. Bagnall (ed.), *MTP International Review of Science, Inorganic Chemistry*, Series 1 (1972) vol. 7; Series 2 (1975) vol. 7.

PIC   S.F.A. Kettle, *Physical Inorganic Chemistry: A Coordination Chemistry Approach*, Spektrum, 1996.

SPL   S.P. Sinha (ed.) *Systematics and Properties of the Lanthanides*, ACS Symposium Series, 1981.

1   Introduction to the Lanthanides

Further Reading

D.M. Yost, H. Russell, Jr and C.S. Garner, *The Rare Earths and Their Compounds*, Wiley, 1947.

T. Moeller, *CIC-1*, **4**, 1.

N.E. Topp, *The Chemistry of the Rare Earth Elements*, Elsevier, 1965.

G.R. Choppin, LPL, 1.

B.T. Kilbourn, *A Lanthanide Lanthology*, Molycorp, 1993 (Parts I and II).

B.T. Kilbourn, *Cerium. A Guide to its Role in Chemical Technology*, Molycorp, 1992.

T. Moeller, *J. Chem. Educ.*, 1970, **47**, 417 (principles).

C.H. Evans (ed), *Episodes from the History of the Rare Earth Elements*, Kluwer, Dordrecht, 1996.

F. Szabadváry, *HRE*, 1988, **11**, 33 (discovery and separation of lanthanides).

R.G. Bautista, *HRE*, 1995, **21**, 1 (separation chemistry).

K.L. Nash, *HRE*, 1991, **18**, 197 (separation chemistry of trivalent lanthanides and actinides).

H.E. Kremers, *J. Chem. Educ.*, 1985, **62**, 445 (extraction).

G.W. de Vore, *LPL*, 321 (geochemistry).

J.-C. Duchesne, *SPL*, 543 (geochemistry).

P. Moller, *SPL*, 561 (geochemistry).

S.R. Taylor and M.M. McLennan, *HRE*, **11**, 485 (geochemistry).

L.A. Haskin and T.P. Paster, *HRE*, **3**, 1 (geochemistry).

R.H. Byrne and E.R. Sholkovitz, *HRE*, 1996, **23**, 497 (marine chemistry and geochemistry).

General References

B.H. Ketelle and G.E. Boyd, *J. Am. Chem. Soc.*, 1947, **69**, 2800 (chromatography).

F.H. Spedding, E.I. Fulmer, J.E. Powell, and J.A. Butler, *J. Am. Chem. Soc.*, 1950, **72**, 2354 (chromatography).

W.B. Jensen, *J. Chem. Educ.*, 1982, **59**, 634 (positions of La and Lu in the Periodic Table).

2   The Lanthanides – Principles and Energetics

Further Reading

S.F.A. Kettle, *PIC*, 240 (f orbitals).

G.T. Seaborg and W.D. Loveland, *The Elements beyond Uranium*, Wiley, 1990, pp. 71–78 (f electrons).

D.R. Lloyd, *J. Chem. Educ.*, 1986, **63**, 502 (lanthanide contraction).

D.A. Johnson, *J. Chem. Educ.*, 1980, **57**, 475 (reactivity and 4f electrons).

L.J. Nugent, *J. Inorg. Nucl. Chem.*, 1975, **37**, 1767; *MTP2*, **7**, 195 (redox potentials, oxidation states).

L.R. Morss, *HRE*, 1991, **18**, 239 (redox potentials).

D.A. Johnson, *Some Thermodynamic Aspects of Inorganic Chemistry*, 2nd edition, CUP, Cambridge, 1982, pp. 50–54, 158–168; *J. Chem. Educ.*, 1986, **63**, 228 (energetics).

D.W. Smith, *Inorganic Substances*, CUP, Cambridge, 1990, pp. 145–149, 163–167 (energetics).

D.A. Johnson., *Adv. Inorg. Chem. Radiochem.*, 1977, **20**, 1 (unusual oxidation states).
F.A. Hart, *CCC-1*, **3**, 1109, 1113 (unusual oxidation states).
T. Moeller, *CIC*, **4**, 75, 97 (unusual oxidation states).

General References

D.A. Johnson, *J. Chem. Educ.*, 1980, **57**, 475 (ionic radii of Lanthanide ions).
D.W. Smith, *J. Chem. Educ.*, 1986, **63**, 228 (enthalpies of formation).
H.G. Friedman, G.R. Choppin, and D.G. Feverbacher, *J. Chem. Educ.*, 1964, **41**, 354 (f orbitals).

3 The Lanthanide Elements and Simple Binary Compounds

Further Reading

B.J. Beaudry and K.A. Gschneider Jr, *HRE*, **1**, 173 (elements).
H.F. Lineberger, T.K. McCluhan, L.A. Luyckx and K.E. Davies, *IAR*, **3**, 19, 167 (applications).
V. Paul-Boncour, L. Hilaire and A. Percheron-Guégan, *HRE*, 2000, **29**, 5 (metals and alloys in catalysis).
GMEL: C1 (oxides), C2 (hydroxides), C3, C4a/b, C6 (halides), C7 (sulfides), C11a/b (borides).
L. Eyring, *HRE*, **3**, 337 (oxides).
R.G. Haire and L. Eyring, *HRE*, 1991, **18**, 413 (oxides).
G. Adachi and N. Imanaka, *Chem. Rev.*, 1998, **98**, 1479 (oxides).
J. Burgess and J. Kijorski, *Adv. Inorg. Chem. Radiochem.*, 1981, **24**, 51 (halides).
J.M. Haschke, *HRE*, **4**, 89 (halides).
H.A. Eick , *HRE*, 1991, **18**, 365 (lanthanide and actinide halides).
G. Meyer and M. Wickleder, *HRE*, 2000, **28**, 53 (simple and complex halides).
A. Simon, Hj. Mattausch, G.J. Miller, W. Bauhofer and R.K. Kremer, *HRE*, 1991, **15**, 191 (metal-rich halides).
G. Adachi, N. Imanaka and Z. Fuzhong, *HRE*, 1991, **15**, 61 (carbides).
E.L. Huston and J.J. Sheridan III, *IAR*, 223 (hydrides).
G.G. Libowitz and A.J. Maeland, *HRE*, **3**, 299 (hydrides).
J.W. Ward and J.M. Haschke, *HRE*, 1991, **18**, 293 (hydrides).
Yu. Kuz'ma and S. Chykhrij, *HRE*, 1996, **23**, 285 (phosphides).
I.G. Vasilyeva, *HRE*, 2001, **32**, 567 (polysulfides).
K. Mitchell and J.A. Ibers, *Chem. Rev.*, 2002, **102**, 1929 (rare-earth and transition-metal chalcogenides).

General References

S.A. Cotton, *Lanthanides and Actinides*, Macmillan, 1991.
R.B. King (ed.), *Encyclopedia of Inorganic Chemistry*, 1st edition, Wiley, Chichester, 1994.

Superconductors

Further Reading

C.N.R. Rao, (ed), *Chemistry of Oxide Superconductors*, Blackwell, Oxford, 1988.
M.B. Maple, M.C. de Andrade, J. Herrmann, R.P. Dickey, N.R. Dilley and S. Han, *J. Alloys Compd.*, 1997, **250**, 585 (superconductors).

Handbook on the Physics and Chemistry of Rare Earths, North Holland, 2000–2001, Volumes 30 and 31, contains several reviews on high-temperature superconductors; the most useful of these are the following:

- M.B. Maple, *HRE*, 2000, **30**, 1 (superconductivity in layered cuprates);
- B. Raveau, C. Michel and M. Hervieu, *HRE*, 2000, **30**, 31 (crystal chemistry of superconducting rare-earth cuprates);
- W.E. Pickett and I.I. Mazin, *HRE*, 2000, **30**, 453 (electronic theory and physical properties of $RBa_2Cu_3O_7$ compounds);
- K. Kobayashi and S. Hirosawa, *HRE*, 2001, **32**, 515 (permanent magnets).

General References

M.K. Wu, J.R. Ashburn, C.J. Torng, P.H. Hor, R.L. Meng, L. Gao, Z.J. Huang, Q. Wang and C.W. Chu, *Phys. Rev. Lett.*, 1987, **58**, 908 ($YBa_2Cu_3O_{7-\delta}$).

K.A. Bednorz and J.G. Muller, *Science*, 1987, **237**, 1133; *Angew. Chem. Int. Ed. Engl.*, 1988, **27**, 735 (superconducting oxides).

4   Coordination Chemistry of the Lanthanides

Further Reading

F.A. Hart, *CCC-1*, **3**, 1087 (complexes).

S.A. Cotton, *CCC-2*, **3**, 93; *EIC-2*, p. 4838 (complexes).

L. Niinsto and M. Leskala, *HRE*, **8**, 203; **9**, 91 (complexes).

L.C. Thompson, *HRE*, **3**, 209 (complexes).

J.F. Desreux, *LPL*, 43 (complexes).

GMEL, D1–D5 (complexes).

S.A. Cotton, *CR Chimie*, 2005, **8**, 129 (establishing coordination numbers for the lanthanides in simple complexes).

G.R. Choppin, *J. Alloys Compd.*, 1997, **249**, 1 (factors involved in lanthanide complexation).

T. Moeller, *CIC*, **4**, 28; *MTP1*, 285 (stability constants).

A.E. Martell and R.M. Smith, *Critical Stability Constants*, Plenum, 1974 onwards, vols 1–6.

H.C. Aspinall, *Chem. Rev.*, 2002, **102**, 1807 (chiral lanthanide complexes).

E.N. Rizkalla and G.R. Choppin, *HRE*, 1991, **15**, 393; *HRE*, 1994, **18**, 529 (hydration and hydrolysis of lanthanides).

N. Marques, A. Sella and J. Takats, *Chem. Rev.*, 2002, **102**, 2137 (pyrazolylborate complexes of the lanthanides).

G.R. Choppin, *J. Alloys Compd.*, 1993, **192**, 256 (aminopolycarboxylate complexes).

G. Bombieri and R. Artali, *J. Alloys Compd.*, 2002, **344**, 9 (DOTA complexes).

J.C.G. Bunzli, *HRE*, **9**, 321; *LPL*, 219 (macrocycle complexes).

A.G. Coutsolelos, *NATO ASI Ser, Ser. C.*, 1995, **459**, 267 (porphyrin complexes).

D.K.P. Ng and J. Jiang, *Chem. Soc. Rev.*, 1997, **26**, 433 (porphyrin complexes).

L.M. Vallarino, *HRE*, 1988, **15**, 4433 (macrocycle complexes from template syntheses).

R.C. Mehrotra, A. Singh and U.M. Tripathi, *Chem. Rev.*, 1991, **91**, 1287 (lanthanide alkoxides).

M.N. Bochkarev, L.N. Zakharov and G.S. Kalinina, *Organoderivatives of Rare Earth Elements*, Kluwer, Dordrecht, 1995 (amides, alkoxides, thiolates).

D.C. Bradley, R.C. Mehrotra, I.P. Rothwell and A. Singh, *Alkoxo and Aryloxo Derivatives of Metals*, Academic Press, San. Diego, 2001.

General References

G.B. Deacon, T. Feng, P.C. Junk, B.W. Skelton, A.N. Sobolev, and A.H. White, *Aurt. J. Chem.*, 1998, **51**, 75 (lanthanide chloride complexes of THF).

G.B. Deacon *et al., J. Chem. Soc., Dalton Trans.*, 2000, 961 (alkoxides).

X.-Z. Feng, A.-L. Guo, Y.-T. Xu, X.-F. Li and P.-N. Sun, *Polyhedron*, 1987, **6**, 1041 (coordination numbers).

J. Marçalo and A. Pires de Matos, *Polyhedron*, 1989, **8**, 2431 (coordination numbers).

B.P. Hay, *Inorg. Chem.*, 1991, **30**, 2881 (nonaqualanthanide ion structure).

A. Habenschuss and F.H. Spedding, *Cryst. Struct. Commun.*, 1982, **7**, 538 (crystal structure Ln cation).

A. Zalkin, D.H. Templeton, and D.G. Karraker, *Inorg. Chem.*, 1969, **8**, 2680 (structure of Ho complex).

T. Phillips, D.E. Sands, and W.F. Wagner, *Inorg. Chem.*, 1968, **7**, 2299 (structure of La complex).

J.D.J. Backer-Dirks, J.E. Cooke, A.M.R. Galas, J.S. Ghotra, C.J. Gray, F.A. Hart, and M.B. Hursthouse, *J. Chem. Soc., Dalton Trans.*, 1980, 2191 (structure of a La crown ether complex).

S.A. Cotton, *Lanthanides and Actinides*, Macmillan, 1991, p. 64 (structure of La EDTA complex).

M. Frechette, I.R. Butler, R. Hynes, and C. Detellièr, *Inorg. Chem*, 1992, **31**, 1650 (structure of La phen complex).

G.H. Frost, F.A. Hart, C. Heath, and M.B. Hursthouse, *Chem. Commun.*, 1969, 1421 (structure of Eu terpy complex).

O. Poncelet, O.W.J. Sartain, L.G. Hubert-Pfalzgraf, K. Folting, and K.G. Caulton, *Inorg. Chem.*, 1989, **28**, 264 [structure of $Y_5O(O Pr^i)_{13}$].

D.L. Clark, J.C. Gordon, P.J. Hay, R.L. Martin, and R. Poli, *Organometallics*, 2002, **21**, 5000 (theoretical calculation).

L. Perrin, L. Maron, O. Eisenstein, and M.F. Lappert, *New J. Chem.*, 2002, **27**, 121 (theoretical calculation).

5 Electronic/Magnetic Properties of the Lanthanides

Magnetism

Further Reading

F.A. Hart, *CCC*, **3**, 1109.

T. Moeller, *CIC*, **4**, 9.

J. Rossat-Mignod, *SPL*, 255.

J.J. Rhyne, *HRE*, 1988, **11**, 293.

S.F.A. Kettle, *PIC*, 265.

Electronic Spectra

Further Reading

F.A. Hart, *CCC-1*, **3**, 1105.

T. Moeller, *CIC*, **4**, 9; *MTP1*, **7**, 275.

W.T. Carnall, J.V Beitz, H. Crosswhite, K. Rajnak and J.B. Mann, *SPL*, 389; *HRE*, **3**, 171.

C.A. Morrison and R.P. Leavitt, *HRE*, **5**, 461.

J.W. O'Laughlin, *HRE*, **4**, 341.

B.R. Judd, *HRE*, **11**, 81 (atomic theory and optical spectroscopy).

B.R. Judd, *LAS*, 267.

J.V. Beitz, *HRE*, 1991, **18**, 159 (absorption and luminescence spectra).

S.F.A. Kettle, *PIC*, 257 (f–f transitions).

C. Görller-Walrand and K. Binnemans, *HRE*, 1996, **23**, 121 (crystal-field parameters).

C. Görller-Walrand and K. Binnemans, *HRE*, 1998, **25**, 101 (intensities of f–f transitions).

N. Sabattini, M. Guardigli and I. Manet, *HRE*, 1996, **23**, 69 (antenna effect in encapsulated complexes).

G.E. Buono-core and H.LiB. Marciniak, *Coord. Chem. Rev.*, 1990, **99**, 55 (quenching of excited states by lanthanide ions and chelates in solution).

GMEL, E1, 1993 (optical spectra of praseodymium compounds).

GMEL, E2a, 1997 (optical spectra of neodymium compounds).

General References

D.G. Karraker, *Inorg. Chem.*, 1968, **7**, 473 (hypersensitivity).

S. Hufner, *SPL*, 313.

Luminescence spectra

Further Reading

C. Fouassier, *EIC-1*, 1984; *EIC-2*, 2730 (luminescence).

K.B. Yatmirskii and N.H. Davidenko, *Coord. Chem. Rev.*, 1979, **27**, 223 (spectra).

J.H. Forsberg, *Coord. Chem. Rev.*, 1973, **10**, 195 (spectra).

S.F.A. Kettle, *PIC*, 263 (luminescence).

D. Parker, *Coord. Chem. Rev.*, 2000, **205**, 109 (luminescent sensors).

Y. Korovin and N. Rusakova, *Rev. Inorg. Chem.*, 2001, **21**, 299 (IR 4f luminescence).

I. Billard, *HRE*, 2003, **33**, 465 (time-resolved, laser-induced emission spectroscopy).

D. Parker and J.A.G. Williams, *J. Chem. Soc., Dalton Trans.*, 1996, 3613 (absorption and luminescence spectra, NMR).

D. Parker, R.S. Dickins, H. Puschmann, C. Crossland and J.A.K. Howard, *Chem. Rev.*, 2002, **102**, 1977 (absorption and luminescence spectra, NMR).

D. Parker, *Chem. Soc. Rev.*, 2004, **33**, 156 (absorption and luminescence spectra, NMR).

T. Gunnlaugsson and J.P. Leonard, *Chem. Commun.*, 2005, 3114 (luminescent sensors).

G.R. Choppin and Z.M. Wang, *Inorg. Chem.*, 1997, **36**, 249 ($^7F_0 \rightarrow {}^5D_0$ transition).

S.T. Frey and W.D. Horrocks, *Inorg. Chim. Acta.*, 1995, **229**, 383 ($^7F_0 \rightarrow {}^5D_0$ transition).

G-y. Adachi and N. Inamaka, *HRE*, 1995, **21**, 179 (chemical sensors).

G. Blasse, *Chem. Mater.*, 1994, **6**, 1465 (scintillator materials).

General References

T. Gunnlaugsson, J.P. Leonard, K. Sénéchal, and A.R. Harte, *J. Am. Chem. Sec.*, 2003, **125**, 12062 (luminescence spectra).

T. Gunnlaugsson, J.P. Leonard, K. Sénéchal, and A.R. Harte, *Chem. Commun.*, 2004, 782 (luminescence spectra).

J.-C.G. Bunzli, B. Klein, G. Chapuis, and K.J. Schenk, *Inorg. Chem.*, 1982, **21**, 808 (luminescence spectra).

T. Yamada, S. Shinoda, and H. Tsukube, *Chem. Commun.*, 2002, 218 (luminescent sensors).

T. Gunnlaugsson, D.A. MacDonail, and D. Parker, *Chem. Commun.*, 2000, 93 (luminescent sensors).

D. Parker and J.A.G. Williams, *Chem. Commun.*, 1998, 245 (luminescent sensors).

D. Parker, K. Senanaayake, and J.A.G. Williams, *Chem. Commun.*, 1997, 1777 (luminescent sensors).

B. Blasse and A. Bril, *J. Chem. Phys.*, 1966, **45**, 3327 (luminescence).

D.A. Durham, G.H. Frost, and F.A. Hart, *J. Inorg. Nucl. Chem.*, 1969, **31**, 833 (luminescence).

R.C. Holz and L.C. Thompson, *Inorg. Chem.*, 1993, **32**, 5251 (luminescence).

D.F. Moser, L.C. Thompson, and V.G. Young, *J. Alloys Compd.*, 2000, **303–304**, 121.

Colour TV

Further Reading

J.R. McColl and F.C. Palilla, *IAR*, 177.

G. Blasse, *HRE*, **4**, 237.

Lasers

Further Reading

M.J. Weber, *HRE*, **4**, 275.

J. Hecht, *The Laser Guidebook*, 2nd edn., McGraw-Hill, New York, 1992.

J. Fready, *Kirk-Othmer Encyclopedia of Chemical Technology, 4th Edition*, 1995, vol. 15, 16.

D.L. Andrews and A.A. Demydov (eds)., *An Introduction to Laser Spectroscopy*, Kluwer, Dordrecht, 2002.

K. Kuriki, Y. Koike, and Y. Okamoto, *Chem. Rev.*, 2002, **102**, 2347 (plastic optical fibre lasers and amplifiers).

ESR

Further Reading

E.M. Stephens, *LPL*, 181 (Gd ESR).

A. Abragam and B. Bleaney, *Electron Paramagnetic Resonance of Transition ions*, Oxford, 1970.

NMR

Further Reading

F.A. Hart, *CCC-1*, **3**, 1100 (NMR).

J. Reuben and G.A. Elgarish, *HRE*, **4**, 483 (NMR).

J.H. Forsberg, *HRE*, 1996, **23,** 1 (NMR of paramagnetic lanthanide compounds).

C. Piguet and C.F.G.C. Geraldes, *HRE*, 2003, **33**, 353 (solution structure of lanthanide-containing coordination compounds by paramagnetic nuclear magnetic resonance).

J. Reuben, *Prog. Nucl. Magn. Reson. Spectrosc.*, 1975, **9**, 1 (shift reagents).

J. Reuben and G.A. Elgarish, *HRE*, **4**, 483 (shift reagents).

A.D. Sherry and C.F.G.C. Geraldes, *LPL*, 219 (shift reagents).

M.R. Peterson Jr and G.H. Wahl Jr, *J. Chem. Educ.*, 1972, **49**, 790 (shift reagents).

R.E. Sievers, *Nuclear Magnetic Shift Reagents*, Academic Press, 1972.

F.A. Hart, *CCC-1*, **3**, 1105 (shift reagents).

M.F. Tweedle, *LPL*, 127 (MRI agents).

R.B. Lauffer, *Chem. Rev.*, 1987, **87**, 901 (MRI agents).

P. Caravan, J.J. Ellison, T.J. McMurry and R.B. Lauffer, *Chem. Rev.*, 1999, **99**, 2293 (MRI agents).

A. Merbach and E. Toth (eds) *The Chemistry of Contrast Agents in Medical Magnetic Resonance Imaging*, Wiley, Chichester, 2001.

M.P. Lowe, *Aust. J. Chem.*, 2002, **55**, 551 ('new generation' MRI contrast agents).

S. Aime, M. Botta and E. Terreno, *Adv. Inorg. Chem. Radiochem.*, 2005, **57**, 173 (MRI agents).

L. Helm, G.M. Nicolle and A.E. Merbach, *Adv. Inorg. Chem. Radiochem.*, 2005, **57**, 327 (water and proton exchange).

General References

T. Viswanathan and A. Toland, *J. Chem. Educ.*, 1995, **72**, 945 (chiral shift reagent).

S.A. Cotton, *Lanthanides and Actinides*, Macmillan, 1991, p. 64 (MRI agents).

J.K.M. Sanders and D.H. Williams, *Chem. Commun.*, 1970, 422 (shift reagents).

Lanthanide Biochemistry

Further Reading

J.-C.G. Bunzli, *LPL*, 219 (bioprobes).

W de W. Horrocks and D.R. Sudnick, *Acc. Chem. Res.*, 1981, **14**, 384 (bioprobes).

J. Reuben, *HRE*, **4**, 515 (bioprobes).

J.R. Ascenso and A.V. Xavier, *SPL*, 501 (bioprobes).

F.S. Richardson, *Chem. Rev.*, 1982, **82**, 541 (bioprobes).

C.H. Evans, *Biochemistry of the Elements, Volume 8; Biochemistry of the Lanthanides*, Plenum, New York, 1990.

J.R. Duffield, D.M. Taylor and D.R. Williams, *HRE*, 1991, **18**, 591 (biochemistry).

J. Burgess, *Chem. Soc. Rev.*, 1996, **25**, 85 (Sc, Y, lanthanides in living systems).

J.M. Harrowfield, in A. Sigel and H. Sigel (eds), *Metal Ions in Biological Systems*, Marcel Dekker, 2003, vol. 40, 106 (interrelations of lanthanides with biosystems).

G.R. Motson, J.S. Fleming and S. Brooker, *Adv. Inorg. Chem.*, 2004, **55**, 361 (Ln complexes as luminescent biolabels).

H. Tsukube, S. Shinoda and H. Tamiaki, *Coord. Chem. Rev.*, 2002, **226**, 227 (chiral recognition and sensing of biological substrates).

6 Organometallic Chemistry of the Lanthanides

Further Reading

F.G.N. Cloke, *COMC-2*, **3**, 1 (zerovalent compounds).

F.T. Edelmann,*COMC-2*, **3**, 11 (all organometallics except zerovalent compounds).

T.J. Marks and R.D. Ernst, *COMC-1*, **3**, 173 (review).

W.J. Evans, *Adv. Organomet. Chem.*, 1985, **24**, 131 (review).

R.D. Kohn, G. Kociok-Kohn and H. Schumann, *EIC-1*, 3618 (review).

H. Schumann and I.L. Fedushkin, *EIC-2*, 4878 (review).

C.J. Schaverien, *Adv. Organomet. Chem.*, 1994, **36**, 283 (review).

M.N. Bochkarev, L.N. Zakharov and G.S. Kalinina, *Organoderivatives of Rare Earth Elements*, Kluwer, Dordrecht, 1995.

W.A. Herrmann (ed.), *Organolanthanoid Chemistry: Synthesis, Structure, Catalysis, in Topics in Current Chemistry*, Springer-Verlag, Berlin, 1996, vol. 179.

H. Schumann, J.A. Meese-Marktscheffel and L. Essar, *Chem. Rev.*, 1995, **95**, 865 (review of organometallic π-complexes).

G. Bombieri and G. Paolucci, *HRE*, 1998, **25**, 265 (organometallic π-complexes).

F.T. Edelmann, *Angew. Chem., Int. Ed. Engl.*, 1995, **34**, 2466 (cyclopentadienyl-free organolanthanide chemistry).

F.T. Edelmann, D.M.M. Freckmann and H. Schumann, *Chem. Rev.*, 2002, **102**, 1851 (non-cyclopentadienyl organolanthanide complexes).

S. Arndt and J. Okuda, *Chem. Rev.*, 2002, **102**, 1953 [mono(cyclopentadienyl) complexes].

M.N. Bochkarev, *Chem. Rev.*, 2002, **102**, 2089 (arene–lanthanide compounds).

F.G.N. Cloke, *Chem. Soc. Rev.*, 1993, **22**, 17 (zerovalent state).

General References

W.J. Evans and B.L. Davis, *Chem. Rev.*, 2002, **102**, 2119 ([(C$_5$Me$_5$)$_3$Ln] systems).

7 The Misfits: Scandium, Yttrium and Promethium

These elements are covered, along with the lanthanides, in Section 39 of the Gmelin Handbook, most of which extends to cover the 1980s.

Further Reading

R.C. Vickery, *The Chemistry of Yttrium and Scandium*, Pergamon, 1960; *CIC*, **3**, 329.

C.T. Horowitz (ed), *Scandium*, Academic Press, 1973 (comprehensive book to that point).

S.A. Cotton, *EIC 2*, p 4838.

F.A. Hart, *CCC-1*, **3**, 1059 (coordination chemistry of Sc and Lanthanides).

S.A. Cotton, *CCC-2*, **3**, 93 (coordination chemistry of Sc and Lanthanides).

S.A. Cotton, *Polyhedron*, 1999, **18**, 1691 (Sc review).

S.A. Cotton, *Comments Inorg. Chem.*, 1999, **21**, 165 [coordination number of Sc^{3+}(aq)].

F. Weigel, *Chem. Zeit.*, 1978, **108**, 339 (comprehensive review of Pm).

G.E. Boyd, *J. Chem. Educ.*, 1959, **36**, 3 (Pm).

E.J. Wheelwright (ed.), *Promethium Technology*, American Nuclear Society, 1973.

S.A. Cotton, *Educ. in Chem.*, 1999, **36**, 96 (discovery of Pm, and its chemistry).

General References

P.S. Coan, L.G. Hubert-Pfalzgraf, and K.G. Caulton, *Inorg. Chem.*, 1992, **31**, 1262 [Y$_5$O(O^{i}P$_r$)$_{13}$ NMR].

W.R. Wilmarth, R.G. Haire, Y.P. Young, D.W. Ramey, and J.R. Peterson, *J. Less Common Metals* (now *J. Alloys Compd*)., 1988, **141**, 275; W.R. Wilmarth, G.M. Bequn, R.G. Haire, and J.R. Peterson, *J. Raman Spectrosc.*, 1988, **19**, 271 (halides and oxide of Pm).

G.R. Willey, M.T. Lakin, N.W. Alcock, *J. Chem. Soc., Chem. Commun.*, 1992, 1619 (Sc crown ether complex).

J.S. Ghotra, M.B. Hursthouse, and A.J. Welch, *J. Chem. Soc., Chem. Commun.*, 1973, 669 (Sc silylamide).

J. Arnold, C.G. Hoffman, D.Y. Dawson, and F.J. Hollander. *Organometallics*, 1993, **12**, 3646 (Sc complexes).

8   The Lanthanides and Scandium in Organic Chemistry

Further Reading

S. Kobayashi, *Eur. J. Org. Chem.*, 1999, 15; S. Kobayashi, M. Sugiura, H. Kitagawa and W.W.-L. Lam, *Chem. Rev.*, 2002, **102**, 2227 (lanthanide triflates in organic synthesis).

S. Kobayashi and K. Manabe, *Pure Appl. Chem.*, 2000, **72**, 1373. ('Green' Lewis Acid catalysis in organic synthesis).

H.B. Kagan, *Tetrahedron*, 2003, **59**, 10351 (SmI$_2$ overview).

G.A. Molander and C.R. Harris, *Chem. Rev.*, 1996, **96**, 307; *Tetrahedron*, 1998, **54**, 3321 (SmI$_2$ in sequencing reactions).

A. Dahlén and G. Hilmersson, *Eur. J. Inorg. Chem.*, 2004, 3393 (SmI$_2$-mediated reductions – influence of various additives).

G.A. Molander and J.A.C. Romero, *Chem. Rev.*, 2002, **102**, 2161 (lanthanocene catalysts in selective organic synthesis).

J. Inanaga, H. Furuno and T. Hayano, *Chem. Rev.*, 2002, **102**, 2211 (asymmetric catalysis and amplification with chiral lanthanide complexes).

K. Mikami, M. Terada and H. Matsuzawa, *Angew. Chem., Int. Ed. Engl.*, 2002, **41**, 3554 (asymmetric catalysis by lanthanide complexes).

H.C. Aspinall and N. Greeves, *J. Organomet. Chem.*, 2002, **647**, 151 (asymmetric synthesis).

M. Shibasaki and N. Yoshikawa, *Chem. Rev.*, 2002, **102**, 2187 (lanthanide complexes in multifunctional asymmetric catalysis).

F. Helion and J.L. Namy, *J. Org. Chem.*, 1999, **64**, 2944; M.-I. Lannou, F. Helion and J.-L. Namy, *Tetrahedron*, 2003, **59**, 10551 (uses of mischmetal in organic synthesis).

D. Klomp, T. Maschmeyer, U. Hanefeld and J.A. Peters, *Chem. Eur. J.*, 2004, **10**, 2088; S.-I. Fukuzawa, N. Nakano and T. Saitoh, *Eur. J. Org. Chem.*, 2004, 2863 (Meerwein–Ponndorf–Verley reaction).

General References

W.J. Evans, J.D. Feldman, and J.W. Ziller, *J. Am. Chem. Soc.*, 1996, **118**, 4581 [CeCl$_3$(H$_2$O)].

T. Imamoto, *Lanthanides in Organic Synthesis*, Academic Press., 1994.

S. Kobayashi (ed.), *Lanthanides: Chemistry and Use in Organic Synthesis*, Springer-Verlag, Heidelberg, 1999.

G.A. Molander, *Chem. Rev.*, 1992, **92**, 29 (application of lanthanide reagents in organic synthesis).

9   Introduction to the Actinides

Further Reading

G.T. Seaborg and W.D. Loveland, *The Elements beyond Uranium*, Wiley, 1990.

G.T. Seaborg, *HRE*, 1991, **18**, 1 (origin of the actinide concept).

M.S. Fred, *ACE*, **2**, 1196 (oxidation states).

J.J. Katz, L.R. Morss and G.T. Seaborg, *ACE*, **2**, 1133, 1142 (oxidation states).

L.J. Nugent, *J. Inorg. Nucl. Chem.*, 1975, **37**, 1767; *MTP2*, **7**, 195 (oxidation states).

L.R. Morss, *HRE*, 1991, **18**, 239 (reduction potentials).

T.W. Newton, *APU*, 24 (redox chemistry of Pu).

Z. Kolaria, *HAC*, **3**, 431 (extraction).

Z. Yong-jun, *HAC*, **3**, 469 (extraction).

L.I. Katzin and D.C. Sonnenberger, *ACE*, **1**, 46 (extraction of thorium).

F. Weigel, *ACE*, **1**, 196 (extraction of uranium).

H.W. Kirby, *ACE*, **1**, 112 (extraction of protactinium).

F. Weigel, *ACE*, **1**, 173 (separation of isotopes).

S. Villani, *Isotope Separation,* American Nuclear Society, 1976; *Uranium Enrichment*, Springer-Verlag, 1979.

P. Pyykko, *Adv. Quantum Chem.*, 1978, **11**, 353 (relativistic effects).

K.S. Pitzer, *Acc. Chem. Res.*, 1979, **12**, 271 (relativistic effects).

P. Pyykko, *Acc. Chem. Res.*, 1979, **12**, 276 (relativistic effects).

P. Pyykko, *Chem. Rev.*, 1988, **88**, 563 (relativistic effects).

K. Balasubramanian, *HRE*, 1991, **18**, 29 (relativistic effects).

N. Kaltsoyannis. *J. Chem. Soc., Dalton Trans.*, 1997, 1 (relativistic effects).

B.A. Hess, *Relativistic Effects in Heavy Element Chemistry and Physics*, Wiley, 2002.

N. Kaltsoyannis, *Chem. Soc. Rev.*, 2003, **32**, 9 (computational actinide chemistry, including relativistic effects).

10   Binary Compounds

Further Reading

For uranium, consult *GMEL,* Supplement No. 55, Main Volume 1936, Supplements from 1975; A1–A7 (element), B2 (alloys), C1–C14 (compounds); D1–D4 (solution), E1–E2 (coordination compounds).

R.G. Haire and L. Eyring, *HRE*, 1991, **18**, 413 (oxides).

D. Brown, *Halides of Lanthanides and Actinides*, Wiley, 1968; *MTP-1*, **1**, 87 (actinide halides).

J.M. Haschke and R.G. Haire, *APU*, 212 (Pu binary compounds).

General References

S.A. Cotton, *Lanthanides and Actinides*, Macmillan, 1991, p.95 (laser separation).

D.D. Ensor, J.R. Peterson, R.G. Haire, and J.P. Young, *J. Inorg. Nucl. Chem.*, 1981, **43**, 2425 (EsF$_3$).

J.P. Young, R.G. Haire, J.R. Peterson, D.D. Ensor, and R.L. Fellows, *Inorg. Chem.*, 1981, **20**, 3979 (EsCl$_3$).

J.H. Burns, J.R. Peterson, and R.D. Baybarz, *J. Inorg. Nucl. Chem.*, 1973, **35**, 1171 (CfCl$_3$).

J.R. Peterson and J.H. Burns, *J. Inorg. Nucl. Chem.*, 1973, **35**, 1525 (CmCl$_3$).

Iu. V. Drobyshevskii, N.N. Prusakov, V.F. Serik, and V.B. Sokolov, *Radiokhimiya*, 1980, **22**, 591 (claim for AmF$_6$).

J.-C. Berthet, M. Nierlich, and M. Ephritikhine, *Chem. Commun.*, 2004, 870 (UO$_2$I$_2$).

J.C. Taylor, *Coord. Chem. Rev.*, 1976, **20**, 197 (uranium halides and oxyhalides).

A.F. Wells, *Structural Inorganic Chemistry*, Clarendon Press, Oxford, 3rd edn, 1962.

11   Co-ordination Chemistry of the Actinides

Further Reading

G.L. Soloveichik, *EIC-1*, 2 (complexes).

D.W. Keogh, *EIC-2*, p. 2 (complexes).

K.W. Bagnall, *CCC-1*, **3**, 1129 (complexes).

C.J. Burns, M.P. Neu, H. Boukhalfa, K.E. Gutowski, N.J. Bridges and R.D. Rogers, CCC-2, **3**, 189 (complexes).

Uranium

Further Reading

E.H.P. Cordfunke, *The Chemistry of Uranium*, Elsevier, 1969.

R.G. Denning *et al.*, *LAS*, 313; *Mol. Phys.*, 1979, **37**, 1109 (uranyl ion).

W.R. Wadt, *J. Am. Chem. Soc.*, 1981, **313**, 6053 (uranyl ion).

P. Pykko, L.J. Laakkonen, and K. Tatsumi, *Inorg. Chem.*, 1989, **28**, 1801 (uranyl ion).

R.G. Denning, *Struct. Bonding.*, 1992, **79**, 215 (review on uranyl ion).

U. Castellato, P.A. Vigato, and M. Vidali, *Coord. Chem. Rev.*, 1981, **36**, 183 (nitrate complexes).

D. Brown, *MTP-2*, **7**, 111 (nitrate complexes).

D. Brown, *MTP-1*, **7**, 87 (halide complexes).

K.W. Bagnall, *MTP-1*, **7**, 139 (thiocyanate complexes).

R.K. Agarwal, H. Agarwal and K. Arora, *Rev. Inorg. Chem.*, 2000, **20**, 1 (thorium complexes of neutral O-donors).

K. Arora, R.C. Goyal, D.D. Agarwal and R.K. Agarwal, *Rev. Inorg. Chem.*, 1999, **18**, 283 (uranium complexes of neutral O-donors).

D.L. Clark, D.E. Hobart and M.P. Neu, *Chem. Rev.*, 1995, **95**, 25 (actinide carbonate complexes).

M.P. Wilkerson, C.J. Burns, H.J. Dewey, J.M. Martin, D.E. Morris, R.T. Paine and B.L. Scott, *Inorg. Chem.*, 2000, **39**, 5277 [uranium(VI) alkoxides and oxo alkoxides].

W.G. Van Der Sluys and A.P. Sattelberger, *Chem. Rev.*, 1990, **90**, 1027 (alkoxides).

W.G. Van Der Sluys, C.J. Burns, J.C. Huffman, and A.P. Sattelberger, *J. Am. Chem. Soc.*, 1988, **110**, 5924; J.M. Berg, D.L. Clark, J.C. Huffman, D.E. Morris, A.P. Sattelberger, W.E. Streib, W.G. Van Der Sluys and J.G. Watkin, *J. Am. Chem. Soc.*, 1992, **114**, 10811 (actinide aryloxides).

D.C. Bradley, R.C. Mehrotra, I.P. Rothwell, and A. Singh, *Alkoxo and Aryloxo Derivatives of Metals*, Academic Press, San Diego, 2001.

J-C. Berthet and M. Ephritikhine, *Coord. Chem. Rev.*, 1998, **178–180**, 83 (amides).

General References

J. Marçalo and A. Pires de Matos, *Polyhedron*, 1989, **8**, 2431 (amides).

Other Elements

Further Reading

GMEL, Supplement No. 40, Main Volume 1940; Supplement 1981 (actinium).

H.W. Kirby, *ACE*, **1**, 14 (actinium).

GMEL, Supplement No. 44, Main Volume 1955; Supplements A1–A4 (element), C1–C7 (compounds), D1–D2 (spectra), E (coordination compounds) (thorium).

L.I. Katzin and D.C. Sonnenberger, *ACE*, **1**, 41 (thorium).

I. Santos, A.P. de Matos, and A.G. Maddox, *Adv. Inorg. Chem. Radiochem.*, 1989, **34**, 65 (thorium).

GMEL System No. 51, Main Volume 1942, 2 supplements 1977 (protactinium).

D. Brown, *AIP*, 343; *Adv. Inorg. Chem. Radiochem.*, 1969, **12**, 1 (protactinium).

J.A. Fahey, *ACE*, **1**, 443 (neptunium).

S.K. Patel, *Coord. Chem. Rev.*, 1978, **25**, 133 (neptunium).

G.T. Seaborg, *AIP*, 1 (plutonium).

F. Weigel, J.J. Katz and G.T. Seaborg, *ACE*, **1**, 499 (plutonium).

J.M. Cleveland, *The Chemistry of Plutonium*, American Nuclear Society, 1979.

W.T. Carnall and G.R. Choppin (eds), *ACS Symp. Series*, 1983, vol. 216 (plutonium).

C.J. Burns and A.P. Sattelberger, *APU*, 61 (organometallic and coordination chemistry of Pu).

M.P. Neu, C.E. Ruggiero and A.J. Francis, *APU*, 170 (bioinorganic chemistry of Pu).

S.B. Clark and C. Delegard, *APU*, 118 (chemistry of Pu in concentrated solutions).

W.W. Schulz and R.A. Pennemann, *ACE*, **2**, 887 (americium).

J.D. Navratil, W.W. Schulz, and G.T. Seaborg, *J. Chem. Educ.*, 1990, **67**, 15 (americium).

N.M. Edelstein, J.D. Navratil and W.W. Schulz, *Americium and Curium Chemistry and Technology*, Reidel, 1985.

E.K. Hulet and D.D. Bode, *MTP-1*, **7**, 1 (nuclear waste processing).

F. Weigel, J.J. Katz and G.T. Seaborg, *ACE*, **1**, 505 (nuclear waste processing).

K. Czerwinski, in *APU*, 77 (Pu separations).

R.J. Silva and H. Nitsche, *APU*, 89 (Pu environmental chemistry).

General References

P.I. Artiukhin, V.I. Medcedovskii, and A.D. Gel'man, *Radiokhimiya*, 1959, **1**, 131; *Zh. Neorg. Khin.*, 1959, **4**, 1324 (plutonium disproportionation) and self-reduction).

A. Von Grosse, *J. Am. Chem. Soc.*, 1934, **56**, 2501 (determination of atomic mass of protactinium).

S.A. Cotton, *Lanthanides and Actinides*, Macmillan, 1991 (plutonium).

12 Electronic/Magnetic Properties of the Actinides

Further Reading

J.L. Ryan, *MTP1*, **7**, 323 (UV–visible spectra).

W.T. Carnall and H.M. Crosswhite, *ACE*, **2**, 1235 (UV–visible spectra).

S. Ahrland, *CIC*, **5**, 473 (UV–visible spectra).

J.P. Hessler and W.T. Carnall, *LAS*, 349 (UV–visible spectra).

N.M. Edelstein and J. Goffart, *ACE*, **2**, 1361 (magnetism).

J.W. Gonsalves, P.J. Steenkamp, and J.G.H. Du Preez, *Inorg. Chim. Acta.*, 1977, **21**, 167; *S. Afr. J. Chem.*, 1977, **30**, 62 (magnetism).

B.N. Figgis and M.A. Hitchman, *Ligand Field Theory and Its Applications*, Wiley, 2000, 311–324 (magnetism and UV–visible spectra).

General References

M. Hirose *et al.*, *Inorg. Chim. Acta*, 1988, **150**, L93 (magnetism).

J.L. Ryan and W.E. Keder, *Adv. Chem. Ser.*, 1967, **71**, 335 (UV–visible spectra).

M.J. Sarsfield, H. Steele, M. Helliwell, and S.J. Teat, *Dalton Trans.*, 2004, 3443 (UV–visible spectra).

B.C. Lane and L.M. Venanzi, *Inorg. Chim. Acta*, 1969, **3**, 239 (magnetism).

J.G.H. DuPreez and B. Zeelie, *Inorg. Chim. Acta*, 1989, **161**, 187 (UV–visible spectra).

D.M. Gruen and R.L. Macbeth, *J. Inorg. Nucl. Chem.*, 1959, **9**, 297 UV–visible spectra).

J.R. Peterson *et al.*, *Inorg. Chem.*, 1986, **25**, 3779 (UV-visible spectra).

13 Organometallic Chemistry of the Actinides

Further Reading

T.J. Marks, *HAC*, **4**, 491; *ACE*, **2**, 1588.

T.J. Marks and A. Streitweiser, *ACE*, **2**, 1547.

T.J. Marks and I.L. Fragala (eds), *Fundamental and Technological Aspects of Organo f-element Chemistry*, Reidel, 1985.

T.J. Marks and R.D. Ernst, *COMC-1*, **3**, 173.

F.T. Edelmann, *COMC-2*, 1995, **4**, 10.

D.L. Clark and A.P. Sattelberger, *EIC-1*, 19.

A.P. Sattelberger and D.L. Clark, *EIC-2*, p. 33.

C.J. Burns and A.P. Sattelberger, *APU*, 61 (organometallic and coordination chemistry of Pu).

General References

T.J. Marks, A.M. Seyam, and J.R. Kolb, *J. Am. Chem. Soc.*, 1973, **95**, 5529 (UCp$_3$X).

T.J. Marks, and W.A. Wachter, *J. Am. Chem. Soc.*, 1976, **98**, 703 (ThCp$_3$X).

A. Streitweiser, Jr, and U. Mueller-Westerhoff, *J. Am. Chem. Soc.*, 1968, **90**, 7364 [U(η^8-C$_8$H$_8$)$_2$].

14 Synthesis of the Transactinides and their Chemistry

Further Reading

B. Fricke, *Structure and Bonding*, 1975, **21**, 89 (predicted properties).

G.T. Seaborg and O.L. Keller Jr, *ACE*, **2**, 1629.

A. Ghiorso, *AIP*, 23.

R.J. Silva, *ACE*, **2**, 1103.

E.K. Hulet *et al.*, *J. Inorg. Nucl. Chem.*, 1980, **42**, 79 (atom-at-a-time chemistry).

K. Kumar, *Superheavy Elements*, Adam Hilger, Bristol and New York, 1989.

G.T. Seaborg and W.D. Loveland, *The Elements Beyond Uranium*, Wiley-Interscience, 1990.

S.A. Cotton, *Chem. Soc. Rev.*, 1996, **25**, 219.

D.M. Hofman and D.M. Lee, *J. Chem. Educ.*, 1999, **76**, 332 (atom-at-a-time chemistry).

S. Hofmann, *On Beyond Uranium*, Taylor and Francis, London, 2002.

M. Schaedel (ed.), *The Chemistry of Superheavy Elements*, Kluwer, 2003.

General References

G.T. Seaborg, *J. Chem. Educ.*, 1969, **46**, 631 (regions of nuclear stability).

G. Herrmann, *Angew. Chem., Int. Ed. Engl.*, 1988, **27**, 1421 (synthesis of element 107, now called bohrium).

J.V. Kratz, *J. Alloys Compd.*, 1994, **213–214**, 22 (apparatus for studying volatile halides).

G.T. Seaborg, *J. Chem. Soc., Dalton Trans.*, 1996, 3904 (Periodic Table).

Element 112

S. Hofmann, V. Ninov, F.P. Hessberger, P. Armbruster, H. Folger, G. Münzenberg, J. Schött, A.G. Popeko, A.V. Yeremin, S. Saro, R. Janik and M. Leino, *Z. Phys. A. Hadrons Nucl.*, 1996, **354**, 229.

Element 113

K. Morita *et al*, *J. Phys. Soc. Jpn.*, 2004, **73**, 2593.

Element 114

Yu.Ts. Oganessian, A.V. Yeremin, A.G. Popeko, S.L. Bogomolov, G.K. Buklanov, N.L. Chelnokov, V.I. Chepigin, B.N. Gikal, V.A. Gorshkov, G.G. Gulbekian, M.G. Itkis, A.P. Kabachenko, A.Y. Lavrentev, O.N. Malyshev, J. Rohac, R.N. Sagadiak, S. Hofmann, S. Saro, G. Giardina and K. Morita, *Nature*, 1999, **400**, 242.

Element 115

Yu.Ts. Oganessian, *et al.*, *Phys. Rev. C.*, 2004, **69**, 021601.

Elements 116 and 118

V. Ninov, K.E. Gregorich, W. Loveland, A. Ghiorso, D.C. Hoffman, D.M. Lee, H. Nitsche, W.J. Swiatecki, U.W. Kirbach, C.A. Laue, J.L. Adams, J.B. Patin, D.A. Shaughnessy, D.A. Strellis and P.A. Wilk, *Phys. Rev. Lett.*, 1999, **83**, 1104; P.J. Karol, H. Nakahara, B.W. Petley and E. Vogt, *Pure Appl. Chem.*, 2001, **73**, 959 (retraction).

Index

Note: Figures and Tables are indicated by *italic page numbers*

With kind thanks to Paul Nash for creation of the index.